Vibration Prediction and Noise Assessment of Building Structures

Vibration Prediction and Noise Assessment of Building Structures

Guest Editors

Chao Zou
Ziyu Tao

MDPI

Basel • Beijing • Wuhan • Barcelona • Belgrade • Novi Sad • Cluj • Manchester

Guest Editors
Chao Zou
School of Civil and Transportation Engineering
Guangdong University of Technology
Guangzhou
China

Ziyu Tao
School of Mechanical and Electrical Engineering
Guangzhou University
Guangzhou
China

Editorial Office
MDPI AG
Grosspeteranlage 5
4052 Basel, Switzerland

This is a reprint of the Special Issue, published open access by the journal *Buildings* (ISSN 2075-5309), freely accessible at: https://www.mdpi.com/journal/buildings/special_issues/XOUNZ3G0HJ.

For citation purposes, cite each article independently as indicated on the article page online and as indicated below:

Lastname, A.A.; Lastname, B.B. Article Title. *Journal Name* **Year**, *Volume Number*, Page Range.

ISBN 978-3-7258-4993-2 (Hbk)
ISBN 978-3-7258-4994-9 (PDF)
https://doi.org/10.3390/books978-3-7258-4994-9

Contents

About the Editors

Chao Zou

Chao Zou is an Associate Professor at Guangdong University. His primary research interests encompass the prediction and mitigation of vibrations and noise generated by rail transit, road traffic, and Transit-Oriented Development (TOD) systems, alongside research into the resource utilization of solid waste. He has led more than 20 significant research projects, including projects funded by grants from the National Natural Science Foundation of China and the Guangdong Natural Science Foundation, as well as various projects commissioned by public and private enterprises. His extensive practical expertise stems from major studies investigating the environmental vibration and noise impacts of metro depots in Guangzhou, Shenzhen, Wuhan, Foshan, Hangzhou, Nanjing, and Changsha; sections of high-speed railway stations; lineside metro environments; urban road traffic noise reduction mechanisms; and solid waste reuse.

Ziyu Tao

Ziyu Tao is affiliated with Guangzhou University. Her research focuses on the prediction and mitigation of vibrations and noise stemming from rail transit, road traffic, and Transit-Oriented Development (TOD) systems, alongside investigations into the thermomechanical coupling effects of high-energy lasers. She has led research projects funded by the National Natural Science Foundation of China and the Hong Kong Polytechnic University Postdoctoral Fellowship Scheme, and has contributed to the successful completion of multiple projects commissioned by public and private enterprises. Building on her successful foundations, which have included receiving a China Scholarship Council grant for a joint PhD training program, Dr. Tao has authored over 30 peer-reviewed journal publications and holds two patents. She actively contributes to the academic community as a Young Editorial Board Member for the SCI-indexed journal *Transportation Safety and Environment*, as a Guest Editor for the SCI-indexed journal *Buildings and Advances in Civil Engineering*, and as a Reviewer for numerous prestigious international journals, including *Structural Control & Health Monitoring*, *Journal of Vibration and Control*, *Machines*, and *Mathematics*.

Preface

Urbanization has intensified the challenges of vibration and noise disturbances in building structures, compromising occupant comfort and critical operations in sensitive facilities like research laboratories and high-tech manufacturing facilities. This Reprint addresses the pressing need for advanced methodologies to predict, assess, and mitigate these disturbances.

Our motivation stems from the complex nature of vibration sources—traffic, construction, and mechanical systems—and their unpredictable propagation through soil structure systems. This collection synthesizes cutting-edge research on analytical modeling, numerical simulations, and probabilistic approaches to overcome limitations in existing prediction frameworks and control strategies.

This work is essential for civil engineers, researchers, and designers developing next-generation urban infrastructure. We thank the contributors for advancing this field, and trust that this Reprint will catalyze the development of innovations that ensure structural integrity and human wellbeing in dynamic environments.

Chao Zou and Ziyu Tao
Guest Editors

buildings

MDPI

ticle

.esearch on a Vibration Model of a Superstructure under the 'ibration Load of Metro Trains

ng Liu [1], Haodong Xu [2], Wei Xia [3], Wenfeng Cai [3] and Senlin Zheng [2,*]

[1] Guangzhou Metro Design & Research Institute Co., Ltd., Guangzhou 510499, China; liuyang@gmdi.cn
[2] School of Water Conservancy and Civil Engineering, South China Agricultural University, Guangzhou 510642, China; xuhd@stu.scau.edu.cn
[3] China Railway Guangzhou Bureau Group Co., Ltd., 151 Zhongshan 1st Road, Yuexiu District, Guangzhou 510088, China; xiawei317@163.com (W.X.); ationfeng@163.com (W.C.)
* Correspondence: zhengsl@scau.edu.cn

Abstract: In view of the problem that vibration of superstructures under vibration loads of metro trains causes, this research used a metro depot and superstructure project as its background and proposed a numerical simulation method based on the impedance analytical model and finite element model to simulate and predict the vibration and secondary noise response of subway trains affecting multi-story buildings at different locations on the ground and in the superstructure. The method's accuracy was verified using real measurement data. The research shows that vibrations generated by subway operations vertically at lower floors remain relatively unchanged, then slowly attenuate before increasing near the top floors. Mitigation measures should primarily address four aspects: rails, fasteners, sleepers, and roadbed. The adverse effects of vibration can be controlled by reducing the excitation intensity of the vibration source, attenuating vibrations along the propagation path, and isolating vibrations in the foundation and interior of the building. This research method can quickly and accurately predict the vibration and noise conditions of superstructure properties and provide support for vibration and noise reduction in practical engineering.

Keywords: transit-oriented development; railway noise and vibration; finite element method; superstructure; secondary noise

1. Introduction

Transit-oriented development (TOD) represents a development paradigm that has seen increasing implementation along railroad lines in China in recent years. Many subway lines produce low-frequency microvibrations attributable to track irregularities, passing curves, train offsets, and wheel–rail abrasion, which are propagated through the tracks, tunnels, and soil strata to adjacent high-rise buildings [1]. Although the amplitude of these low-frequency vibrations is low, prolonged exposure may detrimentally affect the psychological and physiological well-being of residents, and in severe cases, could potentially result in cognitive decline [2–4].

Given the distinct vibration and noise characteristics of the vehicle section and the main line, implementing targeted interventions to attenuate vibrations and noise constitutes a critical technological approach in developing rail-proximate structures. Consequently, it is of substantial importance to simulate, predict, and analyze the vibration response of subway trains in the vehicle section affecting structures at varied ground positions and rooftop levels, in addition to assessing the resultant structural secondary noise [5–7]. Scholars worldwide have developed various methodologies and prediction models to address vehicle-induced environmental vibrations. Rossi F and colleagues established a predictive model for soil vibrations induced by trains, capable of providing root mean square velocities and vibration levels [8]. After validating the feasibility of the impedance method, Masoud Sanayei and colleagues utilized an impedance-based analytical model to simulate the dynamic behavior of a four-story, full-scale building and to predict the floor's response to train-induced vibrations [9]. Wang Yimin and his team conducted empirical analyses on the propagation patterns of environmental and structural vibrations induced by train operations in junction areas [10]. Quanmin Liu and his team employed an FE-SEA hybrid method to investigate the structural noise of steel–concrete composite bridges [11]. Liang Ruihua and colleagues developed a probabilistic prediction model based on Bayesian neural networks (BNNs) for forecasting ground vibrations induced by trains [12]. Considering these research methodologies, this paper utilized a comprehensive combination of numerical simulations and field experiments for the study.

In this research, the vibration response to subway train-induced loads on superstructural elements within a specific section in Wuhan was thoroughly analyzed in the context of an ongoing vehicle section project. The impedance function model [6] and finite element numerical simulation [13] were applied to develop dynamic models encompassing train-track [14,15], roadbed–soil layers, and adjacent buildings [16,17] for the Wuhan section. Subsequently, these computational models were refined using empirical data [18], confirming the appropriateness of the model parameters, boundary conditions, and train loads [19]. This study implemented this method for the detailed predictive analysis of superstructure elements in the sensitive zones of the Ping Sheng vehicle section, assessing both structural vibrations within the subway section [4,20] and the resultant secondary acoustic comfort [20,21]. The closeness of the outcomes to preliminary predictions further corroborates the prediction method's accuracy [22]. The method predicts the vibration and noise conditions of the superstructure under vibration load of metro trains quickly and accurately.

This study not only applies the finite element method but also systematically innovates it to predict subway vibration. Using finite elements, we propose a method that focuses on the vibration and noise of the subway superstructure. This method can accurately and quickly predict the vibration and noise of superstructure properties, providing support for vibration and noise reduction in actual engineering projects.

2. Methodology

The prediction model in this paper consists of two parts: a vehicle–track interaction dynamics model [23] and a moving load–track structure-surrounding strata–buildings structure finite element model [24]. This coupled train-track dynamics model accurately simulates the motion of a six-car subway train running on the track, enabling the calculation of the fastener reaction force, which can then be used as an excitation input for the track–buildings–soil finite element model to determine the building's vibration response. The method has been validated in papers by He L [17], Li H [25], and Qiu Y, Zou C [26].

This paper used a real engineering project as the research background. A finite element model was established that encompassed track structures, surrounding strata, and building structures, reflecting actual operational conditions within a specific section of the Wuhan Metro. Parameters derived from B-type subway cars were integrated into the vehicle–rail dynamics model, allowing for the calculation of wheel–rail forces [27]. These forces were subsequently used in the finite element analysis to compute vibration responses resulting from subway train operations. Finally, the computed vibration responses were compared with measured values obtained from the Wuhan Metro, demonstrating the predictive model's effectiveness and feasibility in capturing environmental vibrations induced by metro operations. Based on this, the impact of subway operations on vibration and secondary noise prediction for the new building above the subway vehicle section was investigated [18], as shown in Figure 1.

This hybrid prediction method, based on ground vibration measurements and predictive modeling of building structures considering soil-structure interactions, calculates the vibration response of a new building using a predictive model of the building structure after obtaining measured ground surface vibrations. This approach has the advantage of not needing to consider tunnels and different soil layers in the model [28,29].

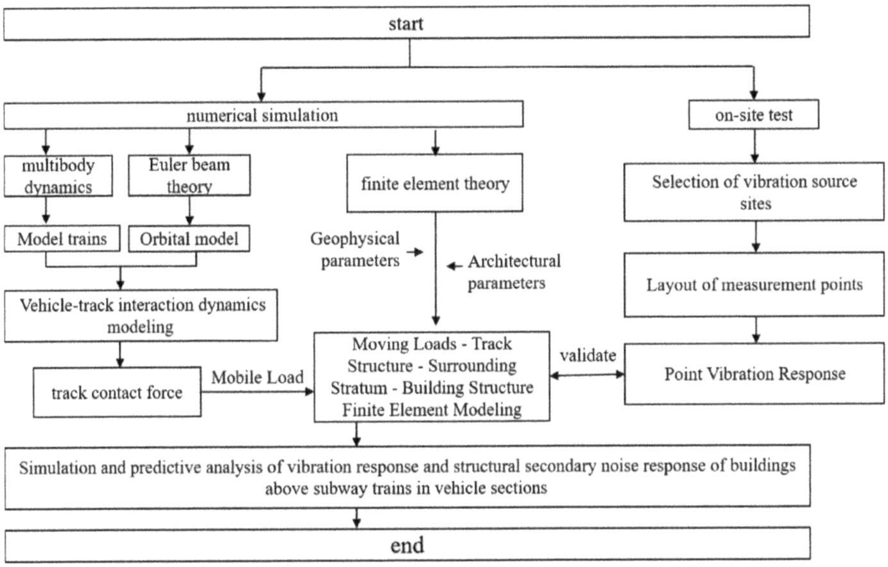

Figure 1. Flowchart of research methodology.

2.1. Modeling of Vehicle–Track Interaction Dynamics

The vehicle–track interaction dynamics model is shown in Figure 2. The model includes a vehicle submodel and a track structure submodel; the x-direction represents the train's running direction, and the z-direction denotes the vertical orientation [30]. Wheel pairs were coupled with the track using Hertzian nonlinear elastic contact theory, accounting for high- and medium-frequency vibrations in the wheel–rail system due to short-wave irregularities. Introducing track irregularities at the wheel–rail contact point allowed the derivation of interaction forces between the wheel and rail, inducing vibrations in both the vehicle and track systems. Subsequently, vehicle–track coupling was achieved through displacement compatibility and force equilibrium between the wheel and rail.

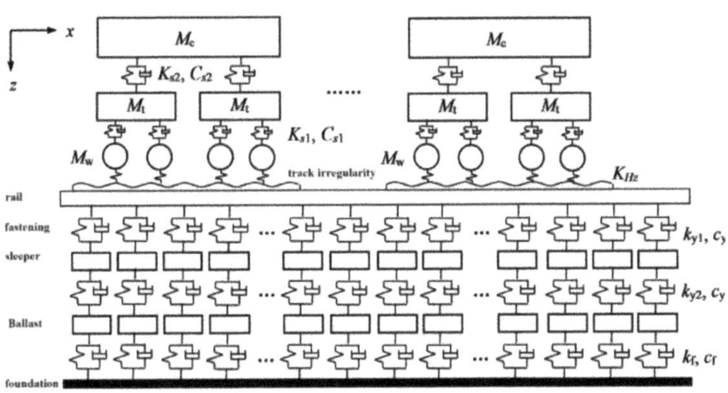

Figure 2. Dynamic interaction model between vehicle and track.

Using multi-rigid body dynamics theory, the subway train dynamics model considered the train as a multi-section vehicle spaced at equal intervals [31]. Each segment was modeled as a multi-rigid body system consisting of a car body, two bogies, and four wheelsets. This model accounted for two degrees of freedom—vertical and pitching movements—for the car body and bogies, while wheelsets were modeled with vertical and lateral movements, resulting in a ten-degree-of-freedom model for each vehicle segment, interconnected by elastic suspension elements [32].

The track structure model included rails, sleepers, and ballast, with the latter representing the overall roadbed in this context. Rails were modeled as finite-length Euler beams, while sleepers and ballast were considered concentrated masses that supported vertical loads only. Connections between rails and sleepers, sleepers and ballast, and ballast and foundation were facilitated through tri-layered equivalent elastic and damping elements. Wheel–rail contact was modeled using the conventional Hertz nonlinear elastic contact theory.

Table 1 displays the dynamic parameters of the subway vehicle model utilized in this study. The excitation of the wheel–rail interaction employed six-level spectral amplitude samples from the United States, incorporating considerations for short-wave disturbances.

Table 1. Parameters of vehicle dynamics.

Project	Vehicle Model
Bogie Wheelbase	2.3 m
Vehicle Spacing	12.6 m
Conductor	19.52 m
Body mass	42,600 kg
Moment of Inertia of the Car Body	1,262,000 kg·m^2
Two-series suspension stiffness	480 kN/m
Two-Series Suspension Damping	60 kN·s/m
Bogie Mass	2550 kg
Moment of Inertia of the Bogie	2410 kg·m^2
A series of suspension stiffness	1500 kN/m
Two-series suspension stiffness	5 kN·s/m
Wheelset Mass	1760 kg

A total of 24 moving wheel loads, representing the dynamic loads of a six-car subway train, were applied to the rails within the track–soil–building model [22]. Figure 3 illustrates the calculated dynamic contact forces between wheel and rail at various locations. The dynamic contact force between wheel and rail serves as the moving load input parameter, introduced into the model via the subroutine VDLOAD to act upon the track structure.

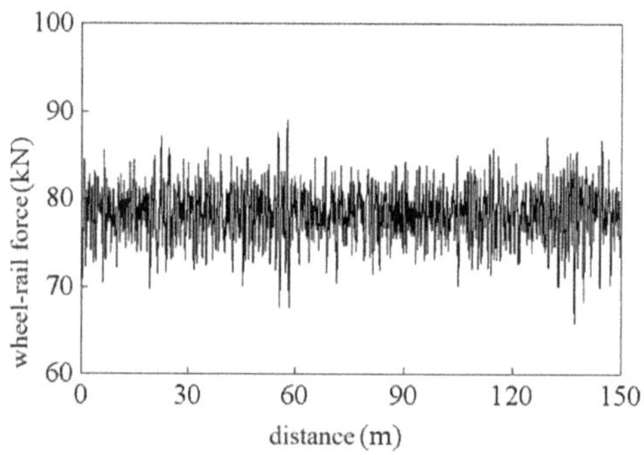

Figure 3. Typical wheel–rail contact force as a function of distance.

2.2. Development of a Three-Dimensional Finite Element Model for Analyzing Track Structures, Surrounding Strata, and Building Interactions

Using the finite element software ABAQUS 6.1, a comprehensive three-dimensional finite element analysis model was developed. This model included the track structure, surrounding soil layers, and the building structure [33].

The vibration-sensitive zone within the Pingsheng vehicle section was chosen as the focus of this study, for which a three-dimensional finite element analysis model encompassing the track structure, surrounding strata, and building structures was developed. Infinite elements were used to model the natural foundation, aiming to negate the effects of artificially truncated boundaries. Calculated dynamic contact forces between wheel and rail served as input parameters for moving loads. These loads were applied using the subroutine VDLOAD to calculate the vibration responses across different building layers. Following this, the secondary noise responses of the structures were calculated [34,35].

The data for the model were derived from actual engineering. Using a ballasted roadbed as an example, the track and roadbed model included rails, fasteners, sleepers, and the roadbed itself (including the surface layer and ballast). The track gauge was set at 1.435 m, featuring a seamless line; the quantitative density of sleepers was 1680 per kilometer; and the roadbed consisted of a gravel layer with a thickness of 0.45 m. Relevant physical property parameters can be found in Table 2.

Table 2. Parameters of the orbital structure model's physical properties.

Model Structure	Modulus of Elasticity E (MPa)	Poisson's Ratio	Density (kg/m^3)
orbit	210,000	0.25	7850
sleeper	30,000	0.2	2400
The upper level of the tunnel bed	300	0.35	1800
Lower level of the tunneling bed	300	0.35	2200

The model's soil strata were simplified into four distinct layers, each simulated via a linear elastic model derived from geological borehole data and shear wave velocity measurements. Considering the influences of weak soil layers, variability among borehole data, and similar shear wave velocities across different strata, the foundation soil was categorized into artificial fill, silty soil, sand layer, and sandstone [36,37]. The sandstone layer's thickness was considered infinite, reflecting the boundary conditions applied within the infinite element simulation model utilized in this study. Soil properties are listed in Table 3.

Table 3. Key physico-mechanical parameters of the foundation soil model.

Soil Layer	Thickness (m)	Density (kg/m^3)	Dynamic Elastic Modulus (MPa)	Dynamic Poisson's Ratio	Shear Wave Velocity (m/s)	Damping Ratio
Artificial soil	1.5	1830	202.5	0.471	190.0	0.03
Mucky soil	23.5	1640	145.8	0.475	162.0	0.03
Sand layer	4	1800	201.5	0.470	225.4	0.03
Sandstone	∞	2200	372.1	0.442	240.2	0.03

Following the specified parameters, individual building structure models were developed for the sensitive area. Beam elements simulated beams and frame columns, while shell elements were used for walls and floor slabs. To accurately capture soil-structure interaction, the embed contact method was employed. Reflecting the vibration propagation mechanism, the numerical model used a foundation separation approach. This method ensured that the track structure's foundation was isolated from the building's foundation, preventing direct vibration wave transfer from the track structure to the superstructure, thereby mitigating significant vibrations. Figure 4 illustrates a representative finite element model.

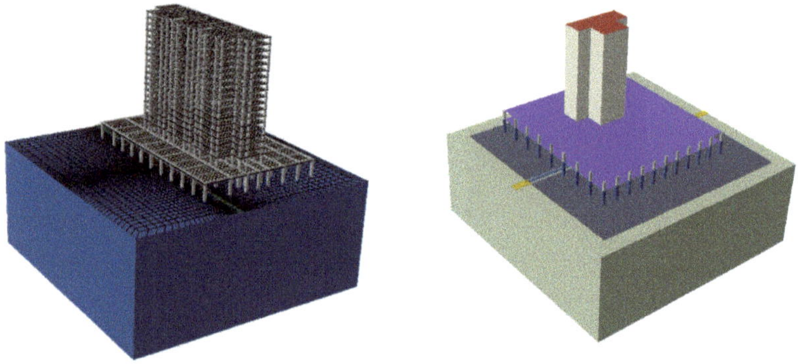

Figure 4. Finite element model of a typical building structure.

Furthermore, the incremental time step was set to 0.002 s to satisfy the computational frequency (200 Hz) requirement. This research method was applied in the actual process of Foshan City Urban Rail Transit Line 4, providing technical support for the vibration and noise reduction in the Ping Sheng Vehicle Section. Two weeks of computational time were required for one analysis, and the method met the engineering needs with higher accuracy and efficiency.

3. Field Testing and Model Verification Processes

3.1. Overview of Test Sites

The total construction area extends over 524,000 m^2, and the land area covers 171,942 m^2, comprising a total of 635 residential units. Currently, this vehicle section is recognized as Wuhan's largest subway vehicle depot, covering an area of 823 acres. It is equipped to facilitate parking for up to 32 trains and includes four maintenance bays. Significantly, the depot's inner areas are equipped with vibration-damping measures, including fasteners and seamless tracks, while its outer areas utilize damping joint cleats as part of their strategy.

3.2. Evaluation Indicators

The indoor vibration of buildings caused by rail transit should be evaluated based on the indoor vibrations detectable by humans and whole-body vibration reflections. Vibration acceleration was evaluated as a source of noise, mainly using the weighted root-mean-square acceleration to assess the impact of vibration on occupants. For residential buildings (including commercial and residential buildings), the single-value evaluation metric for indoor vibration was the Z vibration level. The crossover frequency vibration measurement evaluation metric was the 1/3 octave plumb vibration acceleration level, with a frequency range of 1 to 80 Hz.

The maximum 1/3 octave band sound pressure level within the interior secondary structure noise space of a building when the train passes was used to calculate the model for secondary structural noise, which is defined as:

$$L_{p,i} = L_{Vmid,i} - 22 \tag{1}$$

Equation (1) is applicable to a typical room renovation with a height of around 2.8 m and a reverberation time of approximately 0.8 s, covering an area of about 10 to 12 m^2. Deviation from these conditions necessitates calculations based on Equation (2).

$$L_{p,i} = L_{Vmid,i} + 10\lg\sigma - 10\lg H - 20 + 10\lg T_{60} \tag{2}$$

where $L_{p,i}$ is the maximum 1/3 octave band sound pressure level in the indoor space of a building during the passage of a single train. $L_{Vmid,i}$ is the central vertical 1/3 octave band vibration velocity level in the indoor floor of buildings during the passage of a single train. σ is the sound radiation efficiency. H is the average room height in meters. T_{60} is the reverberation time in seconds.

3.3. Model Validation

Measurements were conducted with the vibration source in its normal operating condition to minimize the influence of external environmental factors on the vibration measurements. These factors included extreme temperature gradients, strong electromagnetic fields, and high winds, as well as avoiding conditions such as rain, snow, thunder, lightning, and wind speeds exceeding 5 m per second at the measurement site.

Vertical accelerations of ground vibration at distances of 6 m, 12 m, 18 m, and 28 m from the test line within the vehicle section were computed. These computed values were then compared with the measured vertical accelerations near the test line. To facilitate comparison, time-domain diagrams were generated from the train passing process curve, with a data length of 9 s specifically selected to encompass the train passing event. Acceleration time-domain data were converted to acceleration spectrograms using the fast Fourier transform (FFT). One-third octave plots for acceleration levels were derived by applying the specific calculations and weighting criteria outlined in the "*Standard for Limits of Vibration and Secondary Radiated Noise of Buildings Caused by Urban Rail Transit and its Measurement Methods*" (JGJ/T 170-2009) [38] and the "*Methods for Measurement of Ambient Vibration in Urban Areas*" (GB 10070-88) [39], as illustrated in Figures 5–8.

Figures 5–8 reveal that both the calculated and measured time-domain curves displayed similar waveforms and overarching trends, accurately capturing the train passing phenomenon. Minimal differences between the calculated and measured acceleration amplitudes were observed at distances of 6 m, 12 m, and 18 m from the track's centerline. However, at a distance of 28 m from the track's centerline, the calculated acceleration amplitudes were slightly larger than those measured. This discrepancy can be attributed to the cell grid division and simplification of soil layers; however, it does not significantly impact the overall findings. As the distance from the track's centerline increased, both the calculated and measured acceleration decay amplitudes aligned closely, exhibiting a consistent decay pattern. The model more effectively simulated the vibration attenuation law caused by subway trains in the time domain.

Analysis of the measured results indicated that soil vibration predominantly occurred at frequencies below 50 Hz, with notable vibration peaks around 20 to 30 Hz. The calculated results demonstrated a high degree of similarity to the measured outcomes, particularly regarding the dominant distribution frequency and peak frequency.

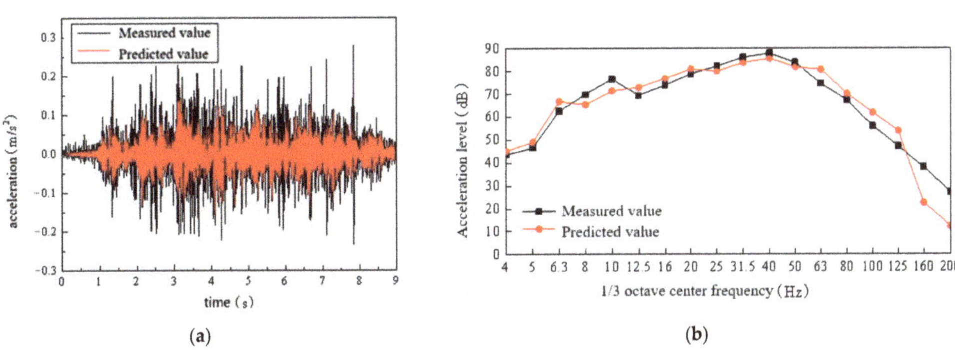

Figure 5. Comparison of measured and calculated acceleration values at 6 m from the track's centerline for both time-domain and frequency-domain analyses: (**a**) time-domain plot of acceleration: measured vs. calculated at 6 m from the track's centerline; (**b**) frequency-domain analysis of acceleration at 6 m from the track's centerline: measured vs. predicted.

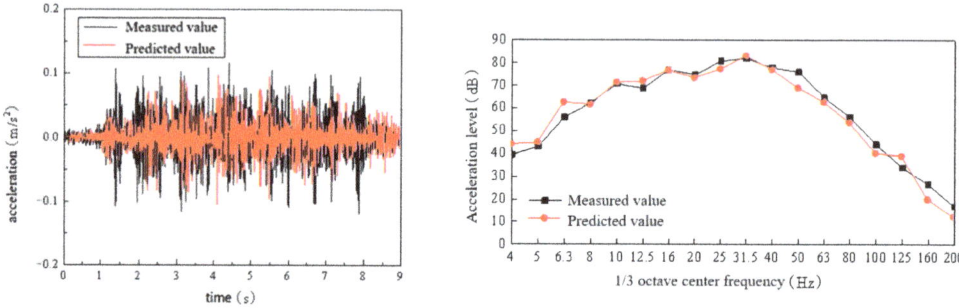

(a) (b)

Figure 6. Comparison of measured and calculated acceleration values at 12 m from the track's centerline for both time-domain and frequency-domain analyses: (**a**) time-domain plot of acceleration: measured vs. calculated at 12 m from the track's centerline; (**b**) frequency-domain analysis of acceleration at 12 m from the track's centerline: measured vs. predicted.

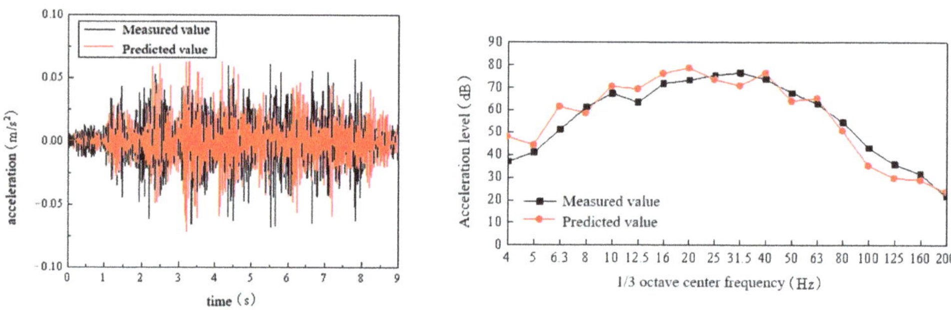

(a) (b)

Figure 7. Comparison of measured and calculated acceleration values at 18 m from the track's centerline for both time-domain and frequency-domain analyses: (**a**) time-domain plot of acceleration: measured vs. calculated at 18 m from the track's centerline; (**b**) frequency-domain analysis of acceleration at 18 m from the track's centerline: measured vs. predicted.

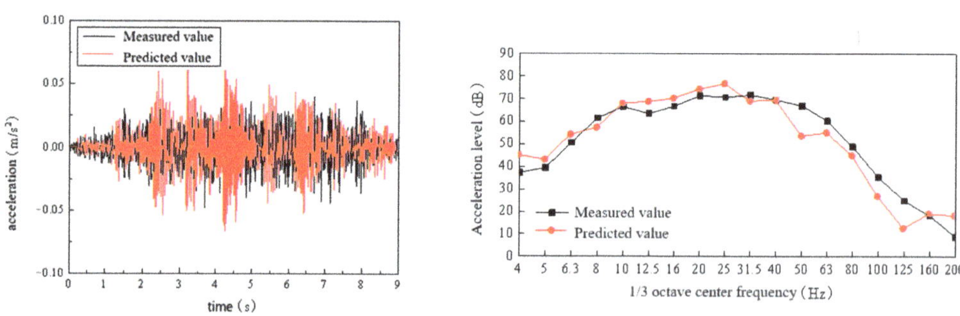

(a) (b)

Figure 8. Comparison of measured and calculated acceleration values at 28 m from the track's centerline for both time-domain and frequency-domain analyses: (**a**) time-domain plot of acceleration: measured vs. calculated at 28 m from the track's centerline; (**b**) frequency-domain analysis of acceleration at 28 m from the track's centerline: measured vs. predicted.

Overall, a pronounced similarity was observed between the model-derived and measured acceleration spectra, particularly in terms of overall trend, dominant distribution frequency, and peak frequency. Generally, the acceleration spectra derived from the model closely matched the observed data, particularly in terms of overall patterns, predominant frequency distributions, and peak frequencies. It was concluded that the developed finite element model effectively replicated the propagation characteristics of ground vibrations adjacent to subway train operations. Therefore, this method can be used in practical engineering to establish a finite element model of the superstructure for accurate prediction and analysis of the vibration response caused by train operations without damping measures.

4. Refined Prediction of Structural Vibrations and Secondary Noise in Building Structures

Utilizing the previously validated model, a finite element model of the superstructure across various sensitive areas within the Ping Sheng vehicle section was developed.

The Ping Sheng vehicle section site is located on the southern side of Ping Sheng Avenue, with a length of approximately 1040 m, a width of approximately 570 m, and a total area of about 27.8 hectares. The Ping Sheng vehicle section is designed to accommodate 46 columns, spanning a length of 297.5 m, a width of 180 m, and covering an area of 49,344 square meters. The depot is equipped with three fixed repair lines, a temporary repair station, a static adjustment line, and a sweeping line. Situated at the northernmost end of the vehicle base, the test line features a gentle slope and extends 888 m to facilitate medium-speed testing.

During the analysis, the maximum vibration level from the least favorable position on each floor was used as the basis for comparison with standard vibration limit values. This helped determine the necessity for additional vibration reduction measures and assessed their effectiveness. Figure 9 presents a representative finite element model illustrating the propagation of train-induced vibrations within the building.

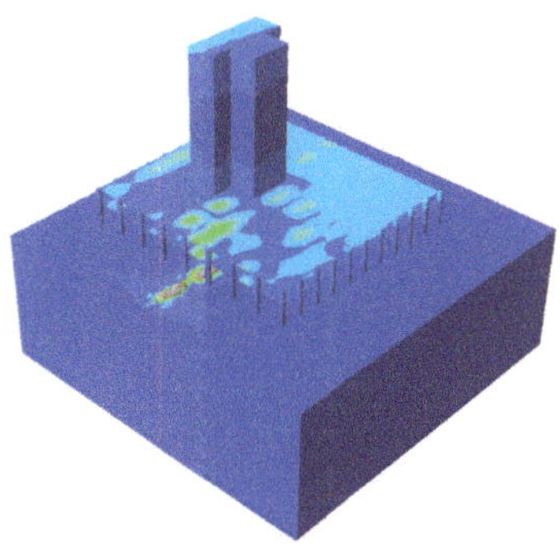

Figure 9. Finite element visualization of train-induced vibration propagation in a building.

Substantial variations in vibration levels and spectral characteristics, as indicated by significant measured data, arose from the operation of different trains on the same line, with Z-axis vibration level differences reaching approximately 15 dB or more. Notably, environmental vibrations exceeding standard limits were not observed during the initial phase of line operation but emerged after a certain period, leading to occasional resident complaints and petitions. This suggests that train-induced vibrations were not constant, exhibiting significant changes over time with a stochastic distribution pattern. Abnormal wheel–rail abrasion, the durability of vibration reduction and isolation measures, and uneven foundation settlement can stimulate the track's overall vertical bending resonance modes, leading to excessive vibration amplitude of the roadbed plate and increased environmental vibration levels.

In China's rail transit environmental vibration impact assessments, the vibration source strength is determined based on the average of the maximum Z-axis vibration levels from analogous tests, in accordance with relevant national standards. However, due to factors such as vehicle type, track structure, and material durability, test results can exhibit dispersion and randomness, displaying distinct distribution characteristics. Furthermore, wheel and rail wear during the rail transportation operation and maintenance phases exhibit significant temporal effects and notable distribution characteristics. Existing research relying on fixed-value predictions often results in insufficient accuracy, overlooking the randomness in vibration source strength due to the operation and maintenance status of rail transit. This limitation affects the accuracy of vibration prediction; thus, finite element calculations incorporate the randomness associated with vibration source strength due to rail transportation's operation and maintenance.

4.1. Influence Zone of the Pharyngeal Area's Upper Lid

The roof of the throat area accommodates an elementary school, a kindergarten, and a commercial building, with the educational facilities consisting of two-story or three-story structures. Located directly above the throat area, this zone primarily experiences vibrations resulting from trains entering and exiting the parking, inspection, and maintenance depots. Given the numerous curves, turnouts, and track irregularities in the throat section, a train speed limit of 25 km/h has been implemented to mitigate their impact. To assess potential vibration effects, the kindergarten building, situated in the most vulnerable zone of this area, was chosen for prediction and analysis.

Figure 10 illustrates the predicted vibration levels resulting from train operations, highlighting variations across different building floors. Additionally, an incremental increase in vibration levels was observed as one ascends through the floors, though at a decelerating rate. The figure shows that each floor of the building exhibits vibration levels ranging from 70 to 79 dB, surpassing the standard daytime limit of 70 dB for vibrations.

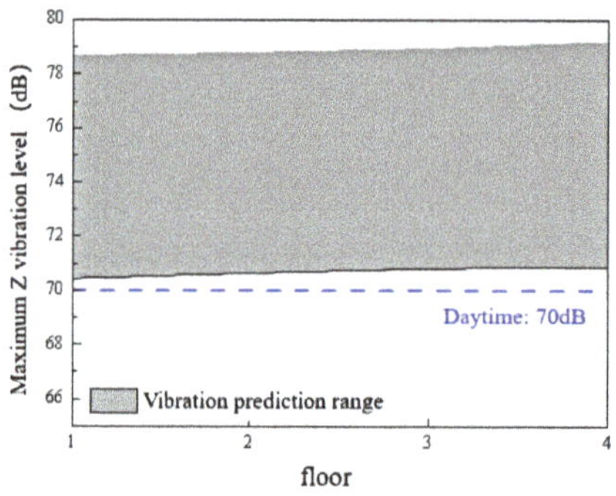

Figure 10. Floor-level vibration variation due to train operations in the throat area.

4.2. Impact Radius of the Test Line

The designated influence area of the test line, accommodating speeds up to 60 km/h, includes five 17F/-2F buildings, an elementary school, a kindergarten, a commercial building, and a gymnasium. Characterized by high population density and diverse buildings, residential structures are positioned at least 30 m from the test line, whereas the school building is located over 25 m away. Consequently, for prediction and analysis, a 17-story residential building and a three-story school building closest to the test line were selected. Figure 11 illustrates the predicted vibrations across different floors of high-rise residential buildings due to train operations.

The figure revealed that vibration levels on each floor ranged from 67.5 to 72.8 dB, exceeding the standard nighttime limit of 67 dB. Notably, floors below the coverage of 15 m were designated for nonresidential purposes, ensuring that the residential areas of the building had a maximum Z-axis vibration level of 71.7 dB.

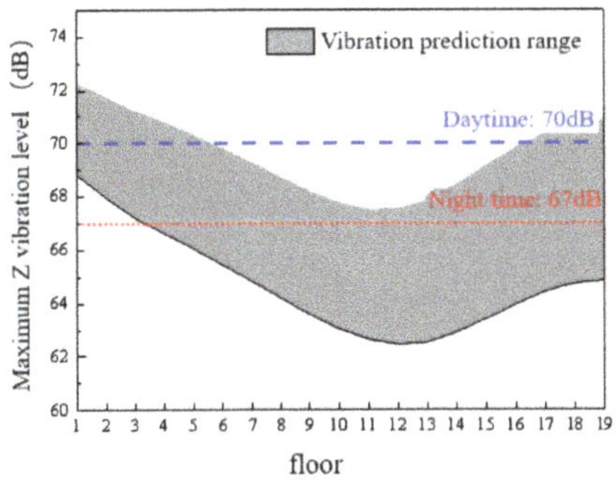

Figure 11. Vibration variation across floors in high-rise buildings due to train operations on the test line.

Figure 12 depicts the predicted range of structural secondary noise due to train operations on the test line. The calculated equivalent sound level revealed that structural secondary noise from train operations ranged from 26.5 to 41.0 dB(A), with the maximum sound level exceeding the standard nighttime limit by 6 dB(A).

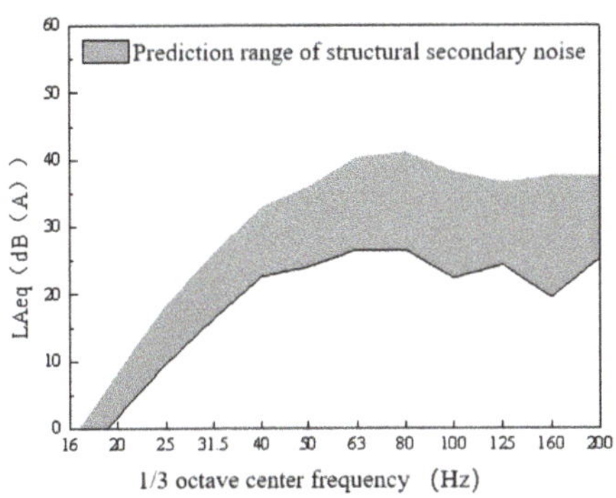

Figure 12. Predicted secondary noise levels in high-rise residential buildings from train operations on the test line.

Figure 13 depicts anticipated variations in building vibrations induced by train operations across different levels of the school building.

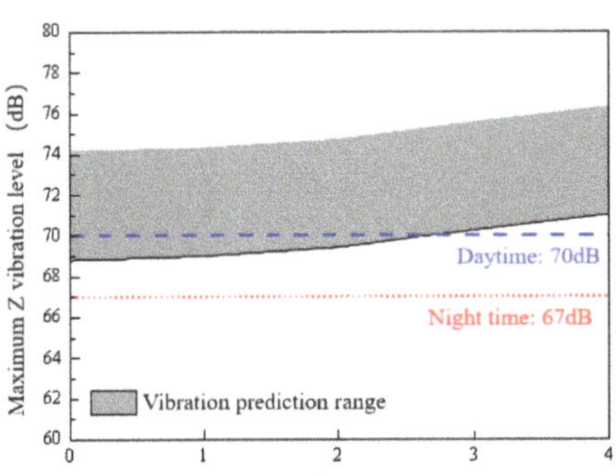

Figure 13. Predicted secondary noise levels in building vibrations induced by train operations across different levels of the school building.

This figure shows that vibration levels on each floor ranged from 68.8 to 76.6 dB, exceeding the standard daytime vibration limit by 6.6 dB.

Figure 14 illustrates the predicted range of structural secondary noise due to train operations on the test line. Calculations of the equivalent sound level revealed that structural secondary noise from train operations ranged from 38.8 to 45.1 dB(A), with the maximum sound level exceeding the specified daytime limit by 7.1 dB(A).

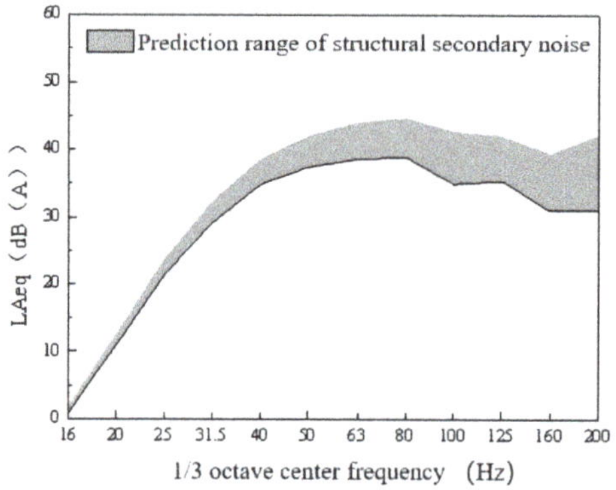

Figure 14. Predicted secondary noise range in high-rise residential buildings due to train operations on the test line.

4.3. Impact Assessment of Maintenance and Material Depots, and Loading/Unloading Lines

The area under consideration encompasses the maintenance warehouse, material warehouse, and material loading/unloading line. It includes 12 buildings on the 17th and 2nd floors above the maintenance warehouse, as well as four buildings on the 17th and 2nd floors above the material warehouse and the material loading/unloading line. These structures are primarily subjected to vibrations resulting from crane operations. Based on preliminary predictions outlined, it has been determined that the crane's vibration impact was more pronounced in the maintenance warehouse compared to the material warehouse and the material loading/unloading line. Consequently, one of the nineteen-story buildings directly situated above the crane (located on its southern side) was selected for finite element calculation and analysis. The results of vibration predictions are presented in Figure 15.

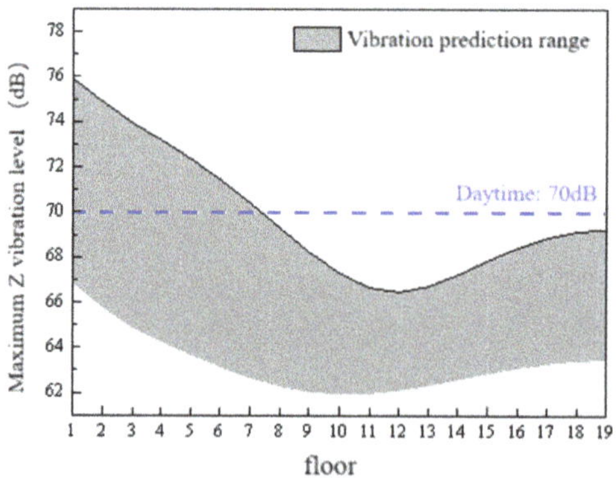

Figure 15. Floor-level vibration variation due to crane operations in the inspection warehouse.

Figure 15 illustrates the structural vibration response of the 17F/-2F building located directly above the service garage on the southern side during crane operations. Vibration levels across each floor ranged from 62 to 76 dB, exceeding the standard daytime limit of 70 dB for vibrations. Given that the upper deck level is not used for residential purposes, a maximum predicted vibration level of 75.8 dB was observed in the residential areas.

Figure 16 depicts the predicted range of structural secondary noise due to crane operations in the service garage. Calculations of the equivalent sound level revealed that structural secondary noise from crane operations ranged from 37.5 to 45.4 dB(A), with the maximum sound level exceeding the daytime limit by 7.4 dB(A).

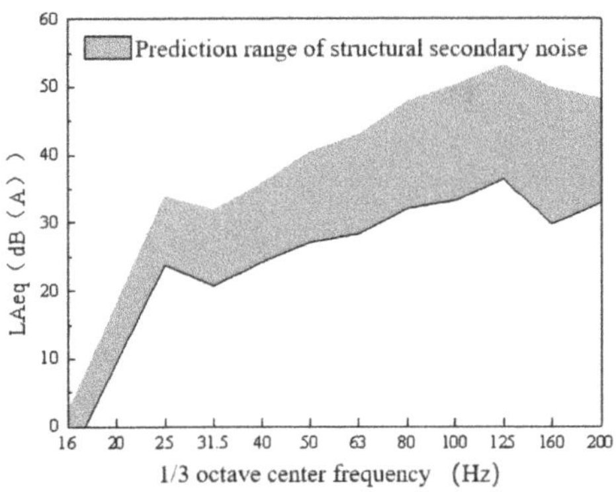

Figure 16. Predicted secondary noise range from crane operations in the inspection depot.

The method has been validated in studies by He J [35] and Li X [40]. After applying the prediction model to a detailed analysis of vehicle-induced vibration in the vibration-sensitive area of the Ping Sheng vehicle section, a comparison was made between the preliminary and refined prediction results. Table 4 summarizes the comparison, highlighting the differences between preliminary predictions and refined finite element predictions for vehicle-induced vibration across various development zones within the Ping Sheng vehicle section. A high level of agreement between the two sets of results is evident, further validating the accuracy and reliability of the proposed model prediction method.

Table 4. Comparative analysis of preliminary and refined finite element predictions for the Pingsheng vehicle section.

Screening and Evaluation of Buildings within the Affected Area: Number and Function	Vibration Source	Preliminary Predictions of Maximum Z-Axis Vibration Levels (dB)	Refined Predictions of Maximum Z-Axis Vibration Levels (dB)
Educational Buildings: 2–3 Story Superstructures, etc. Commercial area: 1~2 story superstructure	Throat area	81.9	79.2
esidential: 17 story superstructure	testing line	70.5	71.7
School: 2~3 story superstructure		75.8	76.6
esidential: 17 story superstructure	Maintenance warehouse	76.2	75.8

5. Conclusions

In this paper, based on the impedance analytical model and the finite element model, a numerical simulation method was proposed for simulating and predicting the vibration and structural secondary noise response of subway trains affecting multi-story buildings at different locations on the ground and within the superstructure. The accuracy of the method was verified via real measurement data. The following conclusions can be drawn from this study:

(1) The study revealed that vibrations generated by subway operations do not follow a simple vertical pattern correlated with floor height, rather, vibrations at lower floors remain relatively unchanged, then gradually attenuate before increasing near the top floors.
(2) Within the superstructure, the impact of vibration on floor slabs exceeds that of columns, leading to a more pronounced vibration sensation for individuals standing at the center of the floor slab.
(3) Vibrations affecting neighboring buildings along the line were predominantly vertical. As the distance from the track increased, vibration attenuation occurred, with high-frequency vibrations diminishing more rapidly than low-frequency vibrations, which remained relatively constant. The impact of vibration also decreased with reduced train speed. Within 40 m of the source, high-frequency vertical vibrations decayed more rapidly than horizontal vibrations, and high-frequency components above 40 Hz diminished even more swiftly.
(4) The comprehensive management of vibration and noise in the vehicle section should target both the sources and causes of vibration and noise and involve studying corresponding management programs. Measures should primarily address four aspects: rails, fasteners, sleepers, and roadbed. Structural damping design should also be implemented to achieve comprehensive management. The adverse effects of vibration can be controlled by reducing the excitation intensity of the vibration source, attenuating vibrations along the propagation path, and isolating vibrations in the foundation and interior of the building.

Author Contributions: Conceptualization: Y.L., H.X. and S.Z.; methodology and formal analysis: Y.L., H.X. and S.Z.; software, resources, and data curation: W.X. and W.C.; validation: Y.L., H.X., S.Z., W.X. and W.C.; investigation and visualization: Y.L., H.X., S.Z., W.X. and W.C.; writing—original draft preparation and writing—review and editing: Y.L., H.X. and W.X. All authors have read and agreed to the published version of the manuscript.

Funding: This research was funded by the National Natural Science Foundation of China (Project No. 52008115), the (Chinese) State Key Laboratory of Subtropical Building Science (Project: No. 2021ZB08).

Data Availability Statement: The data used to support the findings of this study are currently under embargo while the research findings are commercialized. Requests for data, [12 months] after publication of this article, will be considered by the corresponding author.

Conflicts of Interest: Author Yang Liu was employed by the company Guangzhou Metro Design & Research Institute Co., Ltd. Authors Wei Xia and Wenfeng Cai were employed by the company China Railway Guangzhou Bureau Group Co., Ltd. The remaining authors declare that the research was conducted in the absence of any commercial or financial relationships that could be construed as a potential conflict of interest.

References

1. Ma, M.; Liu, W.; Qian, C.; Deng, G.; Li, Y. Study of the train-induced vibration impact on a historic Bell Tower above two spati overlapping metro lines. *Soil Dyn. Earthq. Eng.* **2016**, *81*, 58–74. [CrossRef]
2. Handy, S. Smart growth and the transportation-land use connection: What does the research tell us? *Int. Reg. Sci. Rev.* **2005**, 146–167. [CrossRef]
3. Sheng, X.; Jones, C.J.C.; Thompson, D.J. A theoretical study on the influence of the track on train-induced ground vibrat *J. Sound Vib.* **2004**, *272*, 909–936. [CrossRef]
4. Connolly, D.P.; Kouroussis, G.; Giannopoulos, A.; Verlinden, O.; Woodward, P.K.; Forde, M.C. Assessment of railway vibrati using an efficient scoping model. *Soil Dyn. Earthq. Eng.* **2014**, *58*, 37–47. [CrossRef]
5. Lund, A.; Dahlberg, T. Dynamic train/track interaction including model for track settlement evolvement. *Veh. Syst. Dynan* **2004**, *66*, 667–676.
6. Zou, C.; Moore, J.A.; Sanayei, M.; Tao, Z.; Wang, Y. Impedance Model of Train-Induced Vibration Transmission Across a Tran Structure into an Overtrack Building in a Metro Depot. *J. Struct. Eng.* **2022**, *148*, 1–46. [CrossRef]
7. Lei, X.; Jiang, C. Analysis of vibration reduction effect of steel spring floating slab track with finite elements. *J. Vib. Con* **2014**, *22*, 1462–1471. [CrossRef]

Federico Rossi, A.N. A simple model to predict train-induced vibration: Theoretical formulation and experimental validation. *Environ. Impact Assess. Rev.* **2003**, *23*, 305–322. [CrossRef]

Sanayei, M.; Anish, P.K.; Moore, J.A.; Brett, C.R. Measurement and prediction of train-induced vibrations in a full-scale building. *Eng. Struct.* **2014**, *77*, 119–128. [CrossRef]

Liu, Q.; Li, X.; Zhang, X.; Zhou, Y.; Chen, Y.F. Applying constrained layer damping to reduce vibration and noise from a steel-concrete composite bridge: An experimental and numerical investigation. *J. Sandw. Struct. Mater.* **2018**, *22*, 1743–1769. [CrossRef]

Liu, Q.; Thompson, D.J.; Xu, P.; Feng, Q.; Li, X. Investigation of train-induced vibration and noise from a steel-concrete composite railway bridge using a hybrid finite element-statistical energy analysis method. *J. Sound Vib.* **2020**, *471*, 115197. [CrossRef]

Liang, R.; Liu, W.; Ma, M.; Liu, W. An efficient model for predicting the train-induced ground-borne vibration and un-certainty quantification based on bayesian neural network. *J. Sound Vib.* **2021**, *495*, 115908. [CrossRef]

Connolly, D.; Giannopoulos, A.; Forde, M.C. Numerical modelling of ground borne vibrations from high speed rail lines on embankments. *Soil Dyn. Earthq. Eng.* **2013**, *46*, 13–19. [CrossRef]

Zhai, W.; Wang, K. Lateral hunting stability of railway vehicles running on elastic track structures. *J. Comput. Nonlinear Dynamics* **2010**, *5*, 041009-1–041009-9. [CrossRef]

Torstensson, P.T.; Nielsen, J.C.O. Simulation of dynamic vehicle–track interaction on small radius curves. *Veh. Syst. Dyn.* **2011**, *49*, 1711–1732. [CrossRef]

Kouroussis, G.; Gazetas, G.; Anastasopoulos, I.; Conti, C.; Verlinden, O. Discrete modelling of vertical track–soil coupling for vehicle–track dynamics. *Soil Dyn. Earthq. Eng.* **2011**, *31*, 1711–1723. [CrossRef]

Qiu, Y.; Zou, C.; Hu, J.; Chen, J. Prediction and mitigation of building vibrations caused by train operations on concrete floors. *Appl. Acoust.* **2024**, *219*, 109941. [CrossRef]

Kouroussis, G.; Verlinden, O.; Conti, C. A two-step time simulation of ground vibrations induced by the railway traffic. *Proc. Inst. Mech. Eng. Part C J. Mech. Eng. Sci.* **2011**, *226*, 454–472. [CrossRef]

Pyl, L.; Degrande, G.; Lombaert, G.; Haegeman, W. Validation of a source-receiver model for road traffic-induced vibra-tions in buildings. I: Source model. *J. Eng. Mech.* **2004**, *130*, 1377–1393. [CrossRef]

Galvín, P.; Mendoza, D.L.; Connolly, D.P.; Degrande, G.; Lombaert, G.; Romero, A. Scoping assessment of free_field vibrations due to railway traffic. *Soil Dyn. Earthq. Eng.* **2018**, *114*, 598–614. [CrossRef]

He, L.; Tao, Z. Building vibration measurement and prediction during train operations. *Buildings* **2024**, *14*, 142. [CrossRef]

Zou, C.; Wang, Y.; Zhang, X.; Tao, Z. Vibration isolation of over-track buildings in a metro depot by using trackside wave barriers. *J. Build. Eng.* **2020**, *30*, 101270. [CrossRef]

Zou, C.; Wang, Y.; Moore, J.A.; Sanayei, M. Train-induced field vibration measurements of ground and over-track buildings. *Sci. Total. Environ.* **2017**, *575*, 1339–1351. [CrossRef] [PubMed]

Costa, P.A.; Calçada, R.; Cardoso, A.S. Influence of train dynamic modelling strategy on the prediction of track–ground vibrations induced by railway traffic. *Proc. Inst. Mech. Eng. Part F-J. Rail Rapid Transit* **2012**, *226*, 434–450. [CrossRef]

Li, H.; He, C.; Gong, Q.; Zhou, S.; Li, X.; Zou, C. Tlm-cfspml for 3d dynamic responses of a layered transversely isotropic half-space. *Comput. Geotech.* **2024**, *168*, 106131. [CrossRef]

Ma, M.; Xu, L.; Liu, W.; Tan, X. Semi-analytical solution of a coupled tunnel-soil periodic model with a track slab under a moving train load, Applied Mathematical Modelling. *Simul. Comput. Eng. Environ. Syst.* **2024**, *128*, 588–608.

Zhai, W.; Wang, K.; Cai, C. Fundamentals of vehicle-track coupled dynamics. *Veh. Syst. Dyn.* **2009**, *47*, 1349–1376. [CrossRef]

Sheng, X.; Jones, C.J.C.; Thompson, D.J. A theoretical model for ground vibration from trains generated by vertical track irregularities. *J. Sound Vib.* **2003**, *272*, 937–965. [CrossRef]

Hussein, M.F.M.; Hunt, H.E.M. A numerical model for calculating vibration from a railway tunnel embedded in a full-space. *J. Sound Vib.* **2007**, *305*, 401–431. [CrossRef]

Auersch, L. Wave propagation in the elastic half-space due to an interior load and its application to ground vibration problems and buildings on pile foundations. *Soil Dyn. Earthq. Eng.* **2010**, *30*, 925–936. [CrossRef]

Auersch, L. Dynamic interaction of various beams with the underlying soil–finite and infinite, half–space and Winkler models. *Eur. J. Mech. A/solids* **2008**, *27*, 933–958. [CrossRef]

Xu, L.; Ma, M. Analytical solution of ground-borne vibration due to a spatially periodic harmonic moving load in a tunnel embedded in layered soil. *J. Zhejiang Univ. A* **2023**, *24*, 637–652. [CrossRef]

Auersch, L. Dynamic stiffness of foundations on inhomogeneous soils for a realistic prediction of vertical building resonance. *J. Geotech. Geoenviron. Eng.* **2008**, *134*, 328–340. [CrossRef]

François, S.; Pyl, L.; Masoumi, H.; Degrande, G. The influence of dynamic soil–structure interaction on traffic induced vibrations in buildings. *Soil Dyn. Earthq. Eng.* **2006**, *27*, 655–674. [CrossRef]

Hu, J.; Zou, C.; Liu, Q.; Li, X.; Tao, Z. Floor vibration predictions based on train-track-building coupling model. *J. Build. Eng.* **2024**, *89*, 109340. [CrossRef]

Ditommaso, R.; Mucciarelli, M.; Gallipoli, M.R.; Ponzo, F.C. Effect of a single vibrating building on free-field ground motion: Numerical and experimental evidences. *Bull. Earthq. Eng.* **2009**, *8*, 693–703. [CrossRef]

Iacovino, C.; Ditommaso, R.; Ponzo, F.C.; Limongelli, M.P. The interpolation evolution method for damage localization in structures under seismic excitation. *Earthq. Eng. Struct. Dyn.* **2018**, *47*, 2117–2136. [CrossRef]

38. *JGJ/T 170-2009*; Standard for Limits of Vibration and Secondary Radiated Noise of Buildings Caused by Urban Rail Transit and i Measurement Methods. China Architecture & Building Press: Beijing, China, 2009.
39. *GB 10070-88*; Methods for Measurement of Ambient Vibration in Urban Areas. Standards Press of China: Beijing, China, 1988.
40. Li, X.; Chen, Y.; Zo, C.; Wang, H.; Zheng, B.; Chen, J. Building structure-borne noise measurements and estimation due to trai operations in tunnel. *Sci. Total Environ.* **2024**, *926*, 172080. [CrossRef]

buildings

MDPI

ticle

xperimental Investigation on Building Sound Environment: raffic-Induced Air Noise and Structure-Borne Noise

iang Chen [1,2,3], Lingshan He [4], Xuming Li [5,*], Bokai Zheng [5], Teng Wang [1], Dongyang Wang [1] and Chao Zou [5,*]

[1] Beijing Oriental Yuhong Waterproof Technology Co., Ltd., Beijing 101111, China
[2] Department of Civil Engineering, Tsinghua University, Beijing 100084, China
[3] Beiijao Zane Rail Technology (Beijing) Co., Ltd., Beijing 101111, China
[4] Guangzhou Urban Planning & Design Survey Research Institute Co., Ltd., Guangzhou 510060, China
[5] School of Civil and Transportation, Guangdong University of Technology, Guangzhou 510006, China
* Correspondence: lixuming0216@163.com (X.L.); chao.zou@gdut.edu.cn (C.Z.)

Abstract: The impact of urban traffic on human health is significant. This research conducts field measurements in Guangzhou, China, focusing on a building situated near subgrade roads and viaducts to investigate the characteristics of airborne and structure-borne noise generated by these infrastructures. The analysis involves the use of both sound pressure level and overall sound pressure level, as well as an examination of the transfer function between outdoor and indoor noise levels. The findings indicate that traffic-related airborne noise demonstrates a characteristic frequency at 1000 Hz in this scenario, while viaduct- and building-generated structure-borne noise is predominantly distributed at lower frequencies. Additionally, it is worth noting that structural vibrations generate significantly less energy compared to airborne traffic noise sources. The variation in outdoor road noise across different floors over the entire frequency range demonstrates an initial increase followed by a decrease with rising floor height due to air damping effects as well as sound barriers' attenuation properties. These results enhance engineers' understanding of urban traffic-induced airborne or structure-borne noise while establishing foundational data for designing layouts integrating urban buildings with roads.

Keywords: building environment; structure-borne noise; traffic noise; measurement; air noise

1. Introduction

Currently, individuals are experiencing significant levels of urban environmental pollution in daily life, including air pollution [1–3], light pollution [4,5], environmental vibration [6,7], noise pollution [8], and so on. These effects will impact various parts of the human body and different organs [4]. In urban development, it is almost inevitable that such pollution will occur. While efforts to control other forms of pollution are increasing, the issue of noise pollution is becoming increasingly severe [4]. Research has demonstrated that urban traffic-related activities have a substantial impact on human health [8,9]. As urban roads become increasingly intertwined with buildings in densely populated areas, attention is shifting toward addressing health concerns stemming from road-generated noise [10]. In the past and even now, substantial efforts have been directed toward understanding how road noise impacts individuals at different levels, including annoyance [11], hearing loss [12], sleep disturbance [13], cognitive performance [14], cardiovascular disease [15], etc. However, it is crucial to consider both traffic-induced sound patterns and their distribution when evaluating residents' irritation levels; this serves as a foundation for systematically assessing community annoyance within diverse urban architectural layouts [16].

Urban thoroughfares comprise ground-level streets and elevated highways that generate distinct types of sound as vehicles pass through; these sounds subsequently penetrate nearby structures. Such noises can be categorized as airborne acoustics and structure-borne noise [1,17].

The airborne noise generated from the interaction between vehicle tires [18] and the road surface is transmitted through the air and enters the building interior. Some researchers have conducted studies on traffic road airborne noise, which is caused by multiple generation and amplification phenomena [19], including mechanisms due to tread impact [20], road roughness excitation [21], running deflection, and the acoustic resonances [20] due to pipe resonances, especially for tread pattern. Luca et al. [18] proposed a methodology for simulating tire–road noise through equivalent monopoles, with indoor tests conducted for validation. Additionally, deep-learning-based models are used to predict airborne noise resulting from the tire–road interaction [22]. However, the generation, propagation, and dissipation of sound fields near complex roads present a challenging problem when noise is produced. Various simulation tools were utilized for noise mapping such as CadnA [23], TRANEX [24], SoundPLAN [25], and so on. Hasmaden et al. [25] established noise mapping via SoundPLAN and validated it based on measurements. Nevertheless, discrepancies between the calculated and measured results indicate that analyzing sound fields from neighboring roads poses a complex challenge; conducting field experiments may be more suitable for investigating road noise characteristics.

For structure-borne noise, there are two types of sources near the subgrade road and viaduct road. One type is generated by viaduct vibrations transmitted through the air into buildings, while the other is caused by subgrade road vibrations transmitting into buildings and subsequently transforming into structure-borne noise. The numerical approach is commonly utilized for establishing the room sound field, and the boundary element method (BEM) is widely acknowledged as the most accurate technique for predicting noise from vibrations [17,26,27]. Gu et al. [28] combined the boundary element method and statistical energy analysis (SEA) to predict the full-frequency band noise of the bridge structure; their results indicate that the dominant frequency range of structural vibration and noise of composite bridges is 20~1000 Hz. Sadeghi et al. [29] established a numerical model based on finite element method–SEA theories to simulate building structure-borne noise. However, despite significant progress in acquiring secondary structural noise, there remains a lack of understanding regarding structural vibration noise caused by road traffic.

Given the critical impact of the building sound environment on human health, previous studies have predominantly focused on noise generation. However, urban sound environments are inherently complex, with intricate mechanisms of noise propagation. While previous research has traditionally focused on outdoor noise levels [30–32], or the urban noise map [33,34], the fact remains that a significant portion of individuals spend most of their time indoors. Even when the noise map is utilized for research, it can effectively simulate the outdoor urban noise distribution; however, in complex urban acoustic environments such as indoor spaces, accurately considering noise propagation, coupling loss, and multi-source excitation becomes challenging. In this scenario, conducting field tests proves to be an exceptionally effective method for capturing the intricate acoustic landscape of an actual urban environment. Thus, it is crucial and innovative to conduct a comprehensive investigation of indoor noise levels in urban settings impacted by traffic-induced noise. Therefore, it is important and innovative to conduct a comprehensive study of indoor noise levels in urban environments affected by traffic-induced noise. The authors have previously conducted studies on indoor noise environments [35,36], including measurements taken near a traffic road with open windows and near a construction site with both open and closed windows. In an extension of these earlier studies, this current research also examines indoor and outdoor noise levels with closed windows, along with additional comparative tests. The aim is to investigate the propagation and distribution patterns of indoor sound environments within buildings under the influence of urban traffic noise.

The novelty of this study can be summarized in three main aspects: (1) The primary innovation lies in conducting a comprehensive parametric study on the building noise environment, encompassing traffic-induced airborne noise from subgrade and viaduct roads, as well as structure-borne noise from bridge and building vibrations; (2) In previous research, road noise has primarily been assessed based on the overall sound pressure level.

However, in this study, the author conducted a frequency domain analysis of road noise to investigate the distribution and impact of airborne and structural noise within each frequency band; (3) The third innovation involves the investigation of acoustic propagation characteristics in composite road traffic systems (viaducts and ground road). Both elements are considered simultaneously, leading to more complex conditions for noise transmission.

The structures of this study are shown in Figure 1. This study conducts on-site measurements to investigate the propagation characteristics of air and structural noise generated by road traffic. A series of comparisons has been conducted to examine the characteristics of traffic-induced air noise and structure-borne noise from both the viaduct road and subgrade road. The findings of this study can contribute to engineers' understanding of traffic-induced air and structure-borne noise, while also providing a foundational dataset for urban building and road layout design.

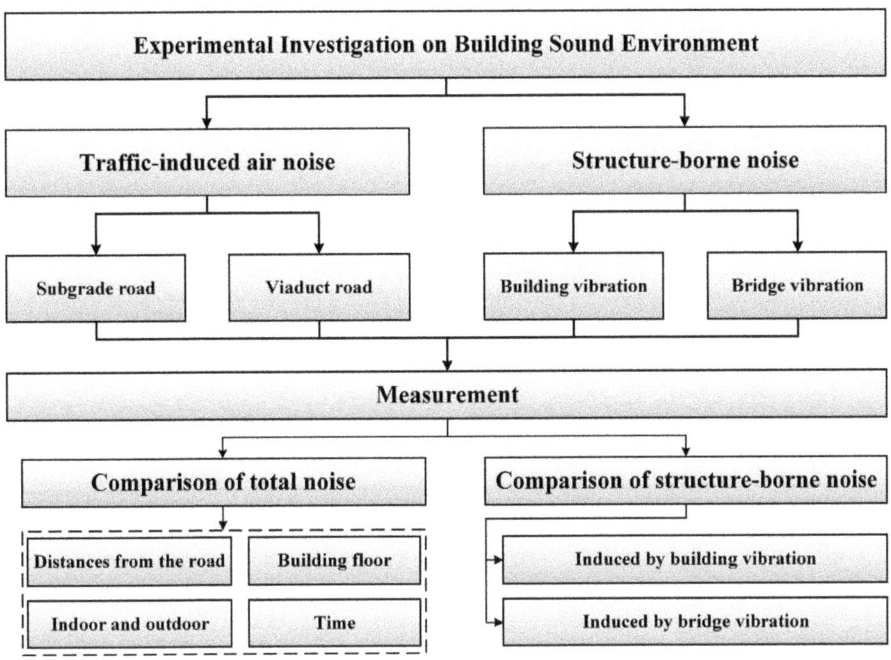

Figure 1. Structures of the study.

2. Measurement

2.1. Measured Site Descriptions

A residential building is chosen for the traffic-induced noise and vibration measurement in Guangzhou, China, which is constructed nearby the subgrade and viaduct roads with a distance of 20 m. The subgrade is a six-lane dual carriageway where trees have been planted on both sides of the road to act as a noise reduction measure. The viaduct road is a four-lane dual carriageway where the sound barrier has been set up on both sides of the road to reduce the noise pollution, which is level with the third floor of the building. The photographs of the buildings during measurement are shown in Figure 2.

Figure 2. Photographs of buildings.

As shown in Figure 3, in the enclosed public corridors of the building, there is a ledge directly facing the road traffic, and no furniture is present due to the nature of it being a public corridor, providing optimal conditions for measurement. The size of the public corridor is 8 m in length, 2 m in width, and 3 m in height, with a volume of 48 m^3. During the measurement period, the door to the public corridor is closed at all times, ensuring that indoor measurements are conducted without potential interference. According to Sabine's formula, the T60 can be expressed empirically as

$$T60 = \frac{0.16 \times V}{S_r \times \alpha_{room}} \tag{1}$$

where S_r and α_{room} are the sound absorption area and average sound absorption coefficient, respectively. By considering the empirical value of the corresponding coefficient, the $T60$ can be subsequently calculated. The general principle postulates a positive correlation between room size and reverberation time, indicating that larger rooms tend to exhibit longer reverberation times. However, given that each room shares the same volume and nearly identical interior configuration, the analysis does not include the measurement and assessment of the room's reverberation time.

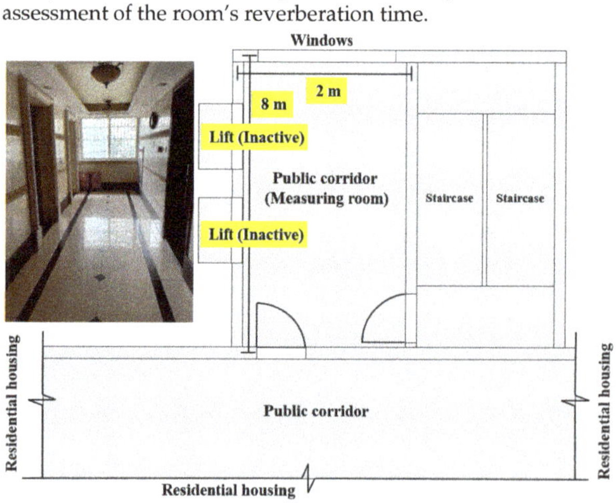

Figure 3. Photograph and spatial layout of the measurement room.

The window is the sliding double-glazed window with an aluminum window frame. The size of the window of each floor is 2 m in width and 1.5 m in height. The window is 14 mm thick double glass with a configuration and thickness of windows 4-6-4 mm and glass-laminated settling with the distance filled by air. Spacers are small pieces of aluminum that separate and support the two panes of glass that comprise the double-glazed window. The *Rw* value of the window can be estimated in the value of 30 dB(A) according to some precious studies [37–40].

2.2. Measurement Program

2.2.1. Setup

The assessment of indoor noise resulting from road traffic is a multifaceted issue, with the indoor noise being comprised of several components under test conditions, as shown in Figure 4. The yellow triangle serves as the observation point. Firstly, the indoor noise originates from the airborne noise (blue wave) generated by the interaction between vehicle tires and the road surface, which is then propagated through the air and infiltrates the interior of buildings. Secondly, when vehicles traverse viaducts, the resulting vibrations produce bridge structure-borne noise (yellow wave), which also penetrates buildings via the air. Lastly, as vehicles travel along roads, their vibrations are transmitted through foundations into buildings, leading to building structure-borne noise (red wave) within them. Collectively, these three components constitute indoor noise caused by road traffic.

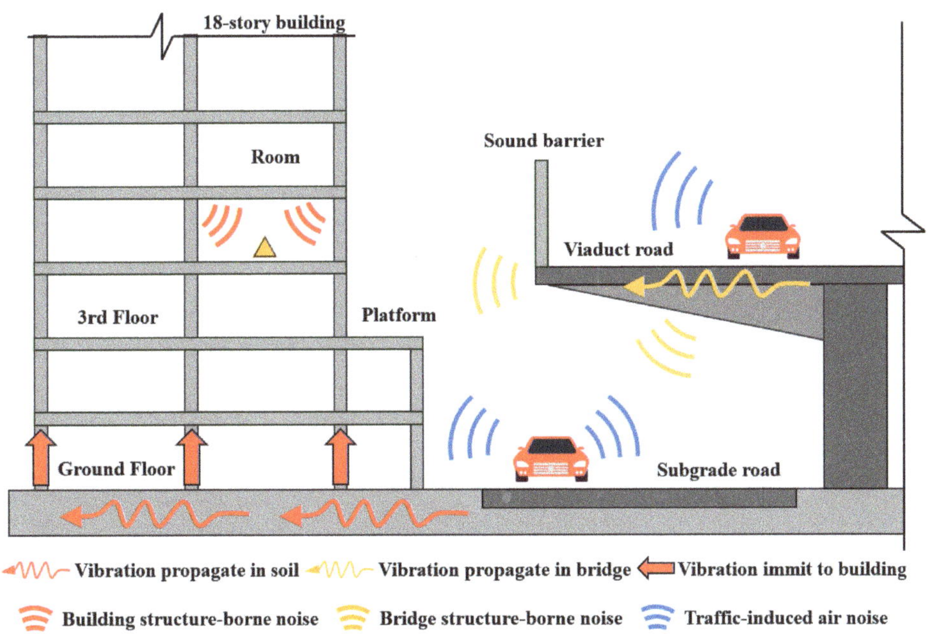

Figure 4. Schematic representation of indoor noise generated by road traffic.

The rooms on the third floor of the building are presently closed and devoid of windows; meanwhile, rooms on upper floors have windows oriented perpendicular to the road. Based on the current test conditions, three sets of control tests were established:

(1) Comparative analysis of road noise levels at different times is conducted at one test point in order to assess changes in the acoustic environment over time and potential impacts on other tests;

(2) Sensors are positioned on the ground floor at varying distances from roads to investigate the attenuation characteristics of road traffic noise within a complex urban acoustic environment;
(3) Indoor and outdoor sensors are installed on the same floor to examine both indoor sound environment characteristics and building sound insulation properties under completely enclosed conditions;
(4) Detection points are established indoors and outdoors at different heights within the building to explore traffic-induced indoor and outdoor noise attenuation characteristics across various floors;
(5) Sensors are deployed at positions closest to both viaduct roads and ground level roads for studying structural noise characteristics specific to viaducts;
(6) Sensors are deployed in the indoor room without windows and corresponding outdoor position for investigating structural noise caused by traffic-induced vibration.

All experiments were carried out within the same facility in order to minimize potential variables that could impact the outcomes. Precision microphones were positioned at various locations inside and outside of the building. This setup and photos taken during measurement are visually depicted in Figure 5. The measurements were carried out on sunny days to minimize the impact of adverse weather conditions, ensuring minimal interference from rain and strong wind. During the building measurements, all doors and windows were kept closed to eliminate potential external factors that could introduce interference.

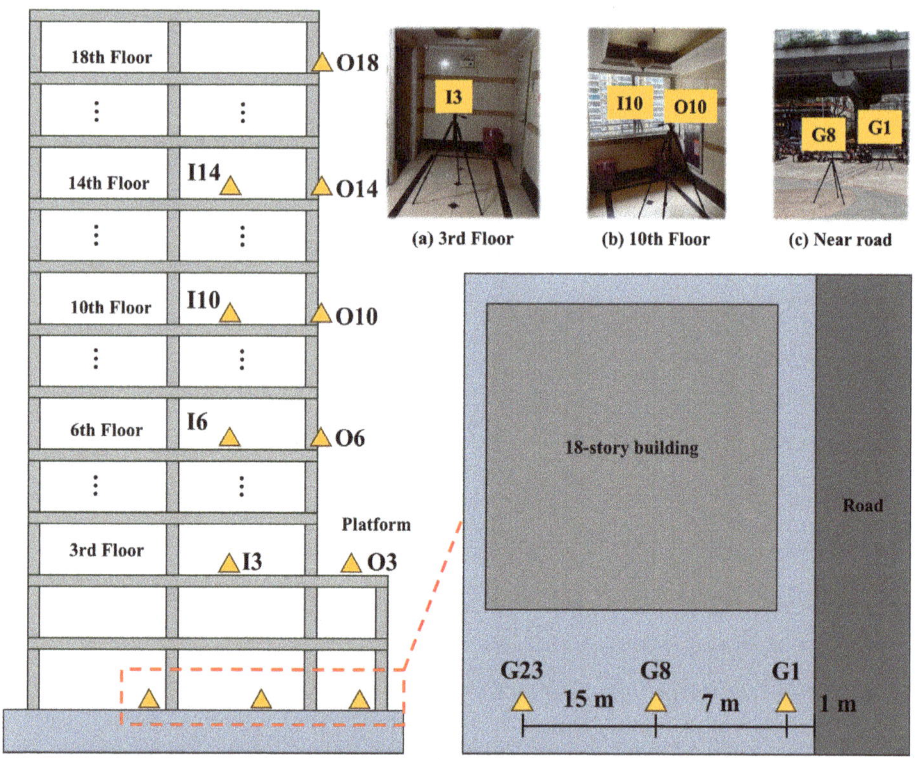

Figure 5. Instrumentation setup and the photos in different positions during measuring.

2.2.2. Instrumentation and Signal Processing

The instrumentation utilized for measurement encompasses a Rion UC-59 microphone and a tripod [8]. The microphone is installed on the preamplifier to obtain a clear noise signal, and the windshield is connected with the head of the microphone to effectively mitigate wind noise interference during the recording process and enhance the clarity and purity of recorded audio, thereby improving the overall recording quality. For the height of the measuring instrumentations, all the assembly instrumentations would be installed on the tripod, ensuring the strict requirement of being 1.5 m away from the floor surface.

Before conducting in situ measurements, a thorough synchronization of the devices was carried out. This involved using a laptop connected to an acquisition instrument. The sampling frequency for each device was set at 10 kHz, chosen to facilitate noise analysis up to 5 kHz, despite such high frequencies not being typically necessary for noise analysis but sufficient for the common frequency of interest at 1~1.5 kHz for traffic noise [41–43]. Some civil engineering machines induce noise and vibrations below 20 Hz, e.g., CAT cars with hammers at 8 and 16 Hz [6]. The frequency band in question should be avoided to prevent potential interference with the analysis. Therefore, in the subsequent discussion, the analysis frequency is set within the range of 20–2000 Hz to facilitate a more precise examination of the propagation and distribution patterns of road noise inside the building.

The acquired raw data were subsequently processed analytically using a custom script developed in MATLAB [44]. This script was specifically tailored to calculate the power spectral density of the signals, focusing on stable intervals coinciding with traffic transit. The data collection was configured for 20 min measurement samples as road noise represents a relatively consistent sound source.

3. Results and Discussion

3.1. Total Noise Due to Traffic Road

3.1.1. Comparison of SPLs at Different Times

Figure 6 shows the A-weight level of the platform on the third floor at different times. The equivalent continuous A sound level in different time periods, as depicted in Figure 6a, demonstrates a consistent trend across the entire frequency spectrum. Additionally, the average total sound pressure level of road noise illustrated in Figure 6b indicates that variations between different time periods do not exceed 2 dB. Based on the average total sound pressure level data, it is evident that road noise remains consistently close to 70 dB throughout the day. This underscores the significant impact of road noise on the community and emphasizes the necessity of implementing noise reduction measures to enhance residents' living environment. On the contrary, it also demonstrates that road noise serves as a relatively consistent source of noise within the building, thus facilitating favorable conditions for comparative analysis and research.

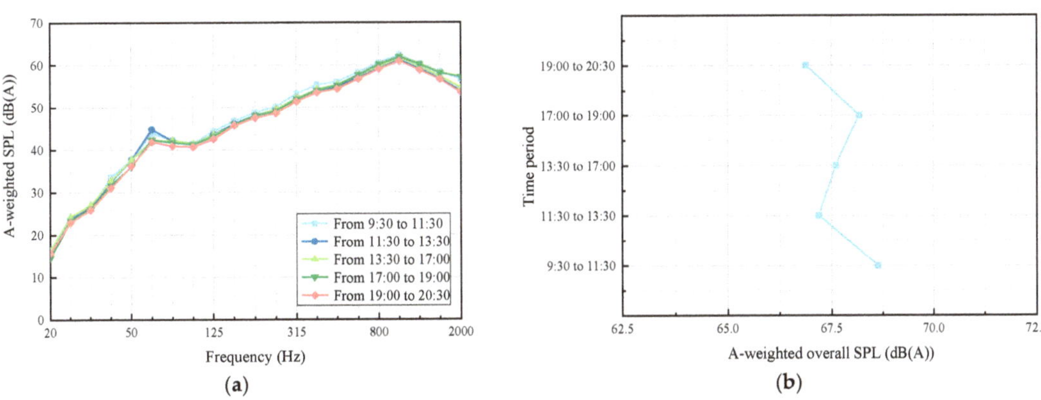

Figure 6. Comparison of sound pressure level at different times. (**a**) A-weighted SPL; (**b**) A-weighted overall SPL.

3.1.2. Comparison of the Noise of Different Distances from the Traffic Road

The comparison of sound pressure levels (SPLs) at varying distances from the traffic road is depicted in Figure 7. The measurements were conducted continuously for one hour under these conditions. To facilitate a clearer comparison, the average SPL in the frequency domain was calculated. For overall SPLs, road noise samples were taken at 20 min intervals to comply with the Chinese code HJ 640-2012 [45]. Figure 7a illustrates that the equivalent continuous A-weighted sound level of road noise initially increases and then decreases with frequency. The peak SPL occurs at 1000 Hz, which is characteristic of air noise. Road noise demonstrates a consistent trend across all frequency bands, decreasing in intensity as the distance from the road increases. The overall SPL of the noise samples in Figure 7b indicate a noticeable decrease in road noise as the distance from the road increases. In the process of sound propagation, as the distance increases, its energy will disperse according to the radiation damping, resulting in a gradual reduction of the sound pressure level. This attenuation is one of the fundamental characteristics of sound propagation.

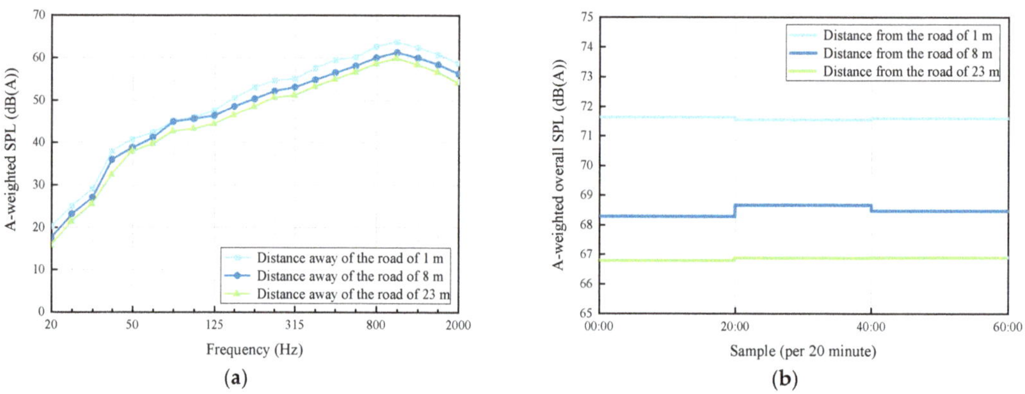

Figure 7. Comparison of the sound pressure levels with different distances from the traffic road. (**a**) A-weighted SPL; (**b**) A-weighted overall SPL.

An interesting phenomenon is observed: SPL decreases non-linearly with increasing distance, possibly due to complex urban noise fields where road noise has a significant impact in close proximity but diminishes during propagation. In contrast, multiple sources contribute to non-linear decreases in noise levels with distance from the road within far-field conditions, which contrasts with the findings of previous studies' noise maps [33,34].

3.1.3. Comparison of SPLs of Indoors and Outdoors

The transmission of road traffic noise from outdoors to indoors primarily involves three main transmission paths: (a) Direct transmission: Sound waves travel directly through air mediums toward exterior walls, windows, and other openings before entering indoor spaces via walls or glass; (b) Penetrate transmission: The sound penetrates through the windows and walls, filling the room; (c) Structural conduction: traffic-induced vibration waves transmission through structures like walls and floors to generate the structure-borne noise. The three transmission paths result in a complex sound environment of the indoor room. Therefore, in the measurement, the sealing of the measurement room should be ensured as far as possible to reduce the influence of direct propagation on the experiment. In the analysis, the indoor sound characteristics are regarded as the acoustic environment containing both structural noise and penetrating noise. The comparison of sound pressure levels (SPLs) between indoor and outdoor environments is illustrated in Figure 8. As depicted in Figure 8a,c,e, the SPLs on the sixth, tenth, and fourteenth floors are higher outdoors than indoors across all frequency bands. This disparity can be attributed to the attenuation effect of window glass soundproofing on road noise propagation from outdoors to indoors, resulting in reduced indoor sound levels. Comparing the total noise pressure level of the samples within the same time period as shown in Figure 8b,d,f, it can be inferred that outdoor noise on the sixth floor exceeds indoor noise by approximately 15 dB, while outdoor noise on the tenth floor surpasses indoor noise by about 18 dB, and outdoor noise on the fourteenth floor exceeds indoor noise by roughly 17 dB. These findings clearly demonstrate that windows play a significant role in reducing indoor noise levels. The findings also indicate that indoor noise exhibits relatively consistent levels, while outdoor noise varies noticeably across different floors. However, once it transmits into the building, the sound pressure level tends to stabilize at around 55–57 dB.

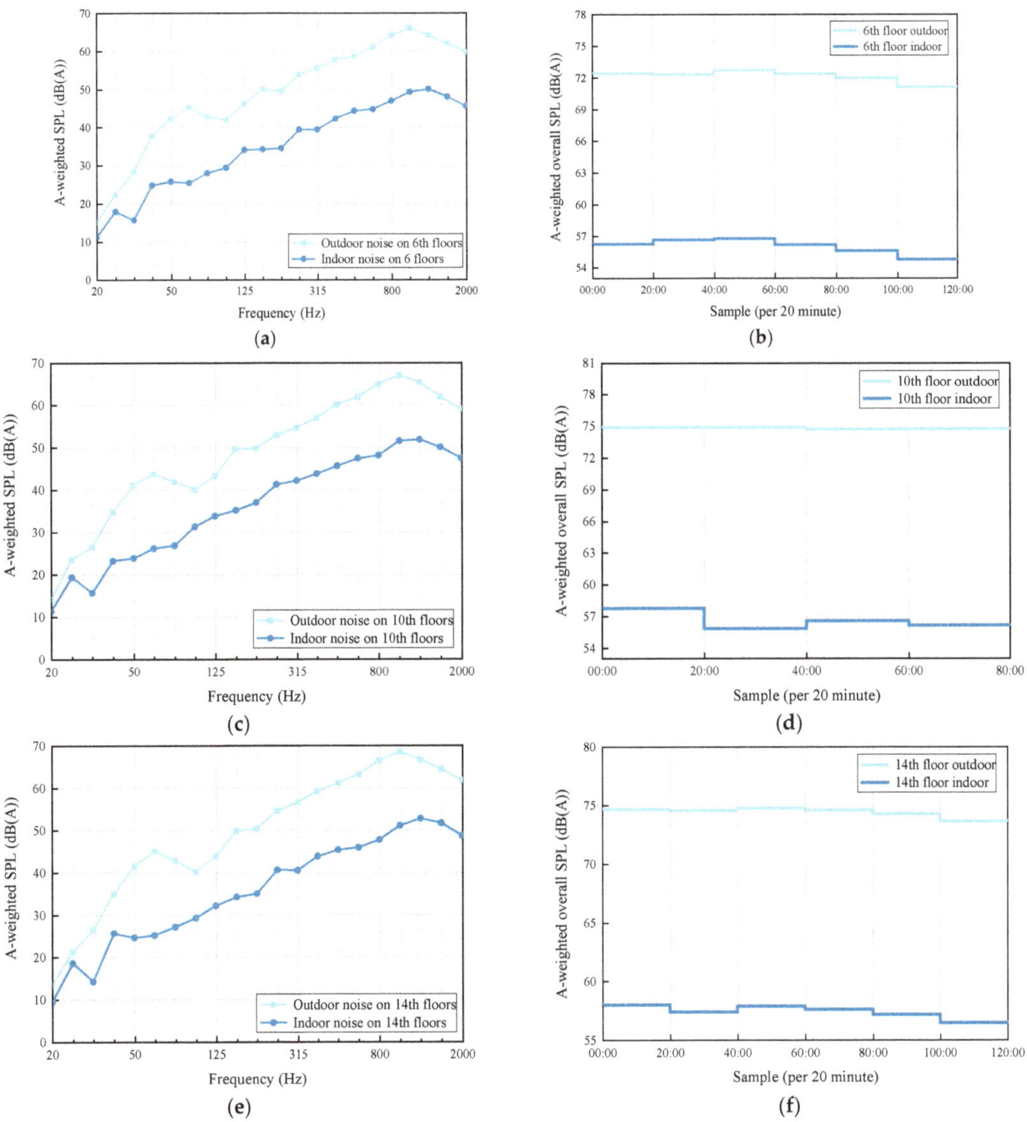

Figure 8. Comparison of the SPLs of indoors and outdoors. A-weighted SPL (**a**) on 6th floor; (**c**) on 10th floor; (**e**) on 14th floor. A-weighted overall SPL (**b**) on 6th floor; (**d**) on 10th floor; (**f**) on 14th floor.

3.1.4. Comparison of SPLs on Different Building Floors

The comparison of sound pressure levels (SPLs) at different floor heights within a building is depicted in Figure 9. As illustrated in Figure 9a, when comparing outdoor noise levels on the third and sixth floors, while maintaining identical distances from the noise receiving points to the noise source on both floors, there is a consistent similarity observed in the low-frequency bands. However, there is a notable reduction in high-frequency noise. This can be attributed to the effectiveness of sound barriers in attenuating mid-to-high-frequency noise due to their susceptibility to physical obstruction and absorption, whereas their insulation effect on low-frequency noise is generally less pronounced owing to longer wavelengths that make them more prone to bypassing the barrier [46]. With increasing floor height, there is an initial increase followed by a decrease in outdoor road noise across the entire frequency range. Similarly, the analysis of average total sound pressure level data depicted in Figure 9b shows a similar propagation pattern with the highest noise occurring on the middle floors.

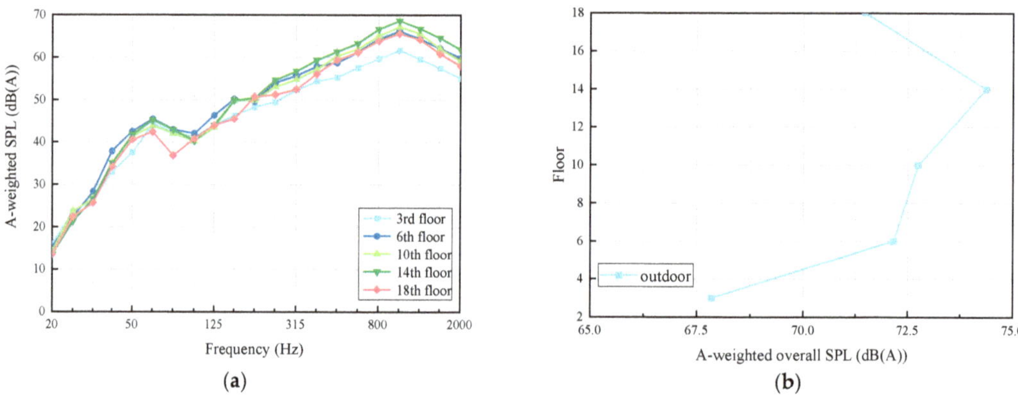

Figure 9. Comparison of the outdoor noise with the different building floor heights; (**a**) A-weighted SPL; (**b**) Overall A-weighted SPL.

The variation in the reduction effect of the sound barrier with building floor height is attributed to the attenuation of noise generated on viaducts, as illustrated in Figure 10. The noise undergoes attenuation through two mechanisms during propagation: firstly, a portion of the noise is blocked by the sound barrier to reduce its level within the building; secondly, air damping contributes to noise attenuation. However, as the building floor height increases, the impact of the sound barrier diminishes because road noise on viaducts can directly reach observation points in buildings without passing through the sound barrier. Consequently, within a specific range, outdoor sound pressure levels inside buildings increase with rising floor heights. Nevertheless, once a certain threshold is reached, the effectiveness of the sound barrier decreases and noise is mainly attenuated via air transmission. Owing to these combined effects of attenuation modes, outdoor noise initially rises and subsequently falls with increasing floor height.

It is evident from Figure 11a that indoor road noise increases across different floors throughout the entire frequency range as floor height rises. This can be attributed to the consistent propagation pattern of outdoor road noise, with indoor detection points not exceeding a certain area above ground level, thus only reflecting an upward trend in noise levels. Furthermore, upon analyzing the average total sound pressure level data in Figure 11b, it becomes apparent that the rate of increase in noise levels gradually diminishes with ascending floor height. As higher floors are more susceptible to distant noise sources, these noises have a relatively minor impact due to their distance, resulting in a slowdown in the growth rate of indoor noise. Furthermore, the attenuation effect of soundproof windows exhibits frequency-dependent characteristics [47], with a tendency to become more pronounced as the frequency increases. The traffic-induced air noise is concentrated in the high-frequency band of interest, while indoor noise experiences significant attenuation when passing through the window glass within that frequency range, resulting that the SPL of indoor noise tends to be consistent.

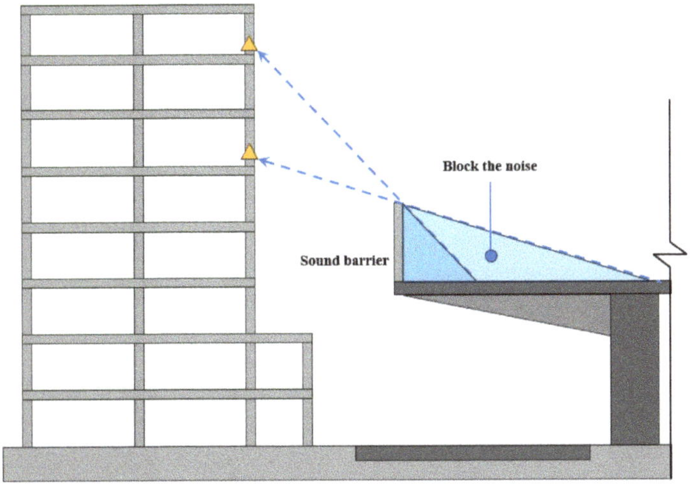

Figure 10. Schematic representation of the reduction effect of the sound barrier.

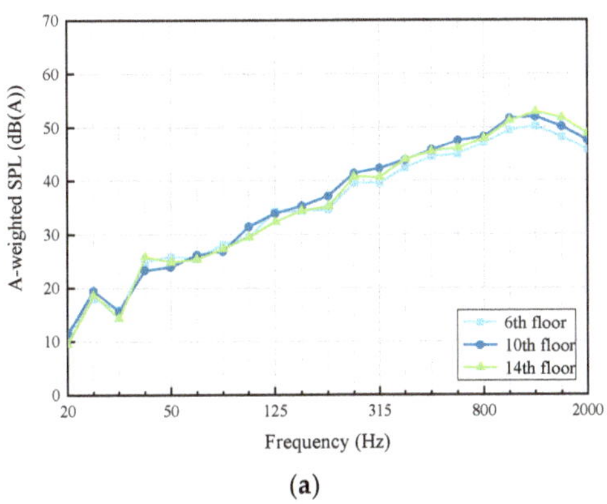

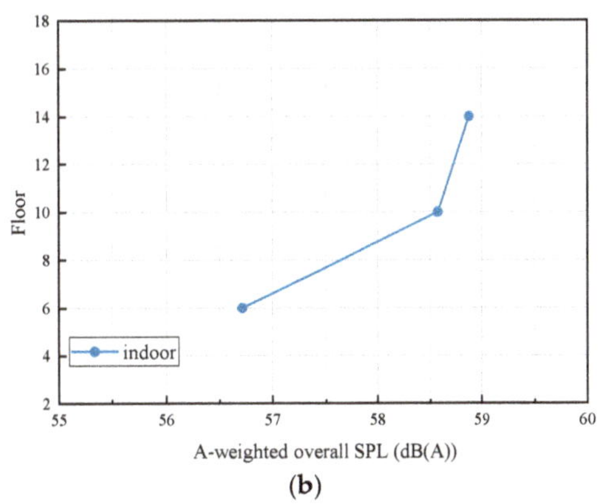

Figure 11. Comparison of indoor noise with different building floor heights; (**a**) A-weighted SPL; (**b**) Overall A-weighted SPL.

3.2. Structure-Borne Noise Due to Traffic Road

3.2.1. Structure-Borne Noise Generated by Bridge

The comparison of sound pressure levels (SPLs) from different types of roads is illustrated in Figure 12. The noise level at the 3-level platform is lower than that at the roadside, exhibiting a similar trend in the frequency domain. However, there is a significant peak at 63 Hz, which may be attributed to structure-borne noise generated by viaduct vibration [17]. Nevertheless, the structure-borne noise resulting from bridge vibration is considerably less than airborne noise caused by traffic operations and can be disregarded.

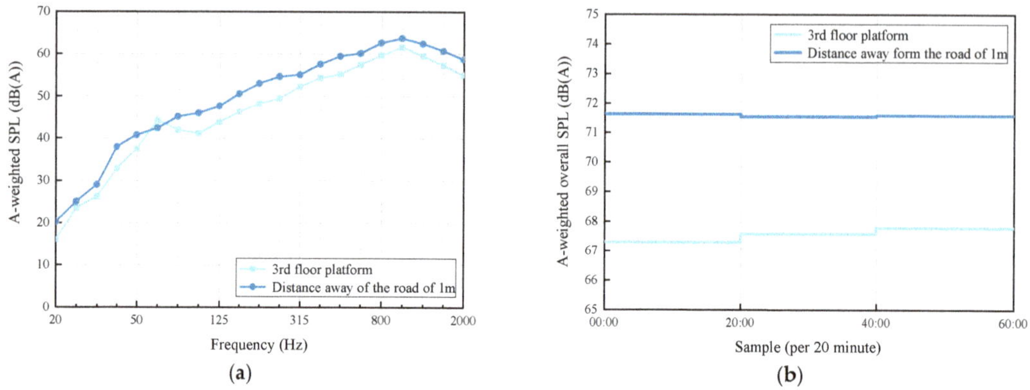

Figure 12. Comparison of the SPLs from different types of road; (**a**) A-weighted SPL; (**b**) Overall A-weighted SPL.

3.2.2. Structure-Borne Noise Generated by Building

A transfer function *TF* from outdoor to indoor is defined to describe the relationship between outdoor and indoor noise, and the following equation is expressed as

$$TF = L_{in}(\omega) - L_{out}(\omega) \tag{2}$$

where $L_{in}(\omega)$ and $L_{out}(\omega)$ represent indoor and outdoor noise, respectively, while ω represents the angular frequency. The A-weighted sound pressure levels of indoor and outdoor noise, as well as their transfer function on the third floor, are illustrated in Figure 13. As depicted in Figure 13a, outdoor noise exceeds indoor noise significantly due to concrete effectively blocking noise energy across all frequencies. Figure 13b displays the transfer function from outdoor noise to indoor noise, revealing two peaks at 25 and 100 Hz possibly attributed to structure-borne noise generated by the building. The vibration induced by traffic transmits into the building and transforms into structure-borne noise, resulting in these two peaks in the transfer function. However, the structure-borne noise generated by building vibration is much less than air traffic-induced noise. The sound energy generated by structures is minimal, which can be disregarded in the analysis or prediction.

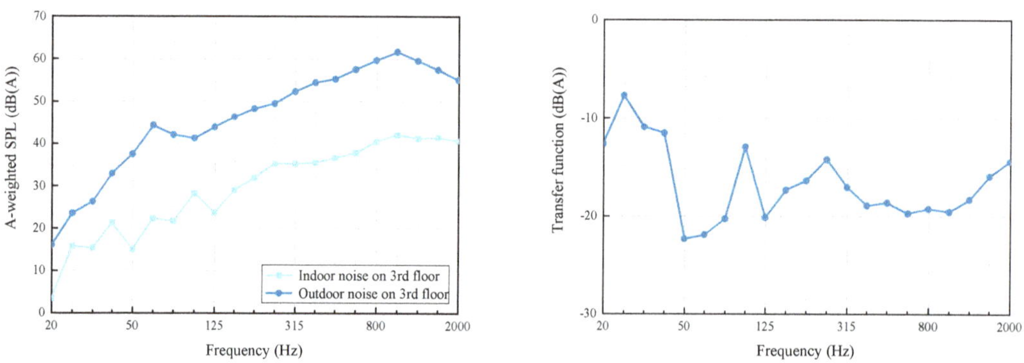

Figure 13. Comparison of indoor and outdoor noise and their transfer function on the third floor; (**a**) A-weighted SPL; (**b**) Transfer function.

4. Conclusions

This paper focus on a building located near both subgrade roads and viaducts to conduct a comprehensive parametric study on the building noise environment, encompassing traffic-induced airborne noise from subgrade and viaduct roads, as well as structure-borne noise from bridge and building vibrations. Road noise has been assessed based on the overall sound pressure level as well as in the frequency domain analysis. The following conclusions are as follows:

(1) Noise from the viaduct road undergoes attenuation through two mechanisms during propagation: firstly, a portion of the noise is blocked by the sound barrier to reduce its level within the building; secondly, air damping contributes to noise attenuation. As the building floor height increases, the impact of the sound barrier diminishes because road noise on viaducts can directly reach observation points in buildings without passing through the sound barrier. Once a certain height is reached, the effectiveness of the sound barrier decreases and noise is mainly attenuated via air transmission. Due to these combined effects of attenuation modes, outdoor noise initially rises and subsequently falls with increasing floor height;

(2) Due to the frequency-dependent characteristics of soundproof windows, the attenuation effect exhibits a tendency to become more pronounced as frequency increases. The traffic-induced air noise is concentrated in the high-frequency band of interest, while indoor noise experiences significant attenuation when passing through the window glass within that frequency range. As a result, the SPL of indoor noise tends to remain consistent;

(3) The non-linear decrease in sound pressure level (SPL) with increasing distance can be attributed to the complex urban noise environment. In close proximity, road noise has a significant impact but diminishes over propagation. However, in the far field, noise from multiple sources leads to a non-linear reduction in noise levels as distance from the road increases;

(4) In comparison to airborne noise, the sound energy generated by bridge and building structures is minimal, and can be disregarded in the analysis or prediction;

(5) The findings of this study can enhance engineers' understanding of traffic-induced air and structure-borne noise, as well as provide a fundamental dataset for urban road layout design. In future research, the model will be developed based on acoustic environment data from the site of measurement, with the aim of establishing an indoor acoustic environment model under the influence of complex sound sources generated by urban road traffic.

Author Contributions: Conceptualization, J.C., X.L. and C.Z.; data curation, X.L. and B.Z.; formal analysis, T.W., D.W. and L.H.; funding acquisition, J.C.; investigation, J.C., X.L. and B.Z.; methodology, C.Z. and L.H.; project administration, J.C.; software, B.Z., T.W. and L.H.; supervision, X.L. and C.Z.; validation, T.W. and D.W.; writing—original draft, J.C., X.L., B.Z. and L.H.; writing—review and editing T.W., D.W. and C.Z. All authors have read and agreed to the published version of the manuscript.

Funding: This research was funded by the project of the Ministry of Housing and Urban-Rural Development of the People's Republic of China, grant number 2022-K-153.

Data Availability Statement: The original contributions presented in the study are included in the article material; further inquiries can be directed to the corresponding author.

Conflicts of Interest: Authors Jialiang Chen, Teng Wang and Dongyang Wang were employed by the company Beijing Oriental Yuhong Waterproof Technology Co., Ltd. Author Jialiang Chen was employed by the company Beiijao Zane Rail Technology (Beijing) Co., Ltd. Author Lingshan He was employed by the company Guangzhou Urban Planning & Design Survey Research Institute Co., Ltd. The remaining authors declare that the research was conducted in the absence of any commercial or financial relationships that could be construed as a potential conflict of interest.

References

1. Szczepanik-Scislo, N.; Scislo, L. Dynamic real-time measurements and a comparison of gas and wood furnaces in a dual-fu heating system in order to evaluate the occupants' safety and indoor air quality. *Buildings* **2023**, *13*, 2125. [CrossRef]
2. Liang, L.; Gong, P. Urban and air pollution: A multi-city study of long-term effects of urban landscape patterns on air quali trends. *Sci. Rep.* **2020**, *10*, 18618. [CrossRef] [PubMed]
3. Scislo, L.; Szczepanik-Scislo, N. Near Real-Time Access Monitoring Based on IoT Dynamic Measurements of Indoor Air Polluta In Proceedings of the 2023 IEEE 12th International Conference on Intelligent Data Acquisition and Advanced Computing Systen Technology and Applications (IDAACS), Dortmund, Germany, 7–9 September 2023; Volume 1, pp. 729–734.
4. Miller, M.R.; Landrigan, P.J.; Arora, M.; Newby, D.E.; Münzel, T.; Kovacic, J.C. Water, Soil, Noise, and Light Pollution: JAC Focus Seminar, Part 2. *J. Am. Coll. Cardiol.* **2024**, *83*, 2308–2323. [CrossRef] [PubMed]
5. Serdar, O.; Erol, N.D.; Aydin, A.N. Effect of light and noise pollution on oxidative stress and proximate composition in *Dreisse polymorpha*. *Ecohydrol. Hydrobiol.* **2024**. Available online: https://www.sciencedirect.com/science/article/pii/S16423593240003 (accessed on 10 July 2024). [CrossRef]
6. Scislo, L.; Szczepanik-Scislo, N. Influence of mechanical ventilation and cooling systems on vibrations of high precision machin In *E3S Web of Conferences*; EDP Sciences: Les Ulis, France, 2019; Volume 100, p. 00080.
7. Hu, J.; Zou, C.; Liu, Q.; Li, X.; Tao, Z. Floor vibration predictions based on train-track-building coupling model. *J. Buil.* **2024**, 109340. [CrossRef]
8. Li, X.; Chen, Y.; Zou, C.; Wang, H.; Zheng, B.; Chen, J. Building structure-borne noise measurements and estimation due to tra operations in tunnel. *Sci. Total Environ.* **2024**, *926*, 172080. [CrossRef] [PubMed]
9. Ma, M.; Xu, L.; Liu, W.; Tan, X. Semi-analytical solution of a coupled tunnel-soil periodic model with a track slab under a movi train load. *Appl. Math. Model.* **2024**, *128*, 588–608. [CrossRef]
10. Das, C.P.; Swain, B.K.; Goswami, S.; Das, M. Prediction of traffic noise induced annoyance: A two-staged SEM-Artificial Neur Network approach. *Transp. Res. Part D Transp. Environ.* **2021**, *100*, 103055.
11. Öhrström, E.; Skånberg, A.; Svensson, H.; Gidlöf-Gunnarsson, A. Effects of road traffic noise and the benefit of access to quietne *J. Sound Vib.* **2006**, *295*, 40–59. [CrossRef]
12. Karpa, M.J.; Gopinath, B.; Beath, K.; Rochtchina, E.; Cumming, R.G.; Wang, J.J.; Mitchell, P. Associations between heari impairment and mortality risk in older persons: The Blue Mountains Hearing Study. *Ann. Epidemiol.* **2010**, *20*, 452–459. [CrossR [PubMed]
13. Muzet, A. Environmental noise, sleep and health. *Sleep Med. Rev.* **2007**, *11*, 135–142. [CrossRef] [PubMed]
14. World Health Organization. *Burden of Disease from Environmental Noise: Quantification of Healthy Life Years Lost in Europe*; Wor Health Organization: Geneva, Switzerland; Regional Office for Europe: Copenhagen, Denmark, 2011.
15. Babisch, W. Cardiovascular effects of noise. *Noise Health* **2011**, *13*, 201–204. [CrossRef] [PubMed]
16. Wang, H.; Wu, Z.; Wu, Z.; Hou, Q. Urban network noise control based on road grade optimization considering comprehensi traffic environment benefit. *J. Environ. Manag.* **2024**, *364*, 121451. [CrossRef] [PubMed]
17. Song, L.; Gao, K.; Liu, Q.; Liu, L.; Feng, Q. Study on the structure-borne noise of U-shaped girder bridges with fully-enclos sound barriers. *Appl. Acoust.* **2023**, *211*, 109497. [CrossRef]
18. Rapino, L.; Ripamonti, F.; Dallasta, S.; Baro, S.; Corradi, R. Synthesis of equivalent sources for tyre/road noise simulation a analysis of the vehicle influence on sound propagation. *Appl. Acoust.* **2024**, *216*, 109751. [CrossRef]
19. Hoever, C.; Tsotras, A.; Pallas, M.A.; Cesbron, J. Factors influencing tyre/road noise under torque. In *INTER-NOISE a NOISE-CON Congress and Conference Proceedings*; Institute of Noise Control Engineering: Wakefield, MA, USA, 2023; Volume 2 pp. 3683–3690.
20. Wang, X. (Ed.) *Automotive Tire Noise and Vibrations: Analysis, Measurement and Simulation*; Butterworth-Heinemann: Oxford, U 2020.
21. Yi, J.; Liu, X.; Shan, Y.; Dong, H. Characteristics of sound pressure in the tire cavity arising from acoustic cavity resonance excit by road roughness. *Appl. Acoust.* **2019**, *146*, 218–226. [CrossRef]
22. Huang, H.; Lim, T.C.; Wu, J.; Ding, W.; Pang, J. Multitarget prediction and optimization of pure electric vehicle tire/road airbor noise sound quality based on a knowledge-and data-driven method. *Mech. Syst. Signal Process.* **2023**, *197*, 110361. [CrossRef]
23. Puckeridge, H.; Braunstein, T.; Weber, C. Comparison of rail noise modelling with CadnaA & SoundPLAN. In Proceedings of t ACOUSTICS 2019, Cape Schanck, VIC, Australia, 10–13 November 2019; Volume 10.
24. Khan, J.; Ketzel, M.; Jensen, S.S.; Gulliver, J.; Thysell, E.; Hertel, O. Comparison of Road Traffic Noise prediction mod CNOSSOS-EU, Nord2000 and TRANEX. *Environ. Pollut.* **2021**, *270*, 116240. [CrossRef] [PubMed]
25. Hasmaden, F.; Zorer Gedik, G.; Yüğrük Akdağ, N. An approach to the design of photovoltaic noise barriers and a case stu from Istanbul, Turkey. *Environ. Sci. Pollut. Res.* **2022**, *29*, 33609–33626. [CrossRef] [PubMed]
26. Ochmann, M. Boundary Element Acoustics Fundamentals and Computer Codes. *J. Acoust. Soc. Am.* **2002**, *111*, 1507–1508.
27. Li, Q.; Xu, Y.L.; Wu, D.J. Concrete bridge-borne low-frequency noise simulation based on train–track–bridge dynamic interacti *J. Sound Vib.* **2012**, *331*, 2457–2470. [CrossRef]
28. Gu, Y.W.; Nie, X.; Yan, A.G.; Zeng, J.H.; Liu, Y.F.; Jiang, Y.X. Experimental and numerical study on vibration and structure-bor noise of high-speed railway composite bridge. *Appl. Acoust.* **2022**, *192*, 108757. [CrossRef]

9. Sadeghi, J.; Vasheghani, M. Improvement of current codes in design of concrete frame buildings: Incorporating train-induced structure borne noise. *J. Build. Eng.* **2022**, *58*, 104955. [CrossRef]

0. Chouksey, A.K.; Kumar, B.; Parida, M.; Pandey, A.D.; Verma, G. Measurement and prediction of road traffic noise at different floor levels of buildings in a mid-sized Indian city. *J. Build. Eng.* **2024**, *92*, 109711. [CrossRef]

1. Huang, B.; Pan, Z.; Liu, Z.; Hou, G.; Yang, H. Acoustic amenity analysis for high-rise building along urban expressway: Modeling traffic noise vertical propagation using neural networks. *Transp. Res. Part D Transp. Environ.* **2017**, *53*, 63–77. [CrossRef]

2. Caldwell, N.J. Study of Traffic Noise bevels at Various Heights of a 39-Story Building. Available online: https://www.nrs.fs.usda.gov/pubs/gtr/gtr_ne25/gtr_ne25_195.pdf (accessed on 10 July 2024).

3. Yang, W.; He, J.; He, C.; Cai, M. Evaluation of urban traffic noise pollution based on noise maps. *Transp. Res. Part D Transp. Environ.* **2020**, *87*, 102516. [CrossRef]

4. Campello-Vicente, H.; Peral-Orts, R.; Campillo-Davo, N.; Velasco-Sanchez, E. The effect of electric vehicles on urban noise maps. *Appl. Acoust.* **2017**, *116*, 59–64. [CrossRef]

5. Wu, J.; Zou, C.; He, S.; Sun, X.; Wang, X.; Yan, Q. Traffic noise exposure of high-rise residential buildings in urban area. *Environ. Sci. Pollut. Res.* **2019**, *26*, 8502–8515. [CrossRef] [PubMed]

6. Zou, C.; Zhu, R.; Tao, Z.; Ouyang, D.; Chen, Y. Evaluation of building construction-induced noise and vibration impact on residents. *Sustainability* **2020**, *12*, 1579. [CrossRef]

7. Balasbaneh, A.T.; Yeoh, D.; Abidin, A.R.Z. Life cycle sustainability assessment of window renovations in schools against noise pollution in tropical climates. *J. Build. Eng.* **2020**, *32*, 101784. [CrossRef]

8. Yang, H.S.; Cho, H.M.; Kim, M.J. On-site measurements for noise reduction through open windows of classrooms with different building dispositions. *Appl. Acoust.* **2018**, *139*, 165–173. [CrossRef]

9. Li, X.; Tang, S.K.; Yim, S.Y.; Lee, R.Y.; Hung, T. Noise reduction of plenum windows on the façade of a high-rise residential building next to heavy road traffic. *Build. Environ.* **2020**, *186*, 107353. [CrossRef]

0. Tan, J.K.A.; Lau, S.K. Experimental study of active noise control for a full-scale plenum window in a domestic apartment. *Appl. Acoust.* **2024**, *224*, 110120. [CrossRef]

1. Lou, K.; Xiao, P.; Kang, A.; Wu, Z.; Dong, X. Effects of asphalt pavement characteristics on traffic noise reduction in different frequencies. *Transp. Res. Part D Transp. Environ.* **2022**, *106*, 103259. [CrossRef]

2. Xu, X.; Cai, J.; Yu, N.; Yang, Y.; Li, X. Effect of loudness and spectral centroid on the music masking of low frequency noise from road traffic. *Appl. Acoust.* **2020**, *166*, 107343. [CrossRef]

3. Chang, T.Y.; Liang, C.H.; Wu, C.F.; Chang, L.T. Application of land-use regression models to estimate sound pressure levels and frequency components of road traffic noise in Taichung, Taiwan. *Environ. Int.* **2019**, *131*, 104959. [CrossRef]

4. Higham, D.J.; Higham, N.J. *MATLAB Guide*; Society for Industrial and Applied Mathematics: Philadelphia, PA, USA, 2016.

5. *HJ 640-2012*; Technical Specifications for Environmental Noise Monitoring Routine Monitoring for Urban Environmental Noise. Chinese code of Ecology Environment: Beijing, China, 2018. (In Chinese)

6. Nieves, J.G.; Picó, R.; Del Rey, R.; Alba, J.; Redondo, J. Assessment of the sound reduction index provided by noise barriers with low sound insulation. *Appl. Acoust.* **2024**, *220*, 109967. [CrossRef]

7. Du, L.; Lau, S.K.; Lee, S.E.; Danzer, M.K. Experimental study on noise reduction and ventilation performances of sound-proofed ventilation window. *Build. Environ.* **2020**, *181*, 107105. [CrossRef]

buildings

MDPI

Article

Train-Induced Vibration and Structure-Borne Noise Measurement and Prediction of Low-Rise Building

Jialiang Chen [1,2,3], Sen Hou [3], Bokai Zheng [4], Xuming Li [4,*], Fangling Peng [1], Yingying Wang [1] and Junjie Chen [3]

1 Beijing Oriental Yuhong Waterproof Technology Co., Ltd., Beijing 101111, China
2 School of Civil Engineering, Tsinghua University, Beijing 100084, China
3 Beiijao Zane Rail Technology (Beijing) Co., Ltd., Beijing 101111, China
4 School of Civil and Transportation, Guangdong University of Technology, Guangzhou 510006, China; 2112309102@mail2.gdut.edu.cn
* Correspondence: lixuming0216@163.com

Abstract: The advancement of urban rail transit is increasingly confronted with environmental challenges related to vibration and noise. To investigate the critical issues surrounding vibration propagation and the generation of structure-borne noise, a two-story frame building was selected for on-site measurements of both vibration and its induced structure-borne noise. The collected data were analyzed in both the time and frequency domains to explore the correlation between these phenomena, leading to the proposal of a hybrid prediction method for structural noise that was subsequently compared with measured results. The findings indicate that the excitation of structure-borne noise produces significant waveforms within sound signals. The characteristic frequency of the structure-borne noise is 25–80 Hz, as well as that of the train-induced vibration. Furthermore, there exists a positive correlation between structural vibration and structure-borne noise, whereby increased levels of vibration correspond to more pronounced structure-borne noise; additionally, indoor distribution patterns of structure-borne noise are non-uniform, with corner wall areas exhibiting greater intensity than central room locations. Finally, a hybrid prediction methodology that is both semi-analytical and semi-empirical is introduced. The approach derives dynamic response predictions of the structure through analytical solutions, subsequently estimating the secondary noise within the building's interior using a newly formulated empirical equation to facilitate rapid predictions regarding indoor building vibrations and structure-borne noises induced by subway train operations.

Keywords: environmental vibration; structure-borne noise; on-site measurement; vibration prediction; noise prediction

1. Introduction

In recent years, due to population growth and increased transportation demand, rail transit systems have emerged as the primary mode of transportation. Subways, trams, high-speed rails, and other modes of transport offer people swift and convenient travel options, effectively mitigating traffic congestion. However, this has also resulted in closer proximity between rail transit and buildings, leading to vibrations and structure-borne noise from train operations that impact the daily lives, education, and scientific activities of nearby residents [1,2]. In the context of train-induced vibrations, the auditory perception of rumbling caused by metro trains is more prominent compared to vibrations. However, the structure-borne noise generated by metro trains is nearly impossible to eliminate, unlike other types of environmental noise due to long-term transportation demand [3].

When a subway operates, the contact between the wheels and rails induces vibrations [4], which then propagate through the soil to the ground, leading to building vibration responses and structure-borne noise.

For environmental vibration, some researchers have conducted extensive studies on the subject. Hu et al. [5] proposed a model that integrates the train, track structure, and building structure into a unified system for trains running on building floors, which can be utilized to predict the base vibrations of buildings and further predict the floor vibration. Auersch [6] proposed a simple and fast prediction of train-induced track forces and ground and building vibrations in which a transfer matrix is utilized for characterizing the dynamic response of individual subsystems as vibration propagates. Sanayei et al. [7] proposed a one-dimensional impedance model in which columns are modeled as rods to account for axial motions only and floor slabs are treated as infinitely long mass-driven points in order to predict floor vibrations caused by trains passing through buildings, and Zou further improved the model and expanded it into two-dimensional [8] and three-dimensional [9] impedance models. Papadopoulos et al. [10] considered how the uncertainty of subsoil conditions could affect the building response by means of Monte Carlo. Edirisinghe et al. [11] modeled a three-dimensional numerical model based on a pipe-in-pipe model that accounts for the through-soil coupling between a tunnel and a piled foundation. Li et al. [12] studied the effect of prolonged train-induced vibrations with different frequencies on loess disintegration, which revealed the macro–micro association of disintegration properties after train-induced vibration.

Compared to the study of train-induced vibration, research on building noise commenced relatively late. Vehicle-induced building structure noise encompasses both air noise and structure-borne noise. Air noise refers to vibrational noise caused by solid structure contact and pneumatic noise induced by gas fluid movement during train operation, primarily occurring at medium and high frequencies [13,14], especially the air noise generated during train operation on ground and elevated lines. Structure-borne noise, which refers to the noise generated by vehicle-induced vibrations propagating through the building structure, is predominantly of medium and low frequency [3,15]. Trains operating on various types of tracks can induce structure-borne noise in conjunction with vehicle-induced vibration, and their acoustic and vibration characteristics can be considered as a one-way coupling [16], where the influence of internal sound pressure on the structure may be disregarded.

In the numerical calculation methods for structural noise, finite elements and boundary elements are commonly employed discrete methods, while the statistical energy method is a frequently used energy method. The statistical energy method demonstrates high computational accuracy and efficiency in dealing with medium- and high-frequency noise [17,18]; however, it is not so suitable for predicting low-frequency noise [19]. Nagy et al. [16] proposed an improved method for predicting indoor structure-borne noise based on global Rayleigh reflectivity and the direct boundary element method. Fiala et al. [20] applied the spectral finite element method to analyze vibration propagation and structure-borne noise in buildings' foundations. Colaço et al. [21] enhanced the calculation efficiency by combining a 2.5-dimensional FEM with the method of fundamental solutions (MFS) model for structure-borne noise analysis. Although both the finite element method and boundary element method offer higher prediction accuracy for low-frequency structure-borne noise compared to empirical formula methods, they are hindered by their high computational cost when predicting indoor noise.

In the majority of application scenarios, the empirical formula method is widely used for noise prediction due to its simplicity and efficiency. Kurzweil et al. [22], Vér [23], and Melke [24] proposed or enhanced empirical prediction methods based on measured data, aiming to predict indoor noise by integrating floor vibration acceleration and relevant parameter correction factors. In these methods, the floor is simplified as the primary excitation source of sound induced by vehicle-induced vibration. Tao et al. [25] employed the empirical formulation in a low-rise over-track building to estimate the structure-borne noise. The findings revealed a notable increase in error at higher frequencies, peaking at almost 10 dB within the 50–80 Hz range. Li et al. [3] combined measurement and deep-learning-based approaches [26–28] to consider the transmission between building vibration and structure-borne noise and compared the predicted results with different empirical models [22–24,29–33]. The findings indicated that various empirical models exhibit distinct fitting effects across different frequency bands.

At present, the majority of research on vibration and structure-borne noise utilizes numerical analysis or empirical formula methods. However, the former requires complex model construction, resulting in low prediction efficiency, while the latter's formulas have a limited application range and insufficient accuracy. The current study involves on-site measurements of vibrations in a two-story frame building and the corresponding structure-borne noise generated by subway trains, with the objective of elucidating the propagation characteristics and response patterns of both vibrations and structural noise, as well as their interrelation. A hybrid prediction methodology that is both semi-analytical and semi-empirical is introduced. The approach derives the dynamic response predictions of the structure through analytical solutions, subsequently estimating the secondary noise within the building's interior using a newly formulated empirical equation. It is important to highlight that this novel empirical formula builds upon existing predictive models and prior research findings. The fundamental component of this empirical formula is the vibration acceleration level, which distinguishes it from previous methodologies that utilized varying reference speeds for vibration velocity levels, thereby establishing a more standardized foundational term.

The contribution of this paper is the proposed fast, hybrid, and user-friendly prediction method utilizing fundamental terms of vibration acceleration levels, which eliminates the need for complex index conversions during the prediction process. This method demonstrates acceptable accuracy and efficiency, with a straightforward and clear formulation.

2. Measurement

2.1. Measured Site Descriptions

A two-story frame structure building located adjacent to a subway line was chosen for testing. Due to the operation of A-type subway cars (6-car organization), the building is positioned approximately 40 m horizontally from the subway tunnel, which has a buried depth of 17 m. The spatial relationship between the building and the tunnel is depicted in Figure 1.

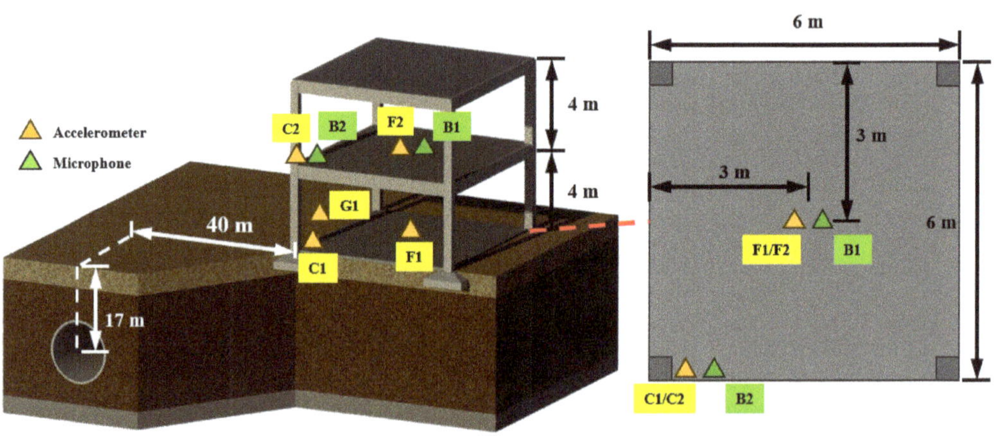

Figure 1. Schematic representation of measurement program and spatial relationship between tunnel and building.

As per the findings of the geological exploration report, the soil dynamic parameters are detailed in Table 1.

Table 1. Dynamic parameters of soils.

Soil Name	Thickness (m)	Density (kg/m^3)	Poisson's Ratio	Young's Modulus (MPa)
Plain fill	2.1	1600	0.45	130
Clay	1.3	1840	0.46	130
Silty soil	8.4	1750	0.47	300
Silty clay	10.6	1930	0.46	380

The building foundation is constructed as an independent foundation with a size of 1.2 × 1.2 m. The structural column size is 0.4 × 0.4 m, and the building has a height of 8 m with 2 layers. The dimensions and dynamic parameters of the building structure are detailed in Tables 2–4, where h, t, w, l, E, and ρ represent the height, thickness, width, length, elastic modulus, and density of the corresponding structural elements. Additionally, the damping ratio and Poisson's ratio of the reinforced concrete structural elements are specified as 0.02 and 0.2, respectively.

Table 2. Parameters of axial load-bearing structure of building.

Floor Number	Load-Bearing Structure	h (m)	t (m)	w (m)	E (Gpa)	P (kg/m^3)
1–2	Structural column	4	0.4	0.4	32.5	2500

Table 3. Parameters of building floor structure.

Floor Number	t (m)	E (Gpa)	l (m)	w (m)	ρ (kg/m^3)
1–3	0.1	30	6	6	2500

Table 4. Parameters of building foundation structure.

Foundation Type	l (m)	w (m)	E (Gpa)	ρ (kg/m^3)
Rigid foundation	1.2	1.2	32.5	2500

In order to study the characteristics of structural noise caused by train operation, microphones were installed in the center of the floor and next to the structural columns (represented by green triangles) on the second floor. Accordingly, accelerometers (represented by yellow triangles) were placed below microphones in their respective locations on the floor as well as at the same location on the 1st floor and outside the building to obtain vibrations.

Figure 1 illustrates the measuring point locations. In a frame building structure, the structural components can be categorized into propagation and response components based on their distinct functions. Among these, columns serve as vertical vibration propagation components due to their significant vertical stiffness, while beams facilitate horizontal vibration transmission. Floor slabs exhibit a rich variety of vibration modes and demonstrate significant compliance, making them serve as vertical vibration response components. Vibrations propagate from the foundation column's base to the structures and are transformed into curved waves at the plate–beam–column junctions, where they propagate, respond, and dissipate towards the far side of the floor. Consequently, considering the response characteristics of various building components and the transmission dynamics of vehicle-induced vibrations within the structure, four measurement points were established: at the column bases and floor center on both the first and second floors. Concurrently, an outdoor ground vibration measurement point was set to facilitate the analysis of train-induced ground vibrations originating from both nearby and distant tunnels, serving as a reference for assessing the energy associated with these vibrations. Testing occurred on a working day with minimal traffic and pedestrian flow, under sunny weather conditions that have little impact on climate.

2.2. Measurement Program

Instrumentation and Signal Processing

The instrumentation utilized for measurement is shown in Figure 2, encompassing a Rion UC-59 microphone (RION Co., Ltd., Tokyo, Japan) and JM 3873 wireless data acquisition system (Jing Ming Technology Co., Ltd., Yangzhou, China). The microphone was installed on the preamplifier to obtain a clear noise signal, and the windshield was connected with the head of the microphone to effectively mitigate wind noise interference during the recording process and enhance the clarity and purity of the recorded audio, thereby improving the overall recording quality. Finally, the instrumentation assembly was installed on the tripod, positioned at a height of 1.5 m above the floor surface. The accelerometer was positioned appropriately on the floor below the microphone. Both the microphone and accelerometer were equipped with time synchronizers and memory cards.

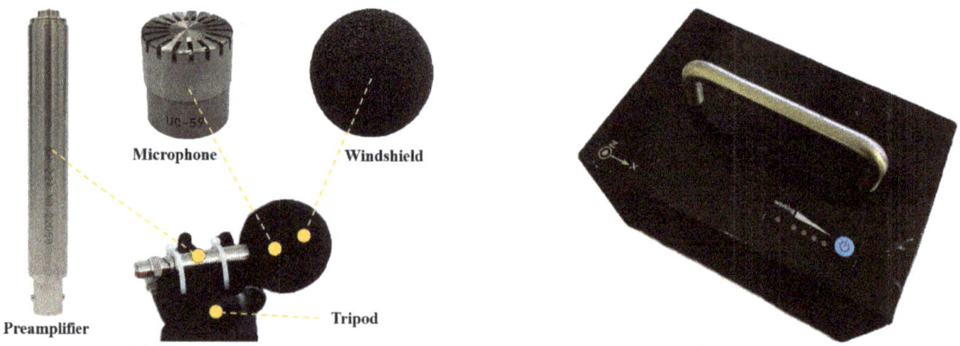

(**a**) Sound level meter (**b**) Accelerometer

Figure 2. Photographs of instrumentation utilized for measurement.

A flowchart that more clearly illustrates the measuring procedures is shown in Figure 3.

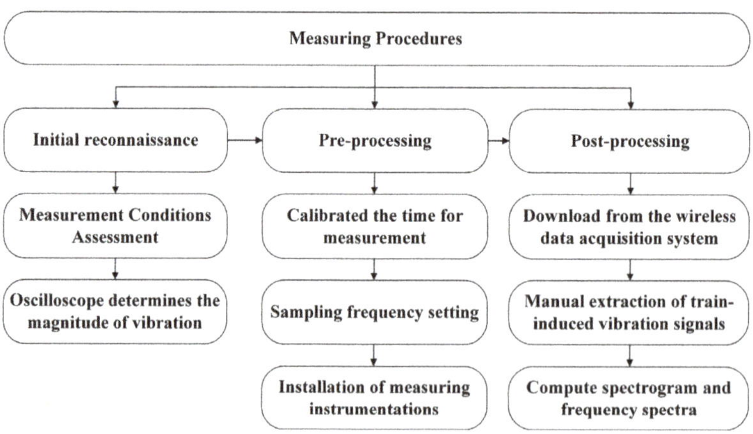

Figure 3. Flowchart of measuring procedures.

The measuring procedures can be categorized into initial reconnaissance, pre-processing, and post-processing.

(1) Initial reconnaissance

During the initial reconnaissance, the measurement site environment was assessed to ascertain its readiness for evaluation and to identify the specific locations of individual test points. Additionally, sensors were positioned on the ground for oscilloscope measurements to quantify the range of vibration acceleration influenced by train-induced vibrations, thereby minimizing potential errors.

(2) Pre-processing

(a) All instruments were calibrated based on the laptop's time prior to measurement initiation, ensuring simultaneous data recording.
(b) According to the Chinese code HJ 453–2018 [30], the dominant frequency range requirement for structure-borne noise analysis is 16–200 Hz. The sampling frequency for each sound level meter was set at 48 kHz, and it was chosen to facilitate noise analysis up to 24 kHz, despite such high frequencies not being typically necessary for noise analysis but sufficient for the common frequency of interest at 16–200 Hz for train-induced structure-borne noise. The sampling frequency was 512 Hz, which captured vibrations resulting from train passage and offered spectral information up to the Nyquist frequency of 256 Hz, allowing for analysis within the frequency range of 16–200 Hz.
(c) After setting the instrumentation and starting the measuring, train pass-by events were simultaneously documented at the metro station, including details of the specific tunnel from which each train originated.

(3) Post-processing

The original time domain signal file was downloaded from the wireless data acquisition system. The acquired raw data were subsequently processed analytically using a custom script. This script was specifically tailored to calculate the power spectral density of the signals, focusing on every train operation signal. Consequently, the signal captured by a sensor must be intercepted in accordance with the vibration waves produced and their spectrogram during the train pass-by events, as shown in Figure 4. Due to the variations in the speed of each train, the time intervals between each intercepted vibration signal differ; however, their spectrograms exhibit similar trends.

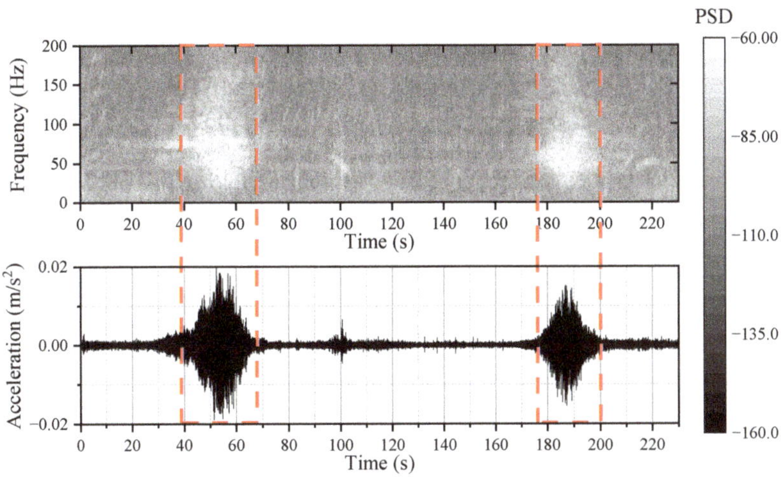

Figure 4. Time history and spectrogram interception of the train pass-by events.

3. Results

A total of 40 trips were collected, and time–frequency conversion processing was conducted. The signals from both the far tunnel and the near tunnel comprised 20 train pass-by events each, thereby providing ample data support for subsequent analysis.

3.1. Vibration in Time Domain

The effective train vibration signal recorded at the scene lasted approximately 14.5 s, with the train running at a speed of about 50 km/h. In Figure 5, the time history signal depicts the vibration acceleration of the outdoor ground measuring points and two-story structural column feet under tunnel vibration. The peak value of soil vibration acceleration on-site was approximately 0.028 m/s^2, while that of two-story structural column feet was about 0.016 m/s^2, indicating a significantly greater soil vibration compared to the building structure. This is because soil–structure coupling loss occurs when a vibration propagates from soil to structures [6]. The underlying cause of train-induced vibrations stems from wheel–rail interaction, with the most significant vibrations occurring at the interface between the wheelset and the rail. Consequently, as an entire train traverses a given point, multiple wheelsets generate a comprehensive vibration wave. This results in an uneven distribution of vibrational energy within the time domain, as shown in Figure 4. Specifically, when a wheelset is perfectly aligned perpendicular to the measurement point, it exhibits heightened vibration energy. During propagation, factors such as soil damping attenuation, soil–structure coupling loss, and building response lead to notable alterations in the frequency domain. However, in terms of time domain analysis, various forms of attenuation diminish the overall wave amplitude without substantially altering its waveform. We have added the corresponding description.

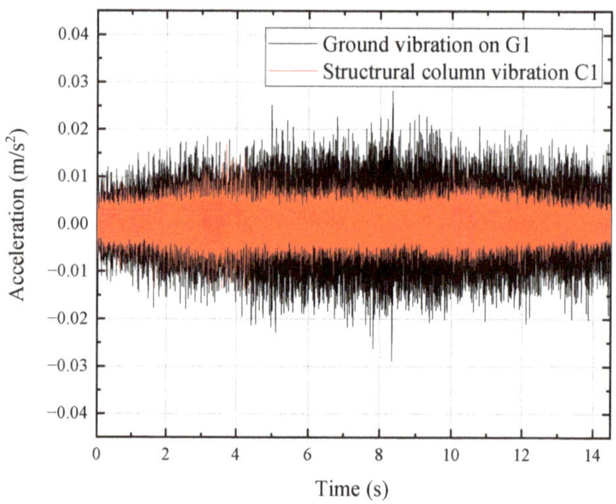

Figure 5. Time domain diagram of vibration acceleration of field soil and building structure.

Figure 6 depicts the time history curve of vibration acceleration at the outdoor ground measurement points under the influence of near-side and far-side tunnel vibrations. The peak soil vibration acceleration in the near-side tunnel measured approximately 0.028 m/s^2, while that in the far-side tunnel was about 0.022 m/s^2. Due to the similarity of the track bed, sleeper, and other tunnel structures, there is minimal distinction between the two when a train passes through. Furthermore, since the distance from the tunnel to the measurement point was approximately 40 m, in this case, the vibrations transmitted by both tunnels to the measurement point were both far-field vibrations, existing in the form of Rayleigh waves. Therefore, these two vibrations had already been subjected to a certain attenuation due to the soil damping effect at high frequencies and therefore have a similar order of magnitude. Nevertheless, it is noteworthy that the soil vibration levels were higher in the vicinity of the near tunnel compared to those near the far tunnel.

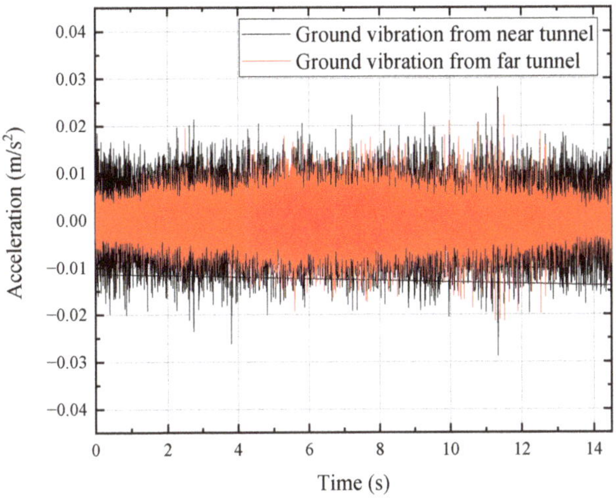

Figure 6. Time domain diagram of vibration acceleration of measuring point G1 on field soil.

3.2. Structure-Borne Noise in Time Domain

Figure 7 shows the time history curve of sound pressure at the central measuring point on the second floor inside the building near the tunnel for a train. As depicted in the figure, the peak sound pressure induced by vehicle-induced vibration from this particular vehicle was approximately 0.04 Pa, exhibiting more distinct acoustic characteristics resembling a vibration waveform. The peak resulting from wheel–rail interaction is clearly discernible.

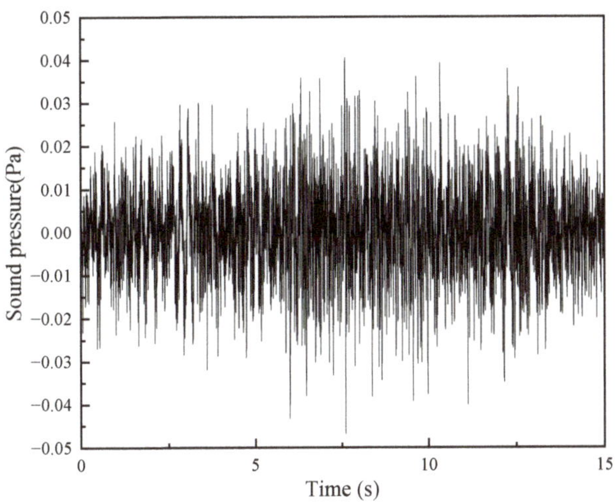

Figure 7. Time domain diagram of sound pressure of near tunnel on floor center.

3.3. Vibration in Frequency Domain

The measured acceleration was processed into vertical vibration acceleration levels through 1/3 octave analysis. Given that the predominant vibration frequencies of the subway fall within the range of 20–80 Hz, a frequency band spanning from 16 to 200 Hz was selected for assessment [3,30]. To enable a clear comparison of vibration propagation in soil and structures, the envelope diagram illustrates the measured ranges of vibration acceleration levels recorded during various train trips. Spectrum analysis was performed by selecting test results with higher magnitudes of vibration to compare acceleration levels near tunnels at different measurement points, as shown in Figure 8.

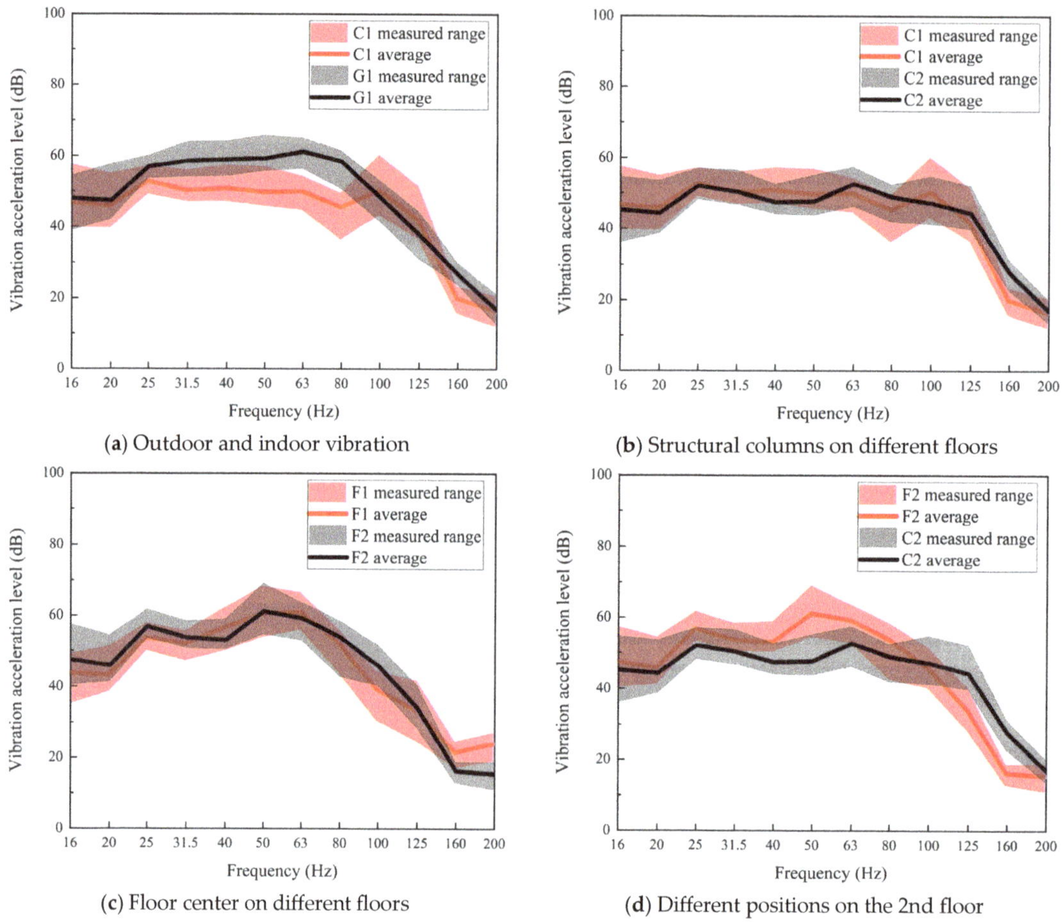

Figure 8. Comparison of vibration acceleration levels in frequency domain at different measuring points of the near tunnel.

Compared to the outdoor vibration G1 and indoor vibration C1, the energy loss of vibration propagating from the soil to the structure was more pronounced. The attenuation tended to be significant at medium and high frequencies, while low-frequency attenuation was less apparent. Vibration acceleration levels between 25 Hz and 100 Hz were notably reduced due to increased attenuation cycles experienced by medium- and high-frequency vibrations at the soil–structure interface, resulting in more pronounced losses. As vehicle-induced vibration primarily distributes its main energy within the range of 25–100 Hz, there were minimal energy components beyond 100 Hz. Consequently, vibration energy experienced only minor losses during propagation, leading to less noticeable attenuation above 100 Hz. An amplification occurred in the frequency band of 100–125 Hz due to elastic building resonance in anti-phase [31].

$$f_a = \frac{v_c}{4H_b} \approx 109 \text{ Hz} \quad (1)$$

Compared to C1 on the first floor and C2 on the second floor, the characteristic frequency of vibration acceleration in the structural columns remained consistent during the propagation process as the building height increased, with minimal variation. This is attributed to the relatively low floor height and the building's two-story configuration. When subjected to train-induced vibrations, the building system tended to vibrate as an entity, resulting in negligible disparity in vibration acceleration between the structural columns on the first and second floors.

Increasing the building height did not significantly affect the difference in vibration acceleration between F1 on the first floor and F2 on the second floor in the frequency domain. Throughout the propagation process, characteristic frequencies remained consistent, leading to overall excitation of the building system and minimal variation in vibration acceleration between floors because the building system tends to vibrate as an entity.

In the two-layer slab, when comparing F2 and C2, the vibration amplitude tended to be greater at the floor center than at the structural column within the frequency range of 20–100 Hz. However, above 100 Hz, it was observed that the vibration amplitude was greater at the structural column than at the floor center. This phenomenon can be attributed to the fact that the structural column serves as a primary conduit for vertical vibration propagation, while the floor primarily responds to and dissipates energy from vertical vibrations. Consequently, in the characteristic frequency band of 20–100 Hz, there is a particularly noticeable vibration response in the floor. In contrast, when examining frequencies that exceed those impacting structural columns, their response diminishes in prominence. Consequently, within the frequency range of 20 Hz to 100 Hz, floor vibrations were more pronounced than those of structural columns; however, this dynamic reversed at elevated frequencies. The vibration exhibited prominent peaks at 50 Hz, attributed to the modal characteristics of the floor, resulting in resonance at this frequency.

To facilitate comparisons between distant and nearby tunnels, spectra of the vibration acceleration levels from similar measurement points for both the near and far tunnels were chosen for comparison, as depicted in Figure 9. In comparison with the outdoor vibration G1 from the near and far tunnels, the amplitude of G1 tends to decrease across the entire frequency spectrum. Furthermore, the disparity at medium to high frequencies is significantly more pronounced than that observed at low frequencies. Due to the shorter period of high-frequency waves, they undergo more cycles than low-frequency waves, resulting in greater dissipation and thus faster attenuation of high-frequency vibrations compared to low-frequency ones. Furthermore, within the same frequency range, a greater distance leads to higher energy consumption, resulting in a larger vibration acceleration level at the near tunnel compared to the far tunnel at identical measurement points.

3.4. Structure-Borne Noise in Frequency Domain

The measured sound pressure was converted into 1/3 octave bands to derive the structural noise sound pressure level. Given that the frequency of structural noise aligns with vibration, the frequency band of 16–200 Hz was selected for analysis. To facilitate a clear comparison of noise propagation within the building, the envelope diagram presents the measured noise pressure level range for different train trips. Furthermore, spectrum analysis was conducted by selecting test results with higher sound pressure levels to obtain a comparison of near-tunnel noise pressure levels at different measuring points, as depicted in Figure 10.

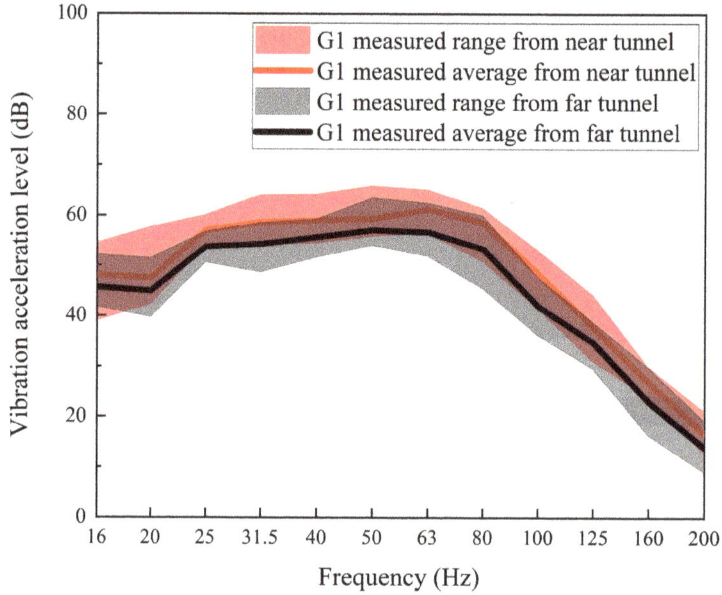

Figure 9. Comparison of vibration acceleration levels in frequency domain of near and far tunnels on field soil.

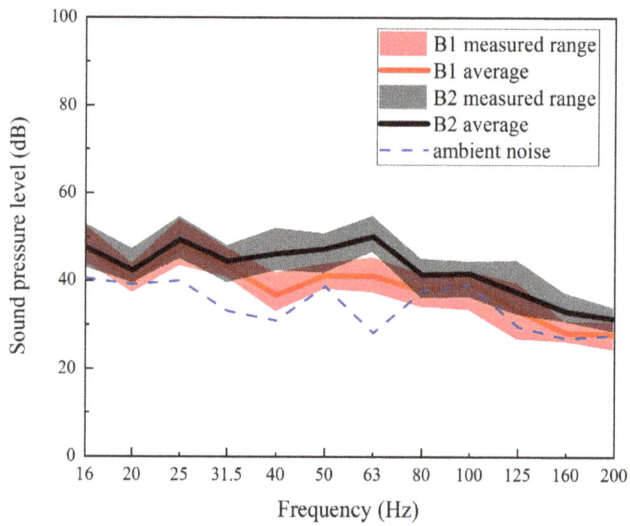

Figure 10. Comparison of noise sound pressure in frequency domain of different measuring points of near tunnel.

The comparison of structural noise between a room panel and its corner revealed that the corner exhibited significantly higher levels of structural noise, particularly above 31.5 Hz. This phenomenon can be attributed to the wall's position, which functions as both a radiant and reflective surface for sound, resulting in the highest observed sound pressure level [3]. In contrast, the center of the floor experiences less structural noise due to its distance from the walls, resulting in reduced reflection superposition effects.

To compare structural noise between the far and near tunnels, respective comparisons were made using the noise pressure level spectra from under the floor center B1 and corner B2 for both tunnels, as illustrated in Figure 11. By comparing the structural noise at the same measuring point in a room caused by trains running in different tunnels, it was observed that as the distance between the tunnel and the building increased, there was no significant difference in the strength of structural noise between the center and corner of the floor in the room within the frequency domain; rather, it tended to be consistent. This can be attributed to structural noise being generated by building vibrations transmitted through solids into rooms. Although this vibration gradually attenuates, it is influenced by factors such as building materials and structural design. Therefore, a slight increase in distance between tracks and buildings may not necessarily have a pronounced impact on structural noise within rooms.

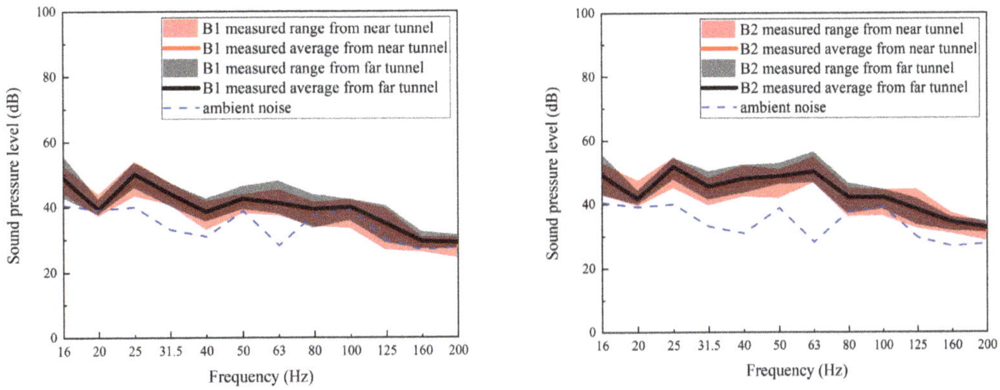

(**a**) Floor center (**b**) Structural column

Figure 11. Comparison of noise sound pressure level in frequency domain of same measurement points of near and far tunnels.

3.5. Transfer Function from Structural Vibration to Structure-Borne Noise

In order to investigate the correlation between structural vibration and structure-borne noise, the transfer difference (the disparity between the sound pressure level and vibration acceleration level) was utilized to quantify the relationship of transmission between structural vibration and structure-borne noise. As the distance of the tunnel does not impact the generation mechanism of structure-borne noise in the room, the transmission difference was computed for all test samples at the center of the floor of the two-story interior, as depicted in Figure 12.

The characteristic frequency band of vibration ranges from 25 Hz to 100 Hz, while that of noise spans from 25 Hz to 80 Hz; both waves share a similar characteristic frequency range. The sound pressure level and vibration acceleration level within this frequency band exhibit a specific decibel difference relationship, leading to a reduction in transmission ratio within the range of 25 Hz to 80 Hz. Frequencies outside this characteristic band become dominant environmental noise and environmental vibrations, causing an overall upward shift in curves above 80 Hz.

The figure reveals a spike at 25 Hz. This may be attributed to resonance phenomena occurring either when the frequency of the vibration wave matches the natural frequency of the building structure or when the sound wave frequency aligns with the acoustic natural frequency of the building room, resulting in an increased vibrational amplitude and consequently a spiking transmission difference. Therefore, there exists a positive correlation between structural vibrations and structure-borne noise; greater structural vibrations lead to more pronounced structure-borne noises. The room's natural frequency [32] can be calculated based on the dimensions of the room, and the minimal natural frequency can be obtained using the following equation:

$$f_r = \frac{c}{2L} \approx 28.4\text{ Hz} \tag{2}$$

where c is the speed of sound and L is the maximum length of the room. The standing wave resonance established at this frequency results in the emergence of a wave peak within the 25–31.5 Hz range of the transfer function.

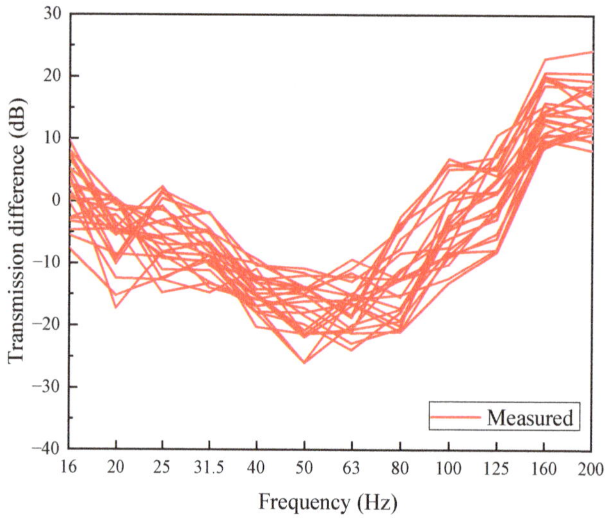

Figure 12. Measured transmission difference.

4. Hybrid Approach for Prediction

4.1. Structural Vibration Estimation Model

Given that the vibration propagation of low-floor buildings tends to affect the overall building movement, we considered the axial wave propagation within the structure. Consequently, a dynamic stiffness model for the structure was developed, treating each column as a single Bernoulli Euler beam and deriving the structural dynamic stiffness matrix for force and displacement:

$$[k_c] = \frac{\overline{E}_c A_c \beta_c}{\sin(\beta_c l_c)} \begin{bmatrix} \cos(\beta_c l_c) & -1 \\ -1 & \cos(\beta_c l_c) \end{bmatrix} \tag{3}$$

where A_c is the cross-section area of the structural column, $\overline{E}_c$ is the Young's modulus of the rod material, $\beta_c = \omega\sqrt{\rho_c/\overline{E}_c}$ is the axial wavenumber, ρ_c is the density of the rod material, ω is the angle frequency, and l_c is the length of the structural column.

The floor and the vertical bearing structure are assumed to be rigidly connected, so when vibration propagates through the floor, it will also propagate to the exterior of the floor, resulting in energy dissipation. Therefore, in this context, the floor can be considered as an element that consumes energy with a fixed value. However, during structural response, it can be viewed as a frequency-dependent mass point; thus, the mass point becomes the primary component of the structural response, and its mass is not fixed during this process. Based on these concepts, the floor was modeled as a mass point primarily responsible for providing energy dissipation:

$$m_s = \frac{8}{i\omega}\sqrt{\frac{\overline{E}_f h_f^3}{12\left(1-v_f^2\right)}\rho_f h_f} \tag{4}$$

where ρ_f, h_f, and v_f are the material density, thickness, and Poisson's ratio of the floor and $\overline{E}_f$ is the complex Young's modulus of the floor.

In accordance with the concept of the structural dynamic stiffness cascade, the system equation in the frequency domain can be formulated as follows:

$$[K_c]\{U\} - \omega^2[M_s]\{U\} = \{F\} \tag{5}$$

where $[K_c]$ is the global stiffness matrix, $[M_s]$ is the global dynamic mass matrix, $\{U\}$ is the displacement vector of the system response, and $\{F\}$ is the input force vector of the system. The structural response can be efficiently obtained by constructing the global matrix and solving the system's dynamic equation through inversion. Numerical integration poses no difficulty in this process, enabling rapid calculation of the structural response. The overall acceleration response {A} of the system can be determined using the following formula:

$$\{A\} = -\omega^2\{U\} \tag{6}$$

4.2. Structure-Borne Noise Estimation Model

Utilizing the aforementioned model for addressing the system's acceleration response, and drawing from a comprehensive review of multiple empirical models, including RIVAS [33], HJ 453–2018 [30], FTA Guidelines [29], Kurzweil [22], and Melke [24], the following corresponding formulations are as shown in Equations (5)–(9), respectively:

$$L_p = L_{V1} - 27 \tag{7}$$

$$L_p = L_{v1} + 10\log_{10}(\sigma_{rad}) - 10\log_{10}(H_r) - 20 + 10\log_{10}(T_r) \tag{8}$$

$$L_A \approx L_{V2} + K_{A-wt} - 5 \tag{9}$$

$$L_p = L_{acc} - 20 \times \log_{10}(f) + 37 \tag{10}$$

$$L_p = L_{v3} + 10\log_{10}(\sigma_{rad}) + 10\log_{10}\left(\frac{4S_{sur}}{S_r}\right) \tag{11}$$

where L_{v1}, L_{v2}, and L_{v3} are the velocity levels, with reference velocities of 1×10^{-9} m/s, 1×10^{-6} in/s, and 5×10^{-8} m/s; L_p is the sound pressure level; L_A is the A-weighting sound pressure level; K_{A-wt} is an A-weighting adjustment at the 1/3-octave band center frequency; L_{acc} is the acceleration level (re: 1×10^{-6} m/s^2); f is the frequency; σ_{rad} is the average radiation ration for the floor bay; S_{sur} is surface area of the room; H_r is the height of the room; and T_r is the reverberation time, which can be calculated according to Sabine's formula [34]:

$$T_r = \frac{0.16 \times V}{S_r \times \alpha_{room}} \tag{12}$$

where S_r, α_{room}, and V represent the sound absorption area, average sound absorption coefficient, and room volume, respectively. Generally, a higher reverberation time is achieved using Saibine's formula; consequently, the methodology for calculating reverberation time is frequently expanded to include Eyring's formula [35]:

$$T_r = \frac{0.16 \times V}{-S_r \times \ln(1 - \alpha_{room})} \tag{13}$$

In most contemporary empirical models, the relationship between velocity and pressure levels is the basis of establishing the model. However, in various environmental vibration specifications, the definition of velocity levels varies, often with distinct reference speeds [29,30,33]. This introduces significant uncertainty to the prompt assessment of structure-borne noise resulting from environmental vibrations. However, in the vast majority of vibration assessments, when using acceleration as an evaluation parameter, the standard reference acceleration is typically set at 1×10^{-6} m/s^2. Taking into account the expediency of rapid forecasting and the consistency of indicators, this paper presents the following empirical formula to forecast unweighted structure-borne noise from vibration acceleration levels based on the Equation (10) proposed by Kurzweil [22]. Other empirical formulas concluded, based on extensive field experiments, according to Equation (8), that an increase in the room's height will lead to a reduction in the sound pressure level of structure-borne noise. Based on this, a correction term for room height has been introduced, and the frequency in the original equation has been adjusted to a circular frequency:

$$L_{\text{p}} = L_{\text{acc}} - 20\log_{10}(\omega) + 37 - 20\log_{10}(H_r) \tag{14}$$

where w is the circular frequency. The dynamic stiffness model of the structure considers the floor as a mass point responsible for damping energy dissipation, resulting in a mass point response. Therefore, obtaining the bottom floor response enables prediction of the superstructure response, and further substitution into the noise model allows for prediction of structure-borne noise in the frequency domain.

4.3. Model Validation

Considering σ_{rad} = 1, S_{sur} = S_r, and α_{room} = 0.1 [36], we compared the measured vibration acceleration level and sound pressure level data in the center of the two-story room with the predicted values calculated using the proposed structural vibration prediction model and structure-borne noise prediction model, as depicted in Figure 13. The results illustrate that the predicted outcomes of the model presented in this paper exhibit a strong fitting performance within the frequency range of 31.5 to 125 Hz, which is recognized as the frequency of interest for structure-borne noise generated from train-induced vibrations. In the 16–25 Hz and 160–200 Hz frequency bands, background noise predominates, leading to a decrease in model accuracy. However, the frequency ranges do not represent the frequencies of interest associated with train-induced vibrations, whose energy levels are minimal and insufficient to significantly impact human comfort. Consequently, the differences can be considered acceptable.

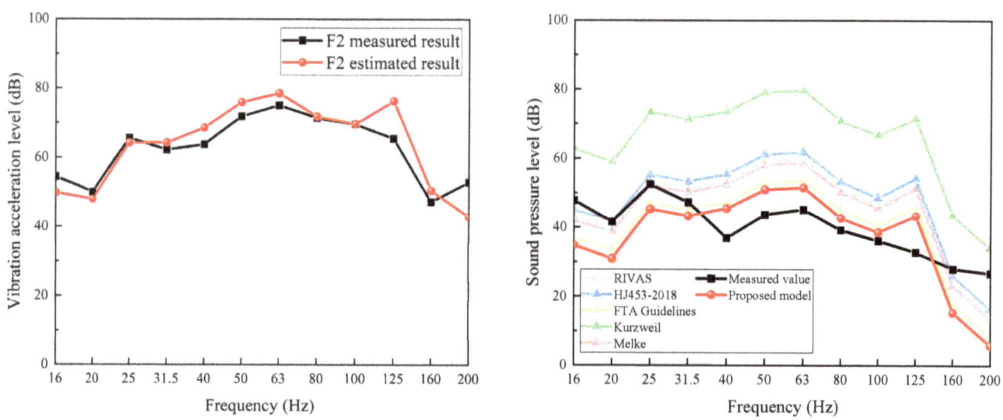

(**a**) Structural vibration (**b**) B1 Structure-borne noise

Figure 13. Comparison of measured and model's predicted results [22,24,29,30,33].

4.4. Error Analysis

Figure 14 illustrates the errors between the measured and predicted results of vibration and noise. As illustrated in Figure 14a, the vibration prediction values generated by the proposed model demonstrate a high degree of accuracy, with errors at each frequency generally remaining within 5 dB. Notably, the errors at 125 Hz and 200 Hz are more pronounced, reaching approximately 10 dB. This discrepancy may be attributed to the predominance of environmental vibrations within these frequency bands, which can adversely affect the model's accuracy. As illustrated in Figure 14b, the predicted structure-borne noise values generated by the proposed model demonstrate a high degree of accuracy, with errors at each frequency generally remaining within 8 dB. In comparison to other empirical models, the proposed model demonstrates superior fitting in the primary frequency bands associated with train-induced vibrations, particularly within the range of 31.5–80 Hz. Although the formulas of RIVAS and Melke exhibit a better fit at 16–31.5 Hz, this frequency band is not of primary concern for train-induced vibrations, which possess alower energy in this frequency band.

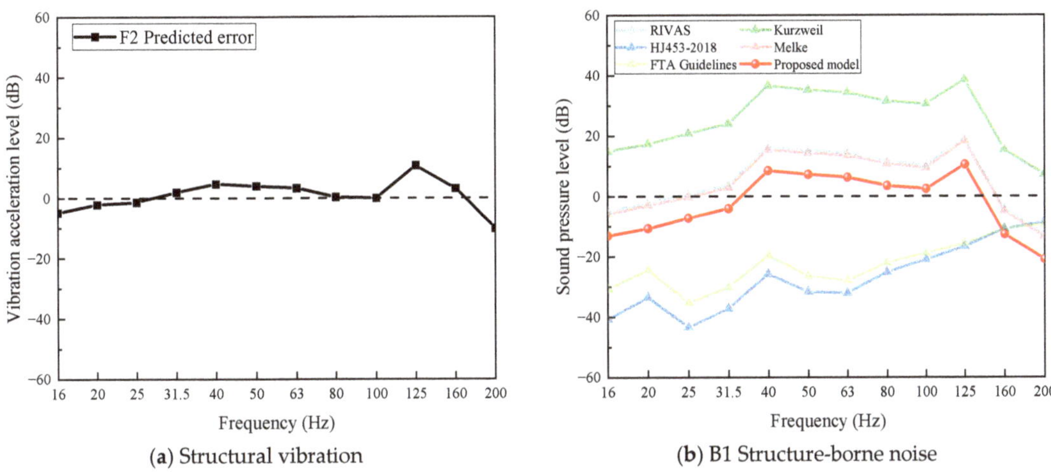

Figure 14. Comparison of errors between measured and model's predicted results [22,24,29,30,33].

4.5. Model Limitation Discussion

A hybrid prediction methodology that is both semi-analytical and semi-empirical has thus been introduced. The approach derives the dynamic response predictions of the structure through analytical solutions and subsequently estimates the secondary noise within the building's interior using a newly formulated empirical equation. This model enables the fast estimation of vibrations and structure-borne noise in low-rise buildings. However, the model is subject to certain limitations.

Firstly, the model's input consists of the foundation basement vibrations, indicating that it is necessary to obtain the vibration values which have undergone soil–structure coupling loss when utilizing this model. By integrating various methodologies, the vibrations are modified during the propagation to enhance both accuracy and efficiency, which aligns with the objectives of the hybrid model proposed in this paper. Consequently, the model presented herein exclusively addresses the final stages of vibration propagation—specifically, propagation and response within the building—thereby excluding any variations in geological conditions. However, this can be accomplished through various soil–structure interaction models, including numerical integral methods [11], simple node-coupled systems [31], and deep learning approaches [37], etc. We posit that it is feasible to conduct vibration measurements for diverse foundations and geological conditions to summarize the law of soil–structure interaction, thereby proposing corresponding coupling loss corrections. Utilizing this idea, the model can maintain both high efficiency and acceptable accuracy. Most importantly, it is user-friendly and can be easily understood and applied by most engineers. We aspire to improve this approach in future work.

What is more, the proposed model is deemed appropriate for frame structures. However, its applicability remains uncertain for superstructure buildings featuring load-bearing wall structures and those with transfer floor configurations. This paper exclusively validated a frame building. Similarly, regarding the proposed structural secondary noise model, only the vibration acceleration levels within the board were considered as inputs for predictions; however, it is important to note that structural noise sources in a room are also generated from walls, which may limit the model's applicability to buildings with special wall structures.

5. Conclusions

A two-story frame structure was subjected to testing for its vibration along a subway line and the resulting structure-borne noise. The study focused on examining the relationship between the vibration and noise caused by train operation in both far and near tunnels. A hybrid prediction method comprising a structural vibration prediction model and a structure-borne noise prediction model was proposed. The results obtained from this method were compared with field measurements. The main conclusions are as follows:

(1) The vibration acceleration characteristics of building components are closely associated with the vibration frequency, showing significant differences in performance across different frequency bands.
(2) Noise within the indoor structure of buildings is not uniformly distributed; corner areas exhibit more pronounced noise than central indoor floor areas.
(3) There exists a positive correlation between structural vibration and structure-borne noise, where greater structural vibrations lead to more noticeable structure-borne noise. The characteristic frequency band of structure-borne noise falls within 25 Hz–80 Hz.
(4) The hybrid prediction method effectively forecasted the impact of subway vibration and structural noise on surrounding buildings under specific working conditions. This method demonstrate high prediction accuracy with a simple and convenient calculation process and is suitable for the preliminary prediction and confirmation stage of specific engineering projects.

Author Contributions: Conceptualization, X.L.; data curation, B.Z.; formal analysis, J.C. (Jialiang Chen), S.H., B.Z., F.P., Y.W. and J.C. (Junjie Chen); funding acquisition, J.C. (Jialiang Chen); investigation, S.H., B.Z., X.L., F.P., Y.W. and J.C. (Junjie Chen); methodology, X.L.; project administration, J.C. (Jialiang Chen); writing—original draft, J.C. (Jialiang Chen) and S.H.; writing—review and editing, B.Z. and X.L. All authors have read and agreed to the published version of the manuscript.

Funding: This research was funded by the project of the Ministry of Housing and Urban-Rural Development of the People's Republic of China, grant number 2022-K-153.

Data Availability Statement: The original contributions presented in the study are included in the article material; further inquiries can be directed to the corresponding author.

Conflicts of Interest: Authors Jialiang Chen, Fangling Peng, and Yingying Wang were employed by the company Beijing Oriental Yuhong Waterproof Technology Co., Ltd. Authors Jialiang Chen, Sen Hou, and Junjie Chen were employed by the company Beijiao Zane Rail Technology (Beijing) Co., Ltd. The remaining authors declare that the research was conducted in the absence of any commercial or financial relationships that could be construed as a potential conflict of interests.

References

1. Ma, M.; Xu, L.; Liu, W.; Tan, X. Semi-analytical solution of a coupled tunnel-soil periodic model with a track slab under a movin train load. *Appl. Math. Model.* **2024**, *128*, 588–608. [CrossRef]
2. He, C.; Jia, Y.; Zhou, S. Semi-analytical method for calculating ground vibrations from a tunnel in a homogeneous half-space wi an irregular surface. *J. Sound Vib.* **2024**, *591*, 118615. [CrossRef]

Li, X.; Chen, Y.; Zou, C.; Wang, H.; Zheng, B.; Chen, J. Building structure-borne noise measurements and estimation due to train operations in tunnel. *Sci. Total Environ.* **2024**, *926*, 172080. [CrossRef] [PubMed]
Sheng, X.; Jones, C.J.C.; Thompson, D.J. A theoretical model for ground vibration from trains generated by vertical track irregularities. *J. Sound Vib.* **2004**, *272*, 937–965. [CrossRef]
Hu, J.; Zou, C.; Liu, Q.; Li, X.; Tao, Z. Floor vibration predictions based on train-track-building coupling model. *J. Build. Eng.* **2024**, *89*, 109340. [CrossRef]
Auersch, L. Simple and fast prediction of train-induced track forces, ground and building vibrations. *Railw. Eng. Sci.* **2020**, *28*, 232–250. [CrossRef]
Sanayei, M.; Zhao, N.; Maurya, P.; Moore, J.A.; Zapfe, J.A.; Hines, E.M. Prediction and mitigation of building floor vibrations using a blocking floor. *J. Struct. Eng.* **2012**, *138*, 1181–1192. [CrossRef]
Zou, C.; Moore, J.A.; Sanayei, M.; Wang, Y. Impedance model for estimating train-induced building vibrations. *Eng. Struct.* **2018**, *172*, 739–750. [CrossRef]
Zou, C.; Moore, J.A.; Sanayei, M.; Tao, Z.; Wang, Y. Impedance model of train-induced vibration transmission across a transfer structure into an overtrack building in a metro depot. *J. Struct. Eng.* **2022**, *148*, 04022187. [CrossRef]
Papadopoulos, M.; François, S.; Degrande, G.; Loimbaert, G. The influence of uncertain local subsoil conditions on the response of buildings to ground vibration. *J. Sound Vib.* **2018**, *418*, 200–220. [CrossRef]
Edirisinghe, T.L.; Talbot, J.P. The significance of source-receiver interaction in the response of piled foundations to ground-borne vibration from underground railways. *J. Sound Vib.* **2021**, *506*, 116178.
Li, S.; Wang, J.; Han, K.; Wang, L.; Zhang, D.; Dong, H.; Cui, B. Promoting effect and microscopic mechanism of train-induced vibration on loess disintegration. *Eng. Geol.* **2024**, *336*, 107559.
Chung, Y.; Gu, K.; Jeon, S.; Park, B.; Park, J. Influence of the Doppler effect and binaural characteristics on perceived annoyance of train pass-by noise. *Appl. Acoust.* **2023**, *211*, 109525. [CrossRef]
He, Y.; Thompson, D.; Hu, Z. Aerodynamic noise from a high-speed train bogie with complex geometry under a leading car. *J. Wind Eng. Ind. Aerodyn.* **2024**, *244*, 105617. [CrossRef]
Sadeghi, J.; Vasheghani, M. Safety of buildings against train induced structure borne noise. *Build. Environ.* **2021**, *197*, 107784. [CrossRef]
Nagy, A.B.; Fiala, P.; Márki, F.; Augusztinovicz, F.; Degrande, G.; Jacobs, S.; Brassenx, D. Prediction of interior noise in buildings generated by underground rail traffic. *J. Sound Vib.* **2006**, *293*, 680–690. [CrossRef]
Liu, Q.; Thompson, D.J.; Xu, P.; Feng, Q.; Li, X. Investigation of train-induced vibration and noise from a steel-concrete composite railway bridge using a hybrid finite element-statistical energy analysis method. *J. Sound Vib.* **2020**, *471*, 115197. [CrossRef]
Fragasso, J.; Helal, K.M.; Moro, L. Transfer-path analysis to estimate underwater radiated noise from onboard structure-borne sources. *Appl. Ocean. Res.* **2024**, *147*, 103979.
Hornikx, M. Acoustic modelling for indoor and outdoor spaces. *J. Build. Perform. Simul.* **2015**, *8*, 1–2. [CrossRef]
Fiala, P.; Degrande, G.; Augusztinovicz, F. Numerical modelling of ground-borne noise and vibration in buildings due to surface rail traffic. *J. Sound Vib.* **2007**, *301*, 718–738.
Colaço, A.; Costa, P.A.; Amado-Mendes, P.; Godinho, L.; Calçada, R. Mitigation of vibrations and re-radiated noise in buildings generated by railway traffic: A parametric study. *Procedia Eng.* **2017**, *199*, 2627–2632. [CrossRef]
Kurzweil, L.G. Ground-borne noise and vibration from underground rail systems. *J. Sound Vib.* **1979**, *66*, 363–370. [CrossRef]
Vér, I.L.; Beranek, L.L. (Eds.) *Noise and Vibration Control Engineering: Principles and Applications*; John Wiley & Sons: Hoboken, NJ, USA, 2005.
Melke, J. Noise and vibration from underground railway lines: Proposals for a prediction procedure. *J. Sound Vib.* **1988**, *120*, 391–406. [CrossRef]
Tao, Z.; Moore, J.A.; Sanayei, M.; Wang, Y.; Zou, C. Train-induced floor vibration and structure-borne noise predictions in a low-rise over-track building. *Eng. Struct.* **2022**, *255*, 113914.
Dong, Y.; Jiang, H.; Mu, M.; Wang, X. Multi-sensor data fusion-enabled lightweight convolutional double regularization contrast transformer for aerospace bearing small samples fault diagnosis. *Adv. Eng. Inform.* **2024**, *62*, 102573.
Dong, Y.; Jiang, H.; Jiang, W.; Xie, L. Dynamic normalization supervised contrastive network with multiscale compound attention mechanism for gearbox imbalanced fault diagnosis. *Eng. Appl. Artif. Intell.* **2024**, *133*, 108098.
Wang, X.; Jiang, H.; Wu, Z.; Yang, Q. Adaptive variational autoencoding generative adversarial networks for rolling bearing fault diagnosis. *Adv. Eng. Inform.* **2023**, *56*, 102027. [CrossRef]
Quagliata, A.; Ahearn, M.; Boeker, E.; Roof, C.; Meister, L.; Singleton, H. *Transit Noise and Vibration Impact Assessment Manual*; No. FTA Report No. 0123; Federal Transit Administration, U.S. Department of Transportation: Washington, DC, USA, 2018.
HJ 453–2018; Technical Guidelines for Environmental Impact Assessment—Urban Rail Transit. China Environmental Science Press: Beijing, China, 2018. (In Chinese)
Auersch, L. Building response due to ground vibration-simple prediction model based on experience with detailed models and measurements. *Int. J. Acoust. Vib.* **2010**, *15*, 101. [CrossRef]
Fahy, F.; Walker, J. (Eds.) *Advanced Applications in Acoustics, Noise and Vibration*; CRC Press: Boca Raton, FL, USA, 2018.

33. Villot, M.; Guigou, C.; Jean, P.; Picard, N. *Report Definition of Appropriate Procedures to Predict Exposure in Buildings and Estima Annoyance*; RIVAS Del.1.6; Railway Induced Vibration Abatement Solutions (RIVAS), European Commission: Brussels, Belgiu 2012.
34. Sabine, W.C. *Collected Papers on Acoustics*; Harvard University Press: Cambridge, MA, USA, 1922.
35. Eyring, C.F. Reverberation time in "dead" rooms. *J. Acoust. Soc. Am.* **1930**, *1*, 217–241. [CrossRef]
36. Zhou, X.; Späh, M.; Hengst, K.; Zhang, T. Predicting the reverberation time in rectangular rooms with non-uniform absorptic distribution. *Appl. Acoust.* **2021**, *171*, 107539. [CrossRef]
37. Li, X.; Chen, Y.; Zou, C.; Wu, J.; Shen, Z.; Chen, Y. Building coupling loss measurement and prediction due to train-induce vertical vibrations. *Soil Dyn. Earthq. Eng.* **2023**, *164*, 107644. [CrossRef]

buildings

MDPI

Article

Probabilistic Prediction and Assessment of Train-Induced Vibrations Based on Mixture Density Model

Ziyu Tao [1,2], Lingshan He [3], Desi Tu [4] and Chao Zou [1,4,*]

1 State Key Laboratory of Performance Monitoring and Protecting of Rail Transit Infrastructure, East China Jiaotong University, Nanchang 330013, China; zi-yu.tao@polyu.edu.hk
2 The Aviation Services Research Centre, Hong Kong Polytechnic University, Hong Kong, China
3 Guangzhou Urban Planning and Design Survey Research Institute Co., Ltd., Guangzhou 510060, China
4 School of Civil and Transportation Engineering, Guangdong University of Technology, Guangzhou 510006, China
* Correspondence: chao.zou@gdut.edu.cn

Abstract: This study presents a probabilistic prediction method for train-induced vibrations by combining a deep neural network (DNN) with the mixture density model in a cascade fashion, referred to as the DNN-RMDN model in this paper. A benchmark example is conducted to demonstrate and evaluate the prediction performance of the DNN-RMDN model. Subsequently, the model is applied to a case study to investigate and compare the uncertainties of train-induced vibrations in the throat area and testing line area of a metro depot. After training, the model is capable of accurately predicting the probability density function (PDF) of train-induced vibrations at different distances from the track and at different frequencies. Utilizing the predicted PDF, probabilistic assessments can be performed to ascertain the likelihood of surpassing predefined limits. By employing a mixture density model instead of a single Gaussian distribution, the DNN-RMDN model achieves more accurate prediction of the PDF for train-induced vibrations. The proposed probabilistic assessment framework can effectively assist in vibration screening during the planning phase and in selecting and designing vibration mitigation measures of appropriate levels.

Keywords: metro depot; train-induced ground vibration; vibration variations; deep learning; residual mixture density network; probabilistic prediction and assessment

1. Introduction

Train-induced vibration has been a topic of concern in research for several decades. Excessive vibrations have the potential to cause discomfort to residents, malfunctions in vibration-sensitive equipment, and even damage to cultural heritage structures located in close proximity to the rail lines. In recent years, there has been an increasing emphasis on assessing the impact of train-induced vibrations on the serviceability of nearby buildings. This focus has been partly driven by the growing prevalence of transit-oriented development (TOD) in mega cities. The efficient prediction and evaluation of train-induced vibrations can support transit-oriented development (TOD) planning and aid in the decision-making process regarding the adoption of vibration control measures.

Train-induced vibration is essentially a random vibration phenomenon. However, most existing prediction methods [1–5] and evaluation standards [6,7] primarily focus on the deterministic behavior of train-induced vibrations, often overlooking their inherent randomness. It is widely recognized that vibration levels are influenced by the characteristics of vibration sources, propagation paths, and the recipients. In the case of train-induced ground-borne vibrations, the dynamics of trains, tracks, and the ground medium play a crucial role. Deterministic methods focus on predicting vibrations from the same train operating on the same track and neglect the uncertainty of vibration propagation caused by spatially varying soil properties. Therefore, for a given vibration source condition, such as the train type, track structures, and operating speed, deterministic methods predict a constant vibration level at a specific location, assuming a simplified soil model, such as the frequently used layered soil model. The predicted constant vibration level generally represents the conditional average of field measurements incorporating uncertainties arising from both vibration source and propagation.

In practice, train-induced ground-borne vibrations often exhibit pronounced variations [8,9]. Consequently, even if the predicted conditional average of train-induced vibrations falls within the limits required by assessment standards, there is still a probability, to some extent, of causing disturbance to the environment. In this context, probabilistic prediction and assessment take precedence over their deterministic counterparts, as they provide a range within which train-induced vibrations are likely to occur. Moreover, probabilistic approaches assist in determining the optimal level of vibration control measures to be implemented.

The variations in field measurements of train-induced ground-borne vibrations stem from factors such as rail unevenness [10], wheel out-of-roundness [11], and the propagation uncertainty caused by spatially varied soil properties [12,13]. Currently, research on the random characteristics of the aforementioned factors primarily focuses on high-speed rail lines or subway mainlines, assuming a constant distance between the track and a fixed receiver location, as well as a consistent track structure. The metro depot is an auxiliary facility of the subway system where metro trains are parked, tested, maintained, and cleaned. Constructing over-track buildings above the metro depot not only effectively utilizes urban land resources but also facilitates the commuting of citizens living or working in the over-track community. Compared to buildings located near the high-speed rail lines or subway mainlines, the over-track buildings of a metro depot are subjected to a more complex vibration field, even though the train operation speed is slower. The variation in train-induced ground-borne vibrations within a metro depot is significant and worthy of investigation due to the following factors:

(1) The large area occupied by a metro depot leads to significant variations in soil properties.
(2) The different functional regions within the metro depot, such as the throat area, testing line area, and parking/maintenance area, showcase distinct train-induced vibration characteristics due to variations in operation speed and track structures.
(3) The concurrent operation of different trains on various rail lines adds extra spatiotemporal complexity to the vibration source.
(4) The irregular distribution of foundations for the over-track platform and buildings results in a complex diffraction field of train-induced vibrations.

Considering the complexity mentioned, the utilization of a combination of the Monte Carlo method and previously introduced deterministic methods to analyze the train-induced uncertainty within a metro depot becomes highly challenging [14] and demands significant computational resources. With the decreasing cost of data acquisition and developments in neural network models [15], as well as computing hardware, data-driven methods are gradually gaining attention for both deterministic predictions [16–22] and probabilistic predictions [23].

This research aims to fill the current research gap by providing probabilistic predictions and assessments of train-induced ground-borne vibrations in the metro depot with the presence of over-track buildings. An efficient probabilistic prediction method based on the deep learning approach and the mixture density model is proposed, which is referred to as the DNN-RMDN model in this paper. In this research, the mixture density model is utilized instead of the Bayesian neural network because the former is adept at capturing intricate, multi-modal output distributions, whereas the latter primarily focuses on estimating uncertainty by modeling the posterior distribution over the network weights. Thus, the proposed method enables the prediction of the probability density function (PDF) of train-induced ground-borne vibrations as a function of distance to the track and frequency. This facilitates probabilistic assessment and assists in making decisions regarding the extent to which vibration control measures should be designed and implemented.

2. Probabilistic Prediction Method Based on Mixture Density Model

This section introduces a mixture density model-based approach, referred to as the DNN-RMDN model in this study, which offers an efficient method for probabilistic prediction and the assessment of train-induced vibrations.

2.1. Methodology

The structure of the DNN-RMDN model is depicted in Figure 1. It consists of two cascaded submodels: a deep neural network (DNN) and a mixture density network (MDN).

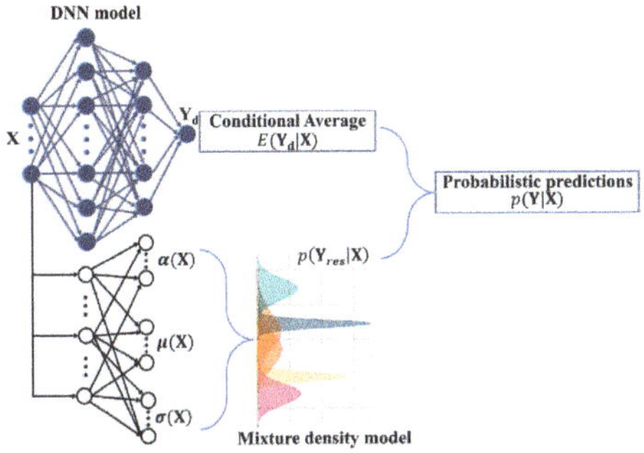

Figure 1. Structure of the DNN-RMDN model.

In the proposed DNN-RMDN model, a DNN is first employed to learn the relationship between the input factors **X** and the conditional mean of random vibrations E($\mathbf{Y_d} \mid \mathbf{X}$). For the application of the DNN model, min–max normalization is utilized on the datasets to aid in the optimization of the model. The rectified linear unit (ReLU) is used as the activation function within the DNN, enabling the introduction of non-linearity and enhancing the model's expressive power. The back-propagation technique, with the Adam algorithm, is utilized to train the DNN model. During training, the mean squared error serves as the loss function, guiding the optimization process.

Following the acquisition of predictions for the conditional average, the mixture density model [24] is employed to learn the probability distribution of the residuals p($\mathbf{Y_{res}} \mid \mathbf{X}$). The p($\mathbf{Y_{res}} \mid \mathbf{X}$) is represented as a linear combination of Gaussian kernels, given by Equation (1) [24,25]:

$$\mathrm{p}(Y_{res}|\mathbf{X}) = \sum_{i=1}^{k} \alpha_i(\mathbf{X}) \cdot \frac{1}{\sqrt{2\pi}\sigma_i(\mathbf{X})} e^{-\frac{1}{2}\left(\frac{Y_{res}-\mu_i(\mathbf{X})}{\sigma_i(\mathbf{X})}\right)^2} \quad (1)$$

where $\alpha_i(\mathbf{X})$, $\mu_i(\mathbf{X})$, and $\sigma_i(\mathbf{X})$ represent the mixing coefficient, mean, and standard deviation of the ith Gaussian kernel, respectively. Within the RMDN model, the logarithmic likelihood loss ($-\log(\mathrm{p}(Y_{res}|\mathbf{X}))$) is employed as the loss function, while the tanh function serves as the activation function [26]. From the outputs of the RMDN model, the mean and standard deviation of input samples can be evaluated as follows [25]:

$$\mu(Y_{res}|\mathbf{X}) = \sum_{i=1}^{k} \alpha_i(\mathbf{X}) \cdot \mu_i(\mathbf{X}) = 0 \quad (2)$$

$$\sigma(Y_{res}|\mathbf{X}) = \sqrt{\sum_{i=1}^{k} \alpha_i(\mathbf{X}) \cdot \left(\sigma_i(\mathbf{X})^2 + \left\| \mu_i(\mathbf{X}) - \sum_{j \neq i} \alpha_j(\mathbf{X}) \cdot \mu_j(\mathbf{X}) \right\|^2\right)} \quad (3)$$

Because the conditional mean is extracted by the DNN model and the dataset residuals are then utilized for training the mixture density model, the mean calculated using Equation (2) equates to zero. This approach helps in simplifying the dataset complexity for the mixture density model through the detrending effect accomplished by the preceding DNN model. The RMDN model can be utilized to capture the inherent variability, known as aleatoric uncertainty, within the residual datasets. By adjusting the hyperparameters of the RMDN model, the epistemic uncertainty can be partially mitigated.

2.2. Benchmark Example

To verify the predictive accuracy of the proposed probabilistic method, a benchmark example is conducted in this section.

The dependent variable Y is composed of a deterministic component Y_d and a random component Y_r, as $Y = Y_d + Y_r$.

The deterministic component Y_d is the function of independent variables x_1 and x_2, and is defined as follows:

$$Y_d(x_1, x_2) = 120 - (x_1 - 5)^2 - 5x_2 \tag{4}$$

The random component Y_r follows a normal distribution, where the mean is 0 and the standard deviation is a function of the independent variable x_1, i.e., $Y_r \sim N(0, \sigma^2(x_1))$. The standard deviation function is defined as follows:

$$\sigma(x_1) = (x_1 + 1.8)/2 \tag{5}$$

In the benchmark example, eleven values of x_1 (equally spaced values from 1 to 11) and three values of x_2 (0, 6, 9) are used to determine the deterministic component Y_d, and a hundred samples for each x_1 are generated to account for the variations in Y, i.e., the random component Y_r. Thus, a total of 3300 discrete data samples of Y are generated.

Using the method described in Section 2.1, a deep neural network (DNN) is first trained to predict the deterministic component Y_d. The adopted DNN model contains two inputs standing for values of independent variables x_1 and x_2, respectively, and one output standing for the value of dependent variable Y. Figure 2a illustrates the predicted conditional mean from the trained DNN and the corresponding training datasets. Figure 2b shows the percentage error between the conditional mean predicted by the trained DNN and the corresponding training datasets. The percentage error is observed to be within 3%, indicating a high level of accuracy. This suggests that the DNN model has been effectively trained and can reliably predict the conditional average of Y for a given pair of input x_1 and x_2.

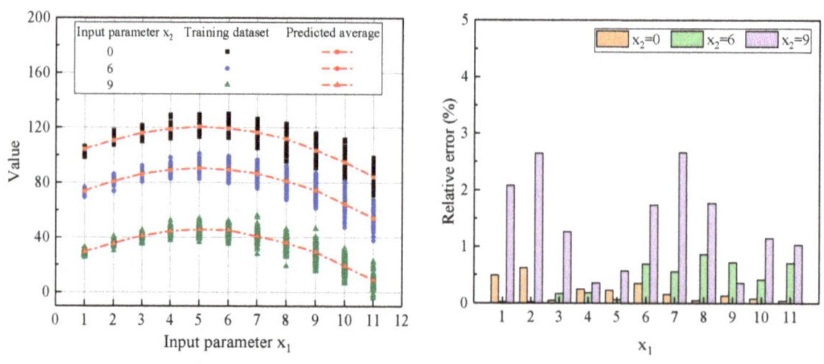

Figure 2. Training results of DNN for predicting the deterministic component. (**a**) Predicted mean and training datasets. (**b**) Percentage error of the trained DNN.

As described in Section 2.1, a followed residual mixture density network was trained to predict the random component Y_r. One hidden layer of 5 neurons and 2 Gaussian kernels was used to build the RMDN. Figure 3 displays the training result of the RMDN. The PDFs of two weighted Gaussian kernels as a function of x_1 are shown, along with their linear combination at each value of x_1. It was observed that kernel 2 outweighs kernel 1, resulting in the linear combination predominantly following the weighted kernel 2. Furthermore, the

PDF of an equivalent single Gaussian distribution, with its mean and standard deviation calculated from Equations (2) and (3), respectively, is superimposed on Figure 3.

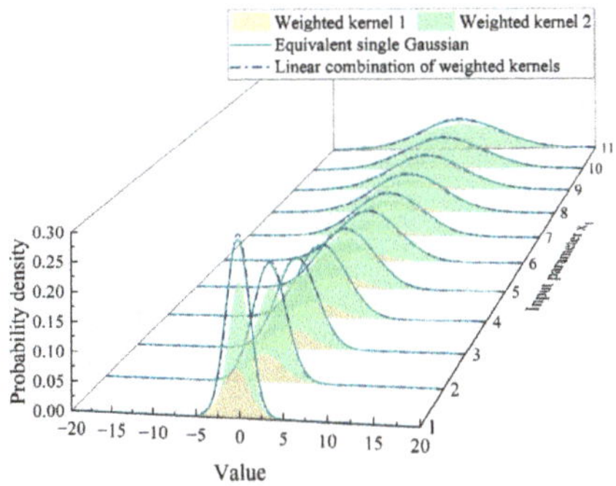

Figure 3. Training results of RMDN for predicting the random component.

Since the samples of the random component in this benchmark example follow a normal distribution as shown in Equation (5), the difference between the actual PDF predicted by the linear combination of Gaussian kernels and the equivalent single Gaussian distribution is negligible. The results also indicate that the inclusion of additional Gaussian kernels in the RMDN will not compromise the model's prediction performance.

Table 1 compares the true standard deviations used in the benchmark example, calculated from Equation (5), with the corresponding predictions from the trained RMDN for each value of x_1. The results indicate that the RMDN is well-trained, with relative errors of standard deviation predictions within 5%.

Table 1. Identification accuracy of the standard deviation.

x_1	TRUE	Neuron5, m = 2	Relative Error (%)
1	1.4	1.4	1.4
2	1.9	2.0	5.8
3	2.4	2.5	5.4
4	2.9	3.0	4.5
5	3.4	3.6	4.7
6	3.9	4.1	4.9
7	4.4	4.6	5.0
8	4.9	5.1	4.1
9	5.4	5.5	2.4
10	5.9	5.9	0.2
11	6.4	6.2	3.4

Figure 4a,b shows the testing results for the deterministic component and random component, respectively. Like the training results, the trained DNN-RMDN well predicted the deterministic and random components of unseen testing data, which implies the proposed method can effectively quantify the inherent data randomness.

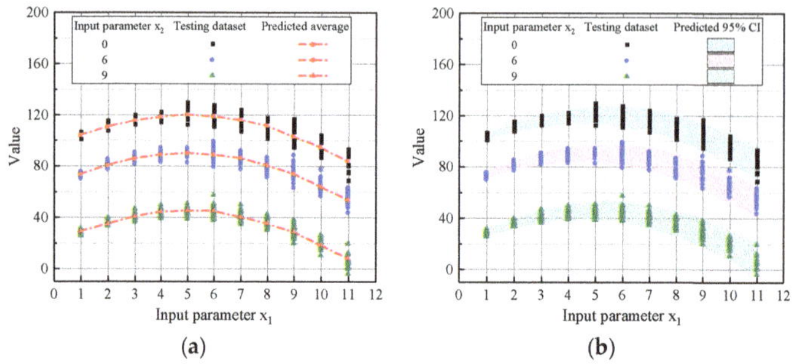

Figure 4. Testing results of the trained DNN-RMDN model. (**a**) Deterministic predictions. (**b**) Probabilistic predictions.

3. Application of the DNN-RMDN Model to Probabilistic Predict Train-Induced Vibrations

As discussed in the Introduction section, the vibration field induced by trains in a metro depot is intricate due to the complex configurations involving track locations and the presence of underground foundation structures, as depicted in Figure 5.

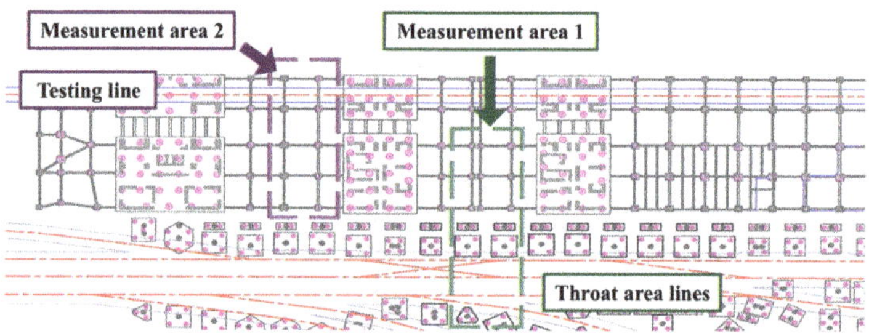

Figure 5. Layout of rail lines and underground foundations in a metro depot with over-track buildings.

Two measurement campaigns were conducted to investigate and quantify the variations in train-induced ground-borne vibrations caused by train passages. These two campaigns specifically focused on the throat area and the testing line, respectively, whose measurement areas are shown in Figure 5. Both measurement campaigns utilized a sampling frequency of 1024 Hz, which is adequate for analyzing frequencies of interest up to 80 Hz. Additional information regarding the measurement equipment, the operating metro train, soil properties, and the track structures can be found in [4,5,9].

For the measurement campaign in the throat area, the ground-borne vibrations were recorded at multiple locations while the investigated metro trains passed through different rail lines in the throat area. For the measurement campaign in the testing area, the ground-borne vibrations were recorded at multiple locations while the same train operated back and forth on the testing line. The variation in train-induced vibrations in the throat area is greater than that in the testing line. This is attributed to the increased uncertainties arising from the vibration source, such as the variation in out-of-roundness among different trains, the uncertainty in the exact distance between the operated rail line in the throat area and the measured location, and the variability in train operation speeds.

3.1. *Vibration Variations in the Throat Area*

3.1.1. Field Measurement Setups

Figure 6 illustrates a schematic diagram of the plan view and cross-sectional view of the measurement setups in the throat area. Five locations on the ground surface, positioned at distances of 4 m, 10 m, 15 m, 25 m, and 35 m from the center line of the nearest track in the throat area, were equipped with accelerometers to record train-induced vibrations during various train passages in the throat area.

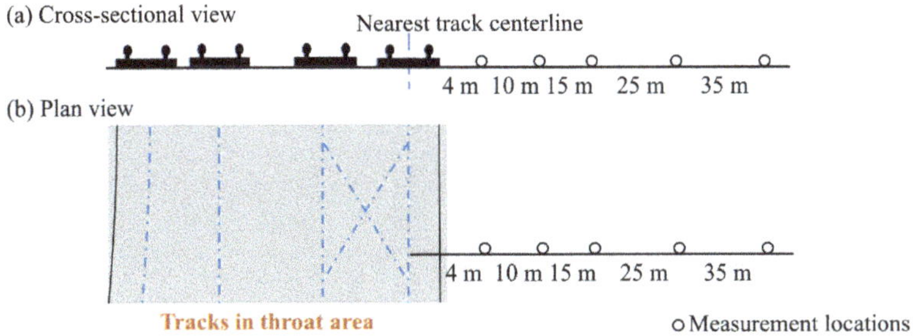

Figure 6. Field measurement setups in the throat area.

Figure 7 shows the one-third octave band spectra of measured train passages at five locations with different distances to the throat area. The analysis was conducted within the frequency range of 8 to 80 Hz, which is of interest for the study [7]. The dominant frequencies were observed to be in the range of 40 to 80 Hz. As the distance from the throat area increases, the dominant frequencies tend to shift towards lower frequency bands. The separation of spectra between different locations becomes more pronounced at frequencies above 25 Hz, particularly for locations that are closer to the throat area. These phenomena can be attributed to the fact that higher frequency components decay more rapidly with distance compared to lower frequency components.

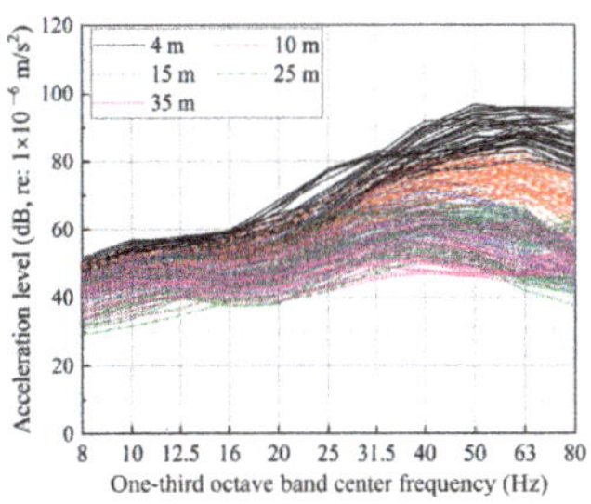

Figure 7. Measured vibration spectra of different train passages at various locations in the throat area.

When examining the vibration spectra at the same location, induced by different train passages in the throat area, it is observed that the acceleration level difference can reach up to 20 dB. Notably, this difference persists even when the vibrations are induced by trains of the same type, passing across the test section at a similar low speed range of 5 to 15 km/h. The variations in measurements at a fixed location are primarily attributed to several factors. Firstly, the wheel out-of-roundness configurations differ among different trains, contributing to measurement variations. Secondly, there is uncertainty regarding the exact distance between the operated rail line and the receiver location in the throat area, as the datasets do not differentiate which train passage operated on which rail line. Additionally, the uncertainty in vibration propagation arises from the spatially varied soil properties and the presence of irregular foundations of the over-track platform and buildings within the soil. These factors collectively contribute to the overall measurement variations observed.

3.1.2. Architecture Determination of the DNN-RMDN Model

This section is dedicated to establishing the architecture of the DNN-RMDN model, as shown in Figure 1. The DNN model is utilized to predict conditional means of vibration levels, while the RMDN model is employed to estimate the uncertainty linked to the predicted mean from the DNN model.

In this research, the DNN model utilizes two inputs: the distance between observation points and the track, and the one-third octave band center frequency. This choice is motivated by two reasons. Firstly, the datasets employed to build the DNN-RMDN model are collected from the same site, featuring the same type of metro trains and negligible variations in train speeds. Consequently, the vibration source and transmission path for the dataset are consistent. The primary factors influencing the vibration levels in the datasets are the distance and the evaluated center frequency, as demonstrated in Figure 7.

On the other hand, one practical application of the DNN-RMDN model is to facilitate the probabilistic assessment of train-induced vibrations, as illustrated in Section 4. When planning and designing railway constructions adjacent to existing buildings or constructing buildings near existing tracks, it is crucial to estimate the vibration levels and the probability of exceeding thresholds based on the distance and the evaluated frequency. These two factors are integral to evaluation standards and are of interest to stakeholders. Therefore, the inputs for the DNN model are chosen as the distance and the center frequency.

In addition, as seen in Figure 7, the variations in vibration levels are not clearly changed with the distance. Thus, the RMDN model's input is simplified to frequency only. Frequency-dependent uncertainty estimations for building serviceability assessments and vibration control measure designs and selection because they are frequency-specific. Furthermore, as depicted in Figure 7, the fluctuations in vibration levels do not exhibit significant changes with distance. Therefore, the input for the RMDN model is simplified to only include frequency. Frequency-dependent uncertainty estimations are crucial for building serviceability assessments and the design and selection of vibration control measures, as they are specific to different frequencies.

A total of 51 synchronized train passages were collected for each of the five locations, with an analysis of 11 center frequencies ranging from 8 to 80 Hz, resulting in a combined total of 2808 samples. Among them, 1966 randomly selected samples were allocated for the training dataset, with 842 samples designated for the testing dataset.

For the DNN model, pyramidal architecture was utilized with three hidden layers: 50 neurons in the first layer, 30 neurons in the second layer, and 20 neurons in the third layer, as detailed in the published work [21]. The learning rate employed was 0.001.

Figure 8a illustrates the mean squared error of the DNN model as the training epochs increase, showing convergence at 400 epochs. Figure 8b compares the predictions from the trained DNN model with the corresponding testing datasets. A linear regression with an R2 value of 0.850 indicates a strong correlation between them, suggesting that the trained DNN model performs effectively with the unseen testing datasets.

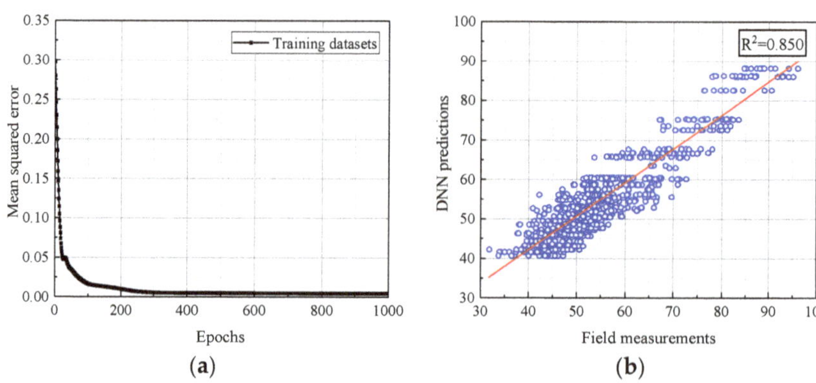

Figure 8. DNN model validation. (**a**) Training process; (**b**) testing results.

A shallow one-layer MLP was employed for the RMDN model since the conditional means were already extracted by the DNN model, and the distributions of residues were simplified through detrending. The hyperparameters of the RMDN model involve the number of neurons in the hidden layer of the MLP and the quantity of Gaussian kernels incorporated in the mixture density model.

The RMDN architecture was chosen based on the standard deviation computed from Equation (3). In Figure 9a, variations in calculated standard deviations at 31.5 Hz are depicted with different numbers of neurons in the hidden layer of the MLP and various numbers of Gaussian kernels within the mixture density model. Notably, the standard deviation of the dataset's residuals shows minimal sensitivity to the number of neurons in the 1-layer MLP, aligning with the detrending process where the means predicted by the DNN model are subtracted from the training dataset. Figure 9b illustrates the variations in calculated standard deviations as the number of Gaussian kernels increases. It is observed that with fewer than 20 Gaussian kernels, there is a decreasing pattern in the estimated standard deviation. This suggests that an order of the mixture density model exceeding 20 is necessary to effectively capture the randomness in the dataset's residuals. Thus, the chosen architecture for the RMDN model comprises a one-layer MLP with 10 neurons and a mixture density model with 20 Gaussian kernels. Figure 9c displays the negative log-likelihood loss function of the RMDN model, converging at 2500 epochs. The learning rate utilized is 0.0001. The error analysis, validation process, and uncertainty quantification utilizing the outcomes of the DNN-RMDN model are discussed in the subsequent section.

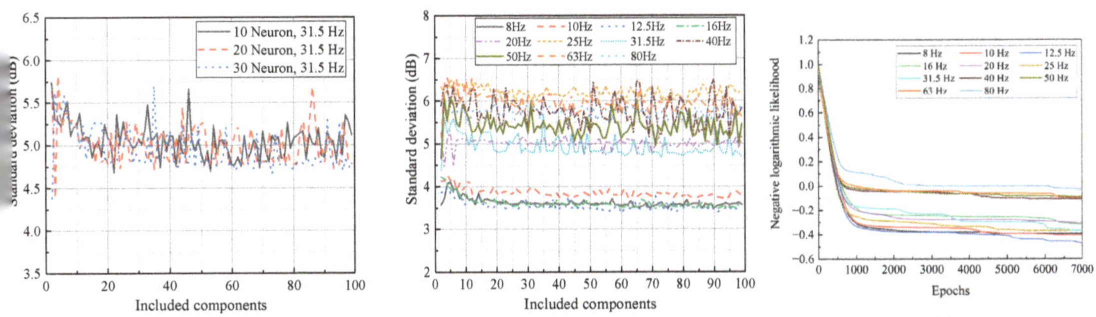

Figure 9. Sensitivity analysis for number of hidden neurons and Gaussian kernels: (**a**) effects of numbers of neurons and Gaussian kernels on σ at 31.5 Hz; (**b**) effects of numbers of Gaussian kernels on σ at different frequencies; (**c**) training process.

3.1.3. Quantifying Uncertainty with PDF

This section presents the training and testing outcomes of the DNN-RMDN model developed in Section 3.1.2. It showcases the predicted PDFs of train-induced vibrations in the throat area at each center frequency of the one-third octave band ranging from 8 to 80 Hz. These PDFs serve as a means to quantify the uncertainty associated with the vibrations.

Figure 10 compares the predicted conditional averages of train-induced vibrations at the five measured locations using the trained DNN model to the training datasets. The percentage error is noted to be below 3.5%, as indicated in Figure 10f. It can be found that the DNN model is effectively trained and capable of accurately capturing the conditional average train-induced acceleration levels at each center frequency and measurement location.

After predicting the conditional averaged vibrations, the residuals of the datasets are calculated by subtracting the train-induced acceleration levels at each location from their corresponding conditional averages. The residual datasets represent the random component of train-induced vibrations and are then utilized for training the RMDN model to predict the PDF which has a zero mean.

Figure 11a illustrates the mixing coefficients of the 20 Gaussian kernels in the mixture density model after training the DNN-RMDN model. Figure 11b displays the PDFs of the kernels with significant mixing coefficients. The PDFs of the kernels with large mixing coefficients are significant because they represent the distributions that contribute the most to the overall PDF estimation. These kernels capture the patterns and characteristics of the residual dataset that have a higher likelihood of occurring.

Figure 12 compares the overall PDF of residual train-induced vibrations at 31.5 Hz, based on the mixture density model consisting of a linear combination of 20 weighted Gaussian kernels as shown in Figure 11, to the equivalent single Gaussian distribution with a mean of zero and standard deviation calculated from the residual datasets using Equation (3). It is evident that the mixture density model has the capability to capture complex PDFs compared to a single Gaussian assumption. As a result, the PDF of train-induced vibrations predicted by the DNN-RMDN model offers greater accuracy in probabilistic assessment. This will be further explored and discussed in Section 4, where a comprehensive probabilistic assessment will be conducted.

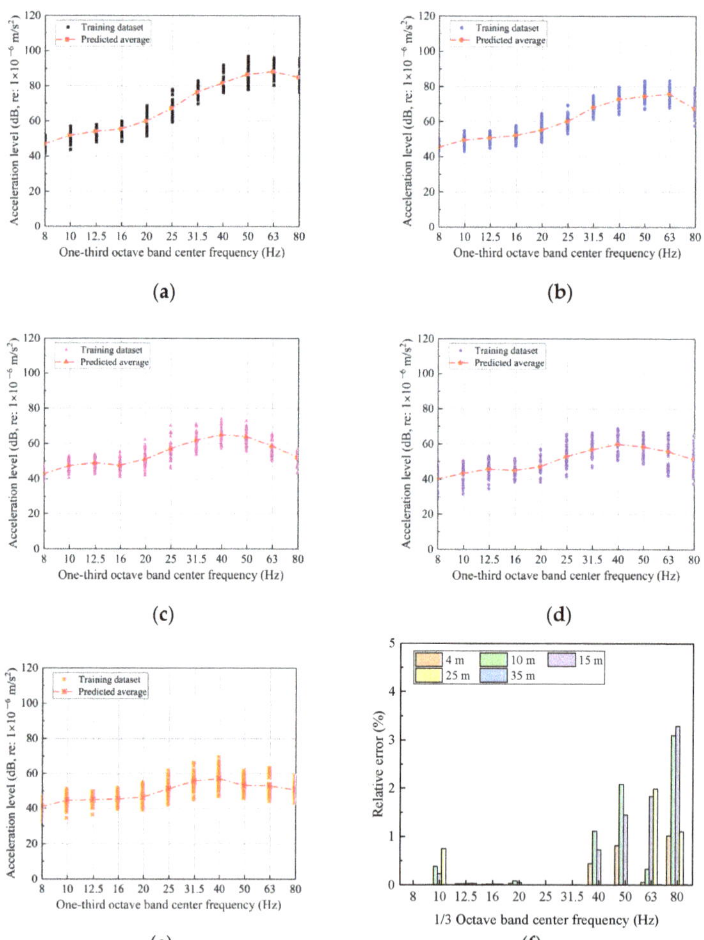

Figure 10. Predictions of conditional averaged vibrations using a trained DNN model: (**a**) 4 m; (**b**) 10 m; (**c**) 15 m; (**d**) 25 m; (**e**) 35 m; (**f**) relative error.

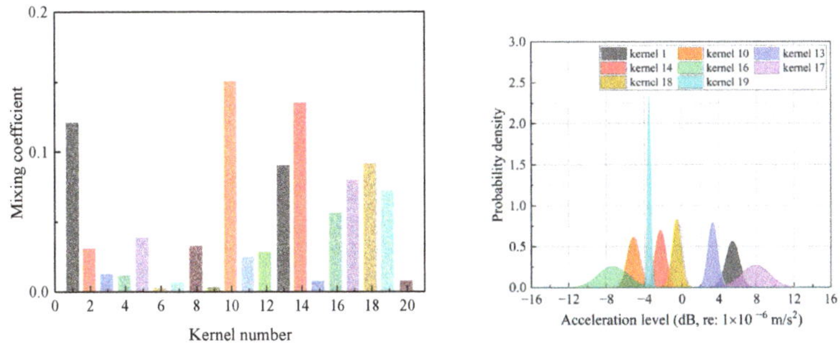

(a) (b)

Figure 11. Mixing coefficients and Gaussian kernels PDF with larger coefficient (31.5 Hz): (**a**) mixing coefficients; (**b**) kernels with large mixing coefficient.

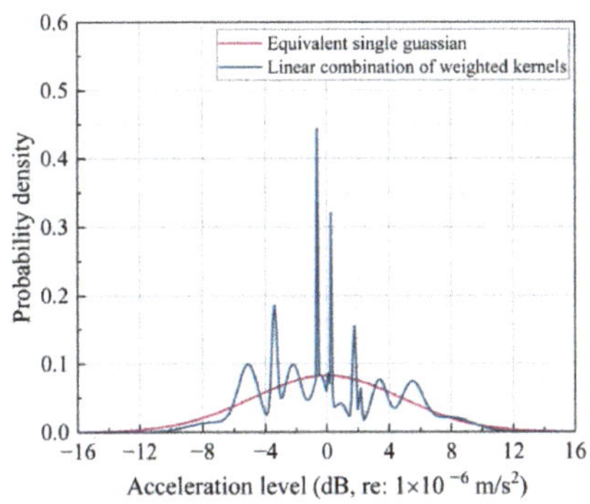

Figure 12. Comparisons of the PDF predicted with the RMDN model and the Gaussian distribution.

Figure 13 presents the predicted PDFs of residual train-induced vibrations by the RMDN model at various center frequencies. It is observed that the RMDN model's predicted PDFs exhibit different shapes at different center frequencies. This finding validates the assumption that the uncertainty of train-induced vibrations is indeed related to the frequency, as the PDF shapes vary accordingly.

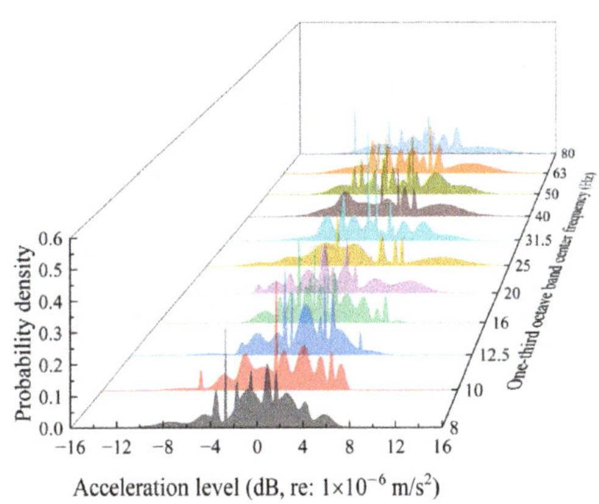

Figure 13. PDFs predicted with RMDN model at different center frequencies.

By utilizing the conditional average acceleration levels predicted by the DNN model at each measurement location, in combination with the RMDN model's predicted residual PDFs at each frequency, it is easy to obtain the PDFs of train-induced vibration levels at each measurement location and frequency. For each measurement location, 100 samples of acceleration-level spectra are generated. In Figure 14, these 100 samples are superimposed with the unseen testing dataset acquired from field measurements. The alignment between the testing dataset and the samples derived from the PDFs predicted by the trained DNN-RMDN model is evident. This validation provides confirmation of the proposed model's probabilistic prediction capability.

3.2. Vibration Variations in the Testing Line Area

3.2.1. Field Measurement Setups

Figure 15 illustrates a schematic diagram of the plan view and cross-sectional view of the measurement setups in the testing line area.

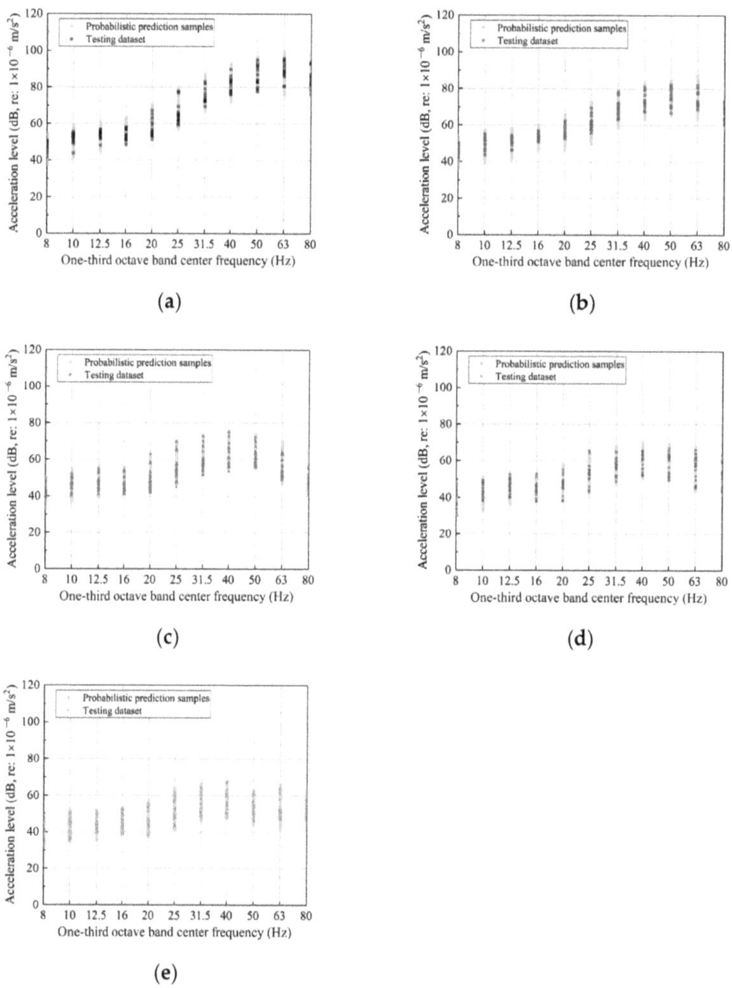

Figure 14. Performance test of the trained DNN-RMDN model: (**a**) 4 m; (**b**) 10 m; (**c**) 15 m; (**d**) 25 m; (**e**) 35 m.

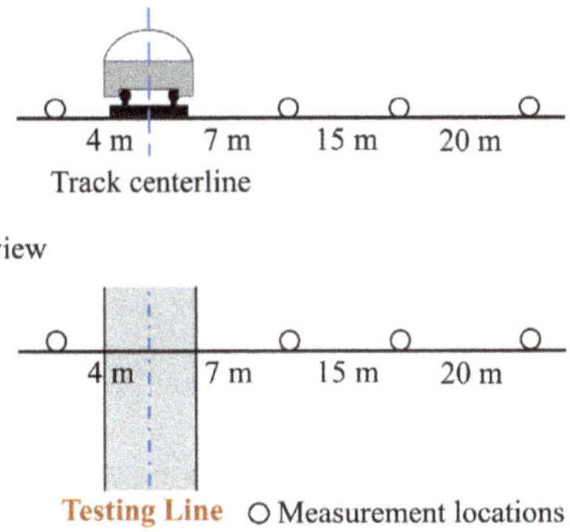

Figure 15. Field measurement setups in the testing line area.

Figure 16 shows the one-third octave band spectra of measured train passages at the four locations with different distances to the track center line of the testing line. The general trend in train-induced ground-borne vibrations in the testing line area mirrors that in the throat area. Specifically, higher-frequency vibration components exhibit a faster decay with distance, and the variation in vibrations is more closely related to frequency rather than the distance from the track. However, there are two distinct differences between train-induced ground-borne vibrations within the testing line area and the throat area. Firstly, when comparing Figure 16 to Figure 7, it is found that the frequency components between 63 Hz and 80 Hz are more prominent in ground-borne vibrations resulting from train passages on the testing line than those in the throat area. Secondly, it is noticeable that the variations in vibrations within the testing line area are significantly reduced compared to those in the throat area. This reduction can be attributed to the field campaign conducted on the testing line, where the same train operated on the same track, thereby minimizing the uncertainties induced by different trains of the same type and the distance between the operating rail line and the receiver location. Furthermore, train passages on the testing line exhibit a smaller variation in operation speeds compared to those in the throat area.

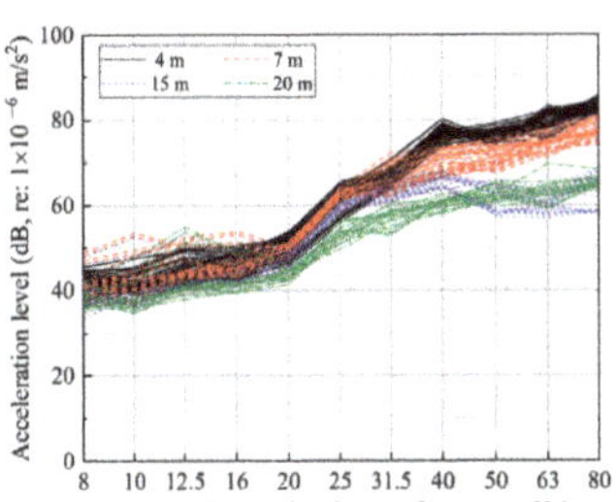

Figure 16. Measured vibration spectra of different train passages at various locations in the testing line area.

3.2.2. Architecture Determination of the DNN-RMDN Model

Similarly to Section 3.1.2, this section is dedicated to establishing the architecture of the DNN-RMDN model, which will serve as the foundation for Section 3.2.3, in which the model is adopted to quantify the uncertainty and predict the PDF of train-induced vibrations in the testing line area.

A total of 45 train passages were recorded at each of the four measurement locations, which were situated 4 m, 7 m, 15 m, and 20 m from the track center line. Consequently, a combined dataset comprising 1980 samples was obtained. Among them, 1386 randomly selected samples were allocated to the training dataset, with 594 samples designated for the testing dataset. The identical DNN model architecture outlined in Section 3.1.2 was employed in this context.

Figure 17a illustrates the mean squared error of the DNN model as the training epochs increase, showing convergence at 400 epochs. Figure 17b compares the predictions from the trained DNN model with the corresponding testing datasets. A linear regression with an R2 value of 0.965 indicates a strong correlation between them, suggesting that the trained DNN model performs effectively with the unseen testing datasets.

In Figure 18a, variations in calculated standard deviations at 31.5 Hz are depicted with different numbers of neurons in the hidden layer of the MLP and various numbers of Gaussian kernels within the mixture density model. Notably, the standard deviation of the dataset's residuals shows minimal sensitivity to the number of neurons in the 1-layer MLP, aligning with the detrending process where means predicted by the DNN model are subtracted from the training dataset. Figure 18b illustrates the variations in calculated standard deviations as the number of Gaussian kernels increases. It has been noted that the estimated standard deviations reached a stable value after incorporating more than 20 Gaussian kernels. This suggests that an order of the mixture density model exceeding 20 is necessary to effectively capture the randomness in the dataset's residuals. Thus, the chosen architecture for the RMDN model comprises a one-layer MLP with 10 neurons and a mixture density model with 20 Gaussian kernels. Figure 18c displays the negative log-likelihood loss function of the RMDN model, converging at 6500 epochs for all 11 investigated center frequencies. The learning rate utilized is 0.0001. The error analysis, validation process, and uncertainty quantification utilizing the outcomes of the DNN-RMDN model are discussed in the subsequent section.

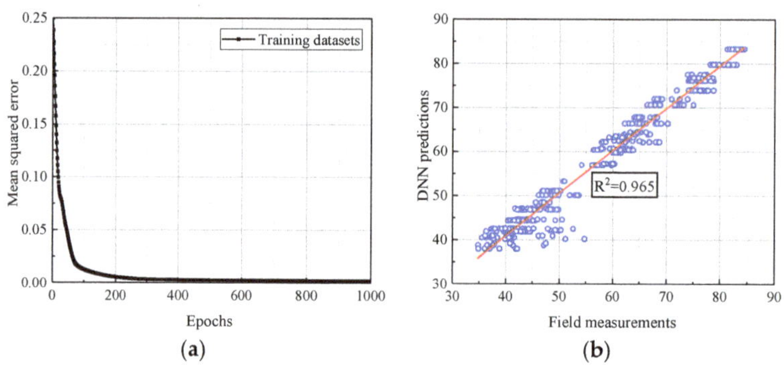

Figure 17. DNN model validation: (**a**) training process; (**b**) testing results.

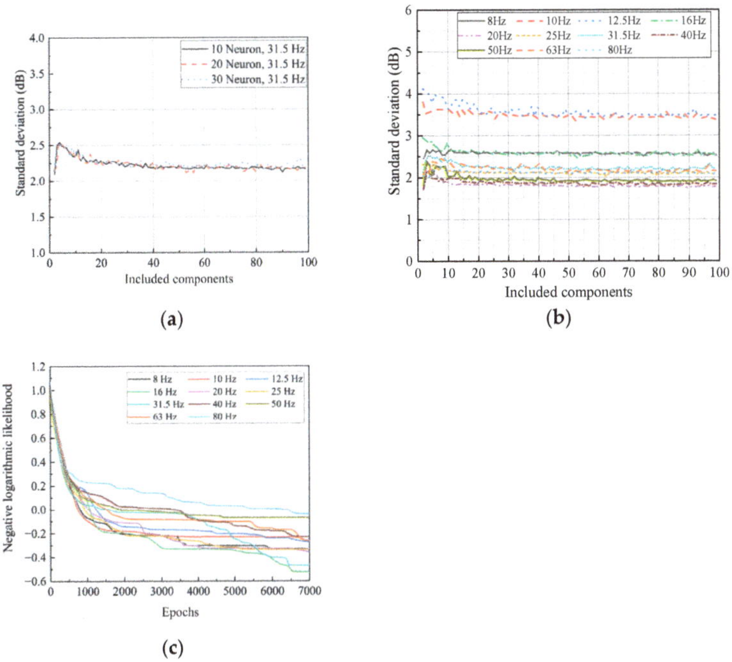

Figure 18. Sensitivity analysis for number of hidden neurons and Gaussian kernels: (**a**) effects of numbers of neurons and Gaussian kernels on σ at 31.5 Hz; (**b**) effects of numbers of Gaussian kernels on σ at different frequencies; (**c**) training process.

3.2.3. Quantifying Uncertainty with PDFs

This section presents the training and testing results of the DNN-RMDN model developed in Section 3.2.2. The predicted PDFs of train-induced vibration levels will be utilized as a foundation for conducting probabilistic assessments in Section 4.

Figure 19 compares the predicted conditional averages of train-induced vibrations at the four measured locations using the trained DNN model to the training dataset. The percentage error is noted to be below 0.8%, as indicated in Figure 19e. It can be found that the DNN model is effectively trained and capable of capturing the conditional average train-induced acceleration levels at each center frequency and measurement location.

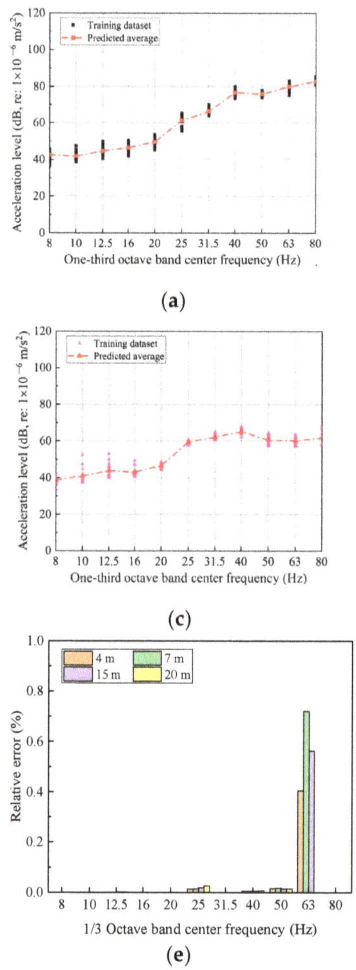

Figure 19. Predictions of conditional averaged vibrations using trained DNN model: (**a**) 4 m; (**b**) 7 m; (**c**) 15 m; (**d**) 20 m; (**e**) relative error.

Following the prediction of conditional average vibrations, the residuals of the dataset are used by the RMDN model to estimate their PDFs as a function of frequency. Figure 20a illustrates the mixing coefficients of the 20 Gaussian kernels in the mixture density model after training the DNN-RMDN model. Figure 20b displays the PDFs of the Gaussian kernels with significant mixing coefficients, which capture the patterns and characteristics of the residual dataset that have a higher likelihood of occurring. By comparing Figure 20a with Figure 11a, it is seen that the distribution of mixing coefficients for ground vibration at 31.5 Hz induced by train passages in the throat area is more uniform than that for ground vibration at the same frequency induced by train passages on the testing line. This indicates that PDF fitting with greater variability requires more kernel components.

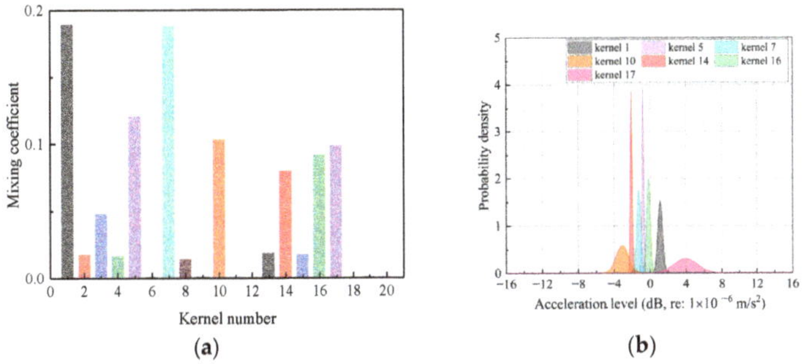

Figure 20. Mixing coefficients and Gaussian kernels PDF with larger coefficient (31.5 Hz): (**a**) mixing coefficients; (**b**) kernels with large mixing coefficient.

Figure 21 compares the overall PDF of residual train-induced vibrations at 31.5 Hz, based on the mixture density model consisting of a linear combination of 20 weighted Gaussian kernels shown in Figure 20, to the equivalent single Gaussian distribution with a mean of zero and standard deviation calculated from the residual dataset using Equation (3). Similarly to Figure 12, it is evident that the mixture density model has the capability to capture complex PDFs compared to a single Gaussian assumption. In addition, the residual PDF of train-induced acceleration levels in the testing line area exhibits a narrower range compared to that in the throat area.

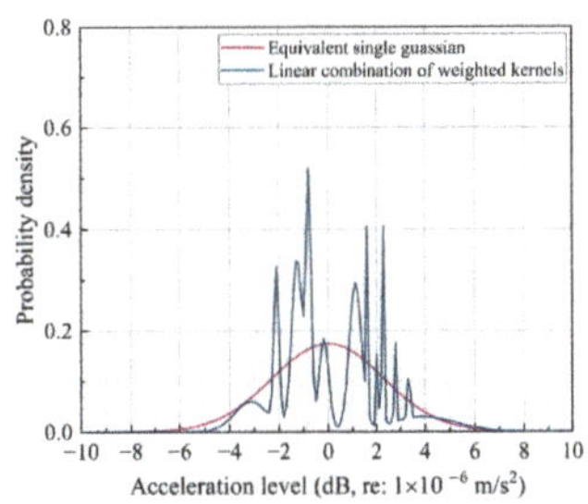

Figure 21. Comparisons of the PDF predicted with RMDN model and the Gaussian distribution.

Figure 22 presents the predicted PDFs of residual train-induced vibrations by the RMDN model at various center frequencies within the one-third octave band. The predicted PDFs of the RMDN model exhibit variations across different frequencies, indicating that the probability of exceeding the limit required by the standard also differs at each frequency. Identifying the frequencies with higher probabilities of exceeding the limit will assist in the design of vibration control measures.

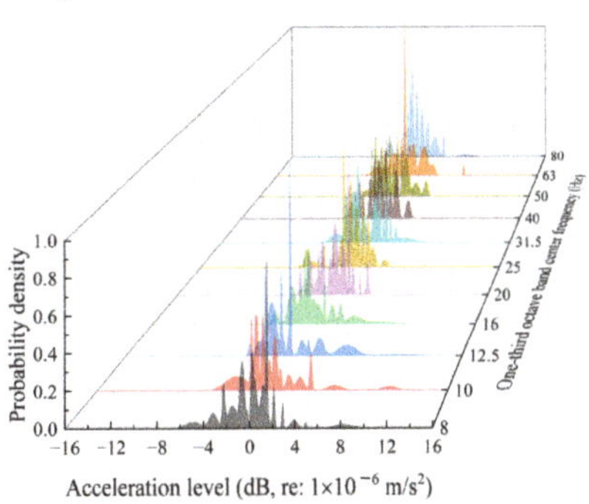

Figure 22. PDFs predicted with RMDN model at different center frequencies.

The PDFs of train-induced vibration levels at each measurement location and frequency are obtained by combining the conditional averages predicted by the DNN model, as shown in Figure 19, with the PDFs of the residual dataset predicted by the RMDN model, as shown in Figure 22. Using the overall PDFs, 100 samples of acceleration level spectra are generated for each distance from the track centerline of the testing line. Figure 23 overlays these samples from the predicted PDF with the unseen testing dataset acquired from field measurements. It is evident that the testing dataset aligns well within the range of samples from the predicted PDF at different distances away from the track, thus validating the performance of the proposed DNN-RMDN method in probabilistic predictions.

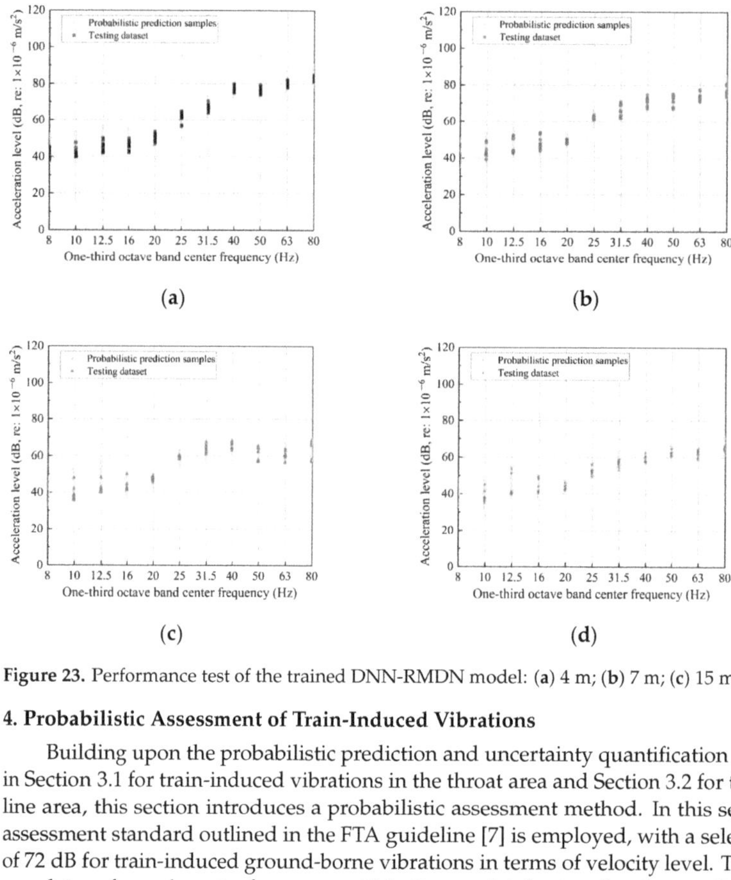

Figure 23. Performance test of the trained DNN-RMDN model: (**a**) 4 m; (**b**) 7 m; (**c**) 15 m; (**d**) 20 m.

4. Probabilistic Assessment of Train-Induced Vibrations

Building upon the probabilistic prediction and uncertainty quantification presented in Section 3.1 for train-induced vibrations in the throat area and Section 3.2 for the testing line area, this section introduces a probabilistic assessment method. In this section, the assessment standard outlined in the FTA guideline [7] is employed, with a selected limit of 72 dB for train-induced ground-borne vibrations in terms of velocity level. The limit is mandatory for each center frequency within the one-third octave band up to 80 Hz.

The predicted PDF p(V | d,f) of train-induced velocity levels using the DNN-RMDN model is expressed by Equation (6). It comprises the conditional average, $\overline{V}_{avg}(d,f)$, which is a function of distance to the track and frequency, predicted by the DNN model, and the residual PDF p(ΔV | f), which is a function of frequency predicted by the cascaded RMDN model:

$$\mathrm{p}(\mathrm{V}|\mathrm{d},\mathrm{f}) = \overline{V}_{avg}(d,f) + \mathrm{p}(\Delta\mathrm{V}|\mathrm{f}) \quad (6)$$

The probability of train-induced velocity levels exceeding the limit at a specific distance from the track and a fixed frequency can be calculated as follows:

$$P(V > V_{lim}|d,f) = 1 - \int_{-\infty}^{V_{lim}} p(V|d,f) \quad (7)$$

The black solid lines in Figure 24 illustrate the PDFs of train-induced vibrations in the throat area at 4 m from the nearest track centerline and frequencies of 31.5 Hz and 50 Hz, respectively. The red dashed lines correspond to the velocity level limit of 72 dB. The shaded areas beneath the PDFs represent the probability of exceeding the limit for each frequency at the specific location. It is evident that, for the given configuration, the probability of exceeding the limit at 50 Hz is greater than that at 31.5 Hz.

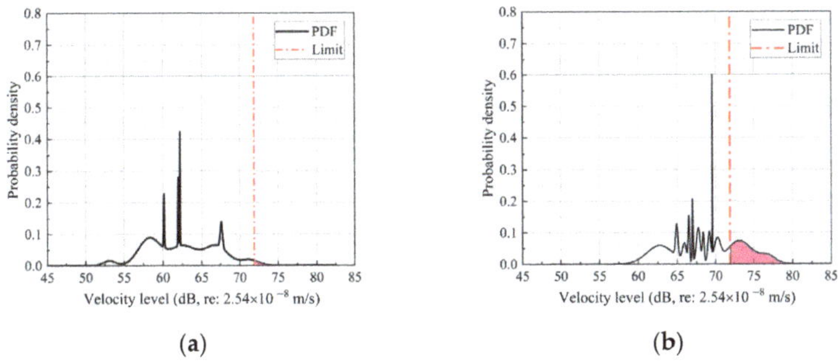

Figure 24. Usage of predicted PDF for calculating exceeding probability: (**a**) 4 m, 31.5 Hz; (**b**) 4 m, 50 Hz.

Figure 25 presents the contour of the probabilistic assessment of train-induced ground-borne vibrations in the throat area at various distances from the tracks and frequencies. It is intuitive to find that train-induced vibrations in the throat area, ranging from 50 to 63 Hz and occurring within 5 m from the nearest track centerline, have a 29% probability of exceeding the limit. The lower probability of exceeding vibration limits can be attributed to the implementation of ballast mattresses on the studied rail lines within the metro depot. These ballast mattresses effectively control and mitigate train-induced vibrations.

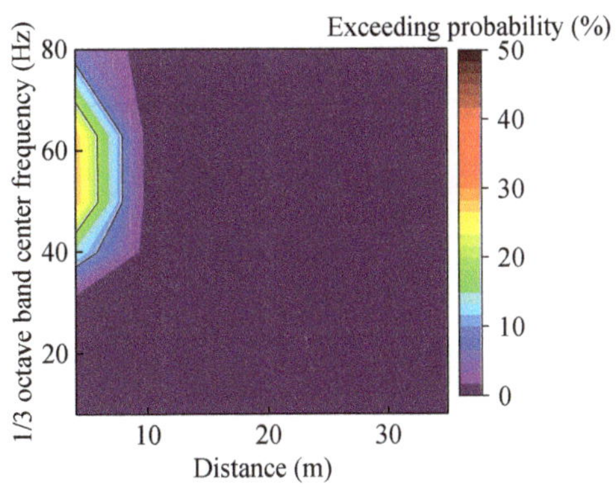

Figure 25. Probabilistic assessment of train-induced ground vibrations in the throat area.

The probabilistic assessment of train-induced vibrations across a wide field range and broad frequency range can be achieved very efficiently in a few seconds based on the predicted PDFs. Thus, the proposed DNN-RMDN method provides a valuable tool in practical applications for planning over-track buildings in complex vibration environments and making informed decisions regarding the design of vibration control measures.

The proposed model is trained and tested using the Python (3.12) language, making it easily integrable with various commercial FEM software like ABAQUS 2017. In practical implementation, only the distance to the track centerline and the investigated center frequency are necessary as data inputs. The model's outputs, which encompass the train-induced ground-borne vibrations and their distributions, can be imported into existing building vibration prediction software as external excitations. Moreover, through the model's implementation, it becomes feasible to predict the probability of ground vibration surpassing the standard at specified distances and frequencies. This capability can assist in selecting building locations with varying vibration requirements and in determining the level and effective frequency range for vibration isolation measures. This approach helps avoid the indiscriminate use of excessively high-level isolation measures or the incorrect selection of the effective frequency range for isolation measures.

5. Conclusions

Field measurements have revealed significant variations in train-induced ground-borne vibrations within a metro depot, even when operating the same type of metro trains at similar low speeds. This research introduces a probabilistic prediction and assessment method, referred to as the DNN-RMDN model, which combines deep learning techniques with a mixture density model. The paper demonstrates the predictive accuracy of the proposed model using a theoretical benchmark example. Subsequently, the DNN-RMDN model is applied to a case study involving two extensive measurement campaigns, aiming to investigate and quantify the random nature of train-induced vibrations in both the throat area and the testing line area of the metro depot. The performance of the DNN-RMDN model remains robust when applied to practical train-induced vibration datasets that exhibit large or medium variations. It demonstrates reliable probabilistic prediction capabilities in such scenarios.

The application of the DNN-RMDN model involves the utilization of a deep neural network (DNN) model to predict the conditional average of train-induced vibrations. Subsequently, a mixture density model with a one-layer shallow multiple layer perceptron is employed to extract the probability density function. The throat area, characterized by a more complex track configuration, variations in measured trains, and varying distances between vibration sources and measurement points, exhibits higher uncertainty in ground vibrations caused by train operations compared to the testing line section. The distribution of weights of the Gaussian kernels in the RMDN model is more evenly distributed in the throat area. By employing a mixture density model instead of a single Gaussian distribution, the DNN-RMDN model achieves a more accurate prediction of the PDF for train-induced vibrations. The predicted PDFs enable efficient probabilistic assessment of train-induced vibrations across a wide field range and broad frequency range, with results obtainable in a matter of seconds. In conclusion, the proposed DNN-RMDN model serves as a practical tool for planning over-track buildings in complex vibration environments and making well-informed decisions regarding the design of vibration control measures.

Due to constraints in the available data, only the distance between the track and observation points along with the center frequency of the 1/3 octave band were considered as inputs in this study. While these limitations in the datasets can help manage uncertainty sources and provide data with increased fidelity and stronger correlations between the data and the analyzed input factors, incorporating other significant uncertainty sources like various rail types and operational conditions as inputs would enhance the model's versatility and scalability.

Given the challenge of controlling uncertainty sources across diverse field measurement datasets, it becomes essential to utilize numerical simulations or surrogate models. To enhance the model's robustness and applicability, future efforts may involve acquiring more extensive datasets covering a range of rail types and operational settings, developing numerical or surrogate models to address the impacts of different uncertainty sources, and strengthening the DNN-RMDN model's adaptability through techniques like data fusion and transfer learning.

Author Contributions: Conceptualization, methodology, writing—original draft preparation, Z.T.; formal analysis, investigation, writing—review and editing, L.H.; software, validation, writing—review and editing, D.T.; resources, data curation, funding acquisition, C.Z. All authors have read and agreed to the published version of the manuscript.

Funding: This research was funded by the open funding project of the State Key Laboratory of Performance Monitoring and Protecting of Rail Transit Infrastructure (Grant No. HJGZ2022110).

Data Availability Statement: The original contributions presented in the study are included in the article, further inquiries can be directed to the corresponding author.

Conflicts of Interest: Author Lingshan He was employed by the Guangzhou Urban Planning and Design Survey Research Institute Co., Ltd. The remaining authors declare that the research was conducted in the absence of any commercial or financial relationships that could be construed as a potential conflict of interest.

References

1. Kouroussis, G.; Verlinden, O. Prediction of railway induced ground vibration through multibody and finite element modelling. *Mech. Sci.* **2013**, *4*, 167–183. [CrossRef]
2. Hu, J.; Zou, C.; Liu, Q.; Li, X.; Tao, Z. Floor vibration predictions based on train-track-building coupling model. *J. Build. Eng.* **2024**, *89*, 109340. [CrossRef]
3. Costa, P.A.; Calçada, R.; Cardoso, A.S. Track–ground vibrations induced by railway traffic: In-situ measurements and validation of a 2.5 D FEM-BEM model. *Soil Dyn. Earthq. Eng.* **2012**, *32*, 111–128. [CrossRef]
4. Zou, C.; Li, X.; He, C.; Zhou, S. An efficient method for estimating building dynamic response due to train operations in tunnel considering transmission path from source to receiver. *Comput. Struct.* **2024**, *305*, 107555. [CrossRef]
5. Tao, Z.Y.; Zou, C.; Yang, G.R.; Wang, Y.M. A semi-analytical method for predicting train-induced vibrations considering train-track-soil and soil-pile-building dynamic interactions. *Soil Dyn. Earthq. Eng.* **2023**, *167*, 107822. [CrossRef]
6. *JGJ/T 170-2009*; Standard for Limit and Measuring Method of Building Vibration and Secondary Noise Caused by Urban Rail Transit. Ministry of Housing and Urban–Rural Development of the People's Republic of China: Beijing, China, 2009.
7. Federal Transit Administration. *Transit Noise and Vibration Impact Assessment Manual*; US Department of Transportation: Washington, DC, USA, 2018.
8. He, C.; Jia, Y.; Zhou, S. Semi-analytical method for calculating ground vibrations from a tunnel in a homogeneous half-space with an irregular surface. *J. Sound Vib.* **2024**, *591*, 118615. [CrossRef]
9. Qiu, Y.; Zou, C.; Hu, J.; Chen, J. Prediction and mitigation of building vibrations caused by train operations on concrete floors. *Appl. Acoust.* **2024**, *219*, 109941. [CrossRef]
10. Lombaert, G.; Galvín, P.; François, S.; Degrande, G. Quantification of uncertainty in the prediction of railway induced ground vibration due to the use of statistical track unevenness data. *J. Sound Vib.* **2014**, *333*, 4232–4253. [CrossRef]
11. Ma, M.; Li, M.; Qu, X.; Zhang, H. Effect of passing metro trains on uncertainty of vibration source intensity: Monitoring tests. *Measurement* **2022**, *193*, 110992. [CrossRef]
12. Sun, H.; Fang, Y.; Yuan, Z.; Weng, Z.; Ni, D. Analytical modeling for the calculation of underground train-induced vibrations in inhomogeneous soils with uncertainty. *AIP Adv.* **2021**, *11*, 115321. [CrossRef]
13. Auersch, L. Train-induced ground vibration due to the irregularities of the soil. *Soil Dyn. Earthq. Eng.* **2021**, *140*, 106438. [CrossRef]
14. Jones, S.; Kuo, K.; Hussein, M.; Hunt, H. Prediction uncertainties and inaccuracies resulting from common assumptions in modelling vibration from underground railways. *Proc. Inst. Mech.Eng. Part F J. Rail Rapid Transit* **2012**, *226*, 501–512. [CrossRef]
15. Jafari, A.; Ma, L.; Shahmansouri, A.A.; Dugnani, R. Quantitative fractography for brittle fracture via multilayer perceptron neural network. *Eng. Fract. Mech.* **2023**, *291*, 109545. [CrossRef]
16. Connolly, D.P.; Kouroussis, G.; Giannopoulos, A.; Verlinden, O.; Woodward, P.K.; Forde, M.C. Assessment of railway vibrations using an efficient scoping model. *Soil Dyn. Earthq. Eng.* **2014**, *58*, 37–47. [CrossRef]
17. Connolly, D.P.; Kouroussis, G.; Woodward, P.K.; Giannopoulos, A.; Verlinden, O.; Forde, M.C. Scoping prediction of re-radiated ground-borne noise and vibration near high speed rail lines with variable soils. *Soil Dyn. Earthq. Eng.* **2014**, *66*, 78–88. [CrossRef]
18. Paneiro, G.; Durão, F.O.; Costa e Silva, M.; Falcão Neves, P. Artificial neural network model for ground vibration amplitudes prediction due to light railway traffic in urban areas. *Neural Comput. Appl.* **2018**, *29*, 1045–1057. [CrossRef]
19. Fang, L.; Yao, J.; Xia, H. Prediction on soil-ground vibration induced by high-speed moving train based on artificial neural network model. *Adv. Mech. Eng.* **2019**, *11*, 1687814019847290. [CrossRef]
20. Tao, Z.; Hu, Z.; Wu, G.; Huang, C.; Zou, C.; Ying, Z. Train-induced vibration predictions based on data-driven cascaded state-space model. *Building* **2022**, *12*, 114. [CrossRef]
21. Liang, R.; Liu, W.; Li, C.; Li, W.; Wu, Z. A novel efficient probabilistic prediction approach for train-induced ground vibrations based on transfer learning. *J. Vib. Control.* **2023**, *30*, 576–587. [CrossRef]
22. Liang, R.; Liu, W.; Ma, M.; Liu, W. An efficient model for predicting the train-induced ground-borne vibration and uncertainty quantification based on Bayesian neural network. *J. Sound Vib.* **2021**, *495*, 115908. [CrossRef]
23. Liu, W.; Liang, R.; Zhang, H.; Wu, Z.; Jiang, B. Deep learning based identification and uncertainty analysis of metro train induced ground-borne vibration. *Mech. Syst. Signal Process.* **2023**, *189*, 110062. [CrossRef]
24. McLachlan, G.J.; Basford, K.E. *Mixture Models: Inference and Applications to Clustering*; M. Dekker: New York, NY, USA, 1988; Volume 38.

25. Bishop, C.M. *Mixture Density Networks*; Aston University: Birmingham, UK, 1994.
26. Guillaumes, A.B. Mixture Density Networks for Distribution and Uncertainty Estimation. Ph.D. Thesis, Universitat Politècnica de Catalunya, Barcelona, Spain, 2017.

buildings

MDPI

rticle

Vibration Measurement and Numerical Simulation of the Effect f Non-Structural Elements on Dynamic Properties of arge-Span Structures

aliang Chen [1,2,3], Wei He [4,5,*], Congbo Sun [4], Sen Hou [3], Junjie Chen [3] and Zhe Wang [6]

[1] Beijing Oriental Yuhong Waterproof Technology Co., Ltd., Beijing 101111, China; chen_galeon@tsinghua.edu.cn
[2] School of Civil Engineering, Tsinghua University, Beijing 100084, China
[3] Beiijao Zane Rail Technology (Beijing) Co., Ltd., Beijing 101111, China; housen@yuhong.com.cn (S.H.); chenjj05@yuhong.com.cn (J.C.)
[4] Faculty of Engineering, China University of Geosciences (Wuhan), 388 Lumo Road, Wuhan 430074, China; cbcbscb@cug.edu.cn
[5] Zhejiang Engineering Center of Road and Bridge Intelligent Operation and Maintenance Technology, Hangzhou 310039, China
[6] Beijing Metro Construction Administration Co., Ltd., Beijing 100068, China; wang305856596@163.com
* Correspondence: hewei2018@cug.edu.cn

Abstract: Non-structural elements have been demonstrated to be essential for the dynamic performance of large-span structures. However, how to quantify their effect has not yet been fully understood. In this study, the contribution of non-structural elements to dynamic properties of large-span structures is systematically investigated via both field measurement and numerical simulation methods. Modal testing of an indoor stadium and an elevated highway bridge was conducted during different construction phases, and the corresponding modal characteristics were identified. Results show that the traditional capacity-based models are incapable of reflecting the actual dynamic characteristics of in-service structures since neglecting the effect of non-structural elements would result in remarkable discrepancies in modal properties. A general modeling framework incorporating the contribution of slab/deck pavement, infill walls (or crash barriers), and joints/connections for large-span structures is developed to quantitatively consider the effect of non-structural elements based on the principle of equivalence of stiffness and mass to the actual structure. The effectiveness of the method is validated by vibration measurement results.

Keywords: non-structural elements (NSEs); serviceability-based modeling; large-span structures; dynamic properties; modal parameter identification; field measurement

1. Introduction

Non-structural elements (NSEs), such as partition walls, ceilings, cladding panels, facades, windows, handrails, etc., are elements that are not designed for load capacity but are used to furnish or for functional purposes [1]. The effect of NSEs has long been ignored in structural design or analysis for many reasons: (i) the unclear mechanism of interaction between main structure and NSEs; (ii) the difficulty in quantifying the mechanical properties of NSEs; (iii) a common, but not always true, belief that a model neglecting NSEs is conservative [2]. However, studies [3,4] show that NSEs significantly modify the response of main structures in terms of stiffness, strength, and ductility, and neglecting the contribution of NSEs may lead to unsafe response predictions. Great research interest has been attracted to the effect of NSCs on dynamic responses of structures in recent years to consider the potential effect of NSEs under ultimate loadings like earthquakes or strong winds [5–10]. All of these involved studies are basically bearing capacity-based since structural safety (strength) is of primary concern, followed by ductility or vulnerability.

Besides safety, vibration serviceability is another important aspect of structural performance, especially for large-span structures. Large-span structures, characterized by low natural frequencies and small damping, are increasingly favored in large public facilities like stadiums or highway bridges. For these structures, steel-concrete composite floor systems or pre-stressed reinforced concrete floor systems are usually employed to realize long spans. Due to the slenderness and low damping, these structures are susceptible to dynamic excitations, such as human activities or vehicle loading, and their performance under dynamic excitation is of great importance for serviceability-based structural design and vibration mitigation, if necessary. Dynamic properties are key parameters for structural analysis. An accurate dynamic model that can properly reflect the dynamic properties (especially the first few modes of interest) of actual in-service structure is of primary importance to structural design and analysis.

Researchers increasingly realize the contribution of NSEs to the dynamic properties of structures [11–22]. Reynolds [11] and Miskovic and Pavic [12] investigated the dynamic properties of the slender floor system and found that NSEs could have a remarkable impact on their dynamic properties. The research by Su et al. [13] revealed that compared with other parameters, NSEs were the largest contributor to structural stiffness. Petrovic and Pavic [14] reported that partition walls would greatly increase structural stiffness and damping. Studies by Devin and Fanning [15] demonstrated that infill walls and cladding panels had a key impact on the dynamic properties of the floor system. The fundamental frequency of the floor system increased by 9% after the addition of the cladding panels. To summarize, it is generally observed that NSEs would alter the natural frequencies of structures [16–19], increase structural damping [20–24], and sometimes even change the order of mode shapes [22].

Although there are several studies concerning the potential influence of NSEs on structural properties, most of them are qualitative evaluations, and the quantification of NSEs is still limited. There have been several attempts to properly model NSEs. The relevant work could be categorized into two types: (i) capacity-based modeling. It falls within the scope of structural safety and refers to establishing a finite element model for structural safety assessment, i.e., to consider the potential effect of NSEs under ultimate loadings like earthquakes or strong winds to ensure structural safety (strength). Telue and Mahendran [5,6] investigated the behavior of both sides of lined steel wall frames using both experimental and FE modeling methods. In the experiment work, they performed the ultimate load-capacity test and load-deformation test and analyzed the potential failure mechanism. Gaiotti [7] modeled the infill walls using the diagonal truss model to consider its contribution to structural responses to wind and seismic loading. The research by Madan et al. [8] also employed the diagonal truss model with hysteric-force deformation to consider the contribution of infill walls under seismic loading. Saifullah et al. [9] investigated the strength and stiffness of plasterboard-lined steel-framed ceiling diaphragms when subject to lateral loading like wind and earthquake loads. They first conducted experimental tests of typical ceiling diaphragms. Then, the FE model was developed using ANSYS software, and the nonlinearity of material, geometric, and element nonlinearity were included in the model. The model was capable of illustrating test results. Brandolese et al. [10] studied the behavior of suspended ceilings using the experimental fragility method. The cyclic quasi-static tests on suspended ceilings were conducted to characterize the damage progression under cyclic displacements and strength degradation. Numerical studies were also performed to validate and expand the experimental work; (ii) serviceability-based modeling. It corresponds to the field of serviceability and refers to establishing a model for vibration serviceability (or comfort) assessment, i.e., to consider the effect of NSEs on structural dynamic properties and responses under in-service or operational loadings, such as human activities or vehicles. Pavic et al. [25] measured the dynamic properties of an office floor and observed a difference of 15% between predicted natural frequencies (i.e., using the initial model developed according to construction drawings without NSEs) and measured results. Based on the initial model, they utilized Shell 63 element to represent the exterior brick walls and achieved satisfactory predictions in natural frequencies to the actual structure. Jiménez-Alonso et al. [23] studied the effect of vinyl flooring on the modal properties of a steel footbridge. They observed an increase in the damping ratio of 2.069% due to the installation of vinyl flooring through a modal test. Then, they set up the FE model of the footbridge using 3D beam elements to model the arches and the deck and 3D cable elements to model the hangers. The non-structural floor (both the glass layer and the vinyl layer) was assumed to have no effect on the structural stiffness and only affect mass. This model was capable of representing the first three modes. Different models should be employed for different calculation purposes. It is evident that the states of structures in (i) and (ii) are apparently different, and the contributions or work status of NSEs are obviously different. Thus, different modeling frameworks and modeling methods for NSEs should be employed according to the actual state (stress state and strain state) of the target structure.

From the above analysis, it is clear that the modeling of NSEs to structural dynamic properties remains limited for serviceability-based analysis and needs further exploration. Moreover, owing to the unclear boundaries between strong vibration and micro-vibration, the different contribution of NSEs in capacity-based modeling and serviceability-based modeling is not clearly understood.

The aim of this study is to properly consider the contribution of NSEs to the dynamic properties of structures within the range of serviceability-based vibration. Field measurements and modal properties identification of various full-scale structures at different construction phases were conducted. A general framework for considering the effect of NSEs on the dynamic properties of large-span structures was proposed and implemented for the structures involved.

The remainder of this paper is structured as follows. Section 2 briefly describes the test structures. Modal testing and parameter identification of test structures under different construction phases are illustrated in Section 3. Section 4 illustrates the basic principles and modeling methods of NSEs. Section 5 verifies the proposed method using full-scale structures. The main conclusions and future work are presented in Section 6.

2. Description of Test Structures

Two full-scale large-span structures, including an indoor stadium and an elevated eight-lane urban highway bridge, are introduced in this section. These structures are all located in China. They generally cover most types of flexible large-span structures favored in engineering practice. Owing to the relatively lightweight, large-span, and low-damping, these structures are susceptible to dynamic loadings like human activity or vehicle load. Thus, their dynamic properties (i.e., modal mass, natural frequency, mode shape, and damping ratio) are of great interest to both engineers and academic societies.

2.1. BU Stadium

BU Stadium, shown in Figure 1, is a large-span indoor stadium for indoor sports like basketball, badminton, and tennis. It also offers a venue for daily training. Compared with the traditional single-story stadium, BU Stadium has two floors, including a large training area of over 2000 m^2 on the first floor and a playing field and grandstands on the second floor. A large-span pre-stressed concrete floor with dimensions of 42 m by 45 m is adopted by the stadium on the second floor. To improve the net height of the first floor, the depth of beams supporting the 2nd floor was designed to be 1.7 m, leading to a depth-span ratio of 1/24.7 of beams. The relatively low depth-span ratio of the second floor makes its dynamic performance under human activity a great concern.

Figure 1. BU Stadium. (a) Elevated view, (b) interior of 2nd floor (before decoration), (c) interior of 2nd floor (after decoration), and (d) impacting test.

Modal tests of the large-span floor were also carried out in two different construction phases, namely, before and after decoration, as shown in Figures 1b and 1c, respectively. The decoration is mainly the pavement of floors, and the construction layers for slab pavement involve a ground leveling layer (mortar), self-leveling mortar, and wooden floor on the structural concrete slab.

2.2. Dongfeng Bridge

Dongfeng Bridge, shown in Figure 2, is an elevated eight-lane urban highway bridge in Wuhan, central China, and it connects the third ring road and the fourth ring road of Wuhan City. The total length of the bridge is 5.808 km. The width of the deck is 33 m, making it one of the widest highway bridges in Wuhan. The whole bridge employs 51 continuous girders, including 16 steel box girders and 35 pre-stressed reinforced concrete box girders. The span varies from 25 m to 89 m. In this study, one three-span continuous steel-box girder (named L7), which had the longest span of 89 m, was selected for modal testing since it is expected to have low natural frequencies and closely spaced modes. Bridge L7 comprises three spans of 57.4 m, 89 m, and 57.4 m and is supported by double piers, as shown in Figure 2a,c. The steel box girder, having single-box cross sections, is divided into six rooms, as shown in Figure 2a. Eight bearings were placed on top of each pier, as illustrated in Figure 2b, including one fixed bearing (B2), four one-way bearings (i.e., B1, B3, and B4 can move freely in the longitudinal direction, and B6 in the transverse direction) and three double-way bearings (i.e., B5, B7, and B8).

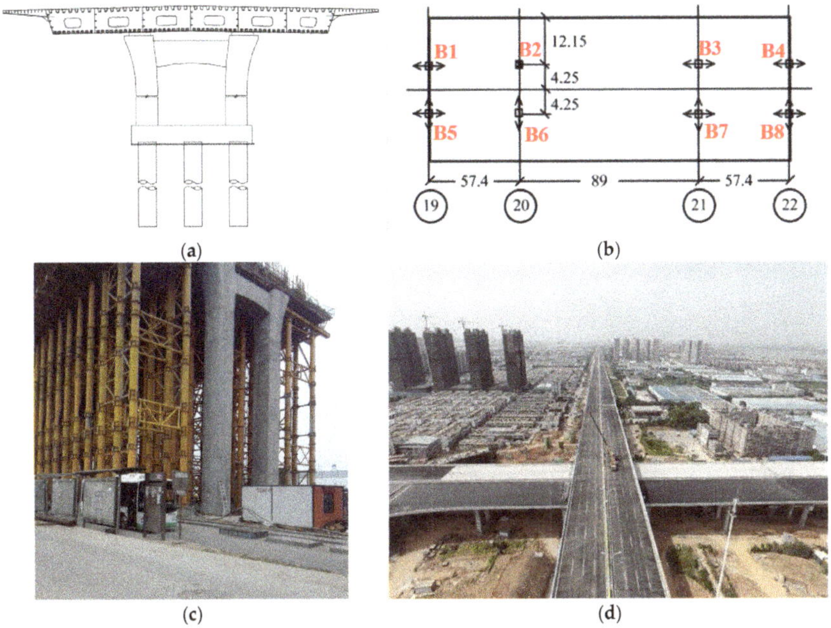

Figure 2. (**a**) Cross section of bridge L7, (**b**) plane view of bridge L7 with the layout of bearings (arrows denote possible movement directions), (**c**) side view (in construction), and (**d**) deck view (deck pavement). All dimensions in m.

The construction details of the L7 Bridge are introduced as follows:

(1) Superstructure

The structural steel utilized is Q345qd. The bridge deck comprises a rectangular orthotropic plate configuration, with the top plate typically exhibiting a thickness of 16 mm, which is increased to 40 mm in proximity to both the middle and end supports. Longitudinal reinforcement of the top plate is achieved through U-shaped stiffeners spaced approximately 550 mm apart. The bottom plate generally has a thickness of 14 mm, is also thickened to 40 mm near the middle and end supports, and features T-shaped stiffeners in its longitudinal direction with spacing around 400 mm. Cross girders are positioned at intervals of 3 m, maintaining a thickness of 14 mm while being reinforced near support locations and at cantilever ends. A total volume of 437 m^3 of weighted concrete is poured at both the middle and end supports. The surface treatment for the bridge deck consists sequentially, from top to bottom, of a layer comprising 9 cm SBS modified asphalt followed by an additional layer of 8 cm C50 steel fiber concrete. To ensure robust adhesion between the bridge deck surface and underlying concrete, shear pins with a diameter of 20 mm along with 10 mm welded steel mesh are strategically placed atop the steel box girder. Furthermore, the box girder incorporates a bidirectional structural slope set at an inclination rate of 2%, aligned with the centerline trajectory.

(2) Substructure

The bridge features an H-shaped double-column solid pier, with heights measuring 9.894 m, 8.491 m, 9.081 m, and 11.106 m (corresponding to bridge piers 19–22), and a center-to-center spacing of 8.5 m between the columns. The concrete used for the piers is classified as C40. A low-profile abutment is implemented, with dimensions of $10 \times 8.5 \times 3$ m constructed from C30 concrete. The pile foundation consists of drilled and cast-in-place concrete piles designed for end bearing, approximately 50 m in length, and also utilizes C30 concrete.

(3) Other Structures

The expansion joint on both sides of the beam is of type SF320, and the support is a JQGZ universal ball-type support. The adjacent L6 beam (19th pier) is a four-span continuous concrete box girder (4 × 30 m), while the L8 beam is a variable-width-and-height four-span continuous steel box girder (34 m + 36 m + 40 m + 44 m).

Three construction phases of bridge L7 were considered: (i) Phase 1, the structural construction of the bridge has been completed, but without deck pavement; (ii) Phase 2, 8 cm thick steel fiber reinforced concrete was paved on the steel deck; (iii) Phase 3, another 9 cm thick modified asphalt concrete was paved on Phase 2 structure, and the bridge is ready for in-service use. The construction method of the bridge deck pavement is shown in Figure 3.

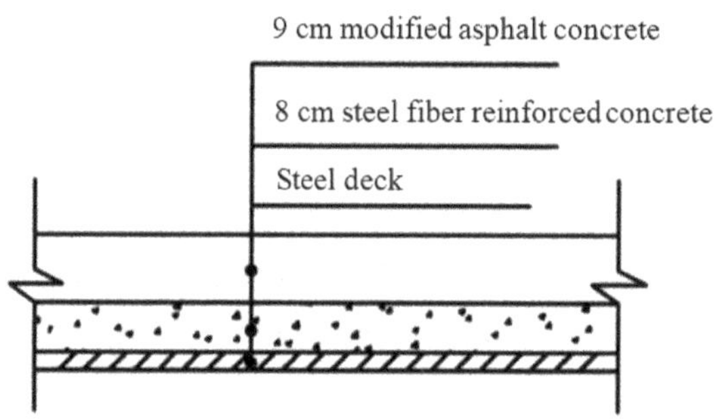

Figure 3. Construction method for bridge deck pavement.

3. Modal Testing and Parameter Identification of Test Structures Under Different Construction Phases

Table 1 summarizes the dynamic testing and analysis details of all involved structures. The ambient vibration (AV) testing method [26–30] was employed in modal testing of all involved structures owing to their huge masses or flexibility. The heel impact (HI) excitation method [31,32] was also introduced in the modal testing of BU Stadium to check the performance of AV testing. During the experiment, the test subject (Gender: male; Age: 24 years; Body mass: 71.3 kg) was instructed to jump off the chair and impact the floor with heels, as shown in Figure 1d.

Table 1. Modal testing and analysis methods for the tested structures.

Test Structure	Testing Method	Accelerometer	Sampling Frequency (Hz)	Identification Method
BU Stadium	AV + HI	VSE-15-D1	200	SSI + PP
Dongfeng Bridge	AV	B&K 4507B	256	SSI + PP

HI: heel impact testing; AV: ambient vibration testing; ID: initial displacement method; SSI: stochastic subspace identification; PP: peak-picking.

Two types of accelerometers were used for data acquisition, including Bruel & Kjaer 4507B—piezoelectric accelerometers (nominal sensitivity of 1 mV/g and frequency range from 0.3 Hz to 6000 Hz) and VSE-15-D1 accelerometers (nominal sensitivity of 5 mV/gal and frequency range from 0.1 Hz to 200 Hz). Data were acquired using PULSE data acquisition software. The data acquisition system and test instruments are depicted in Figure 4.

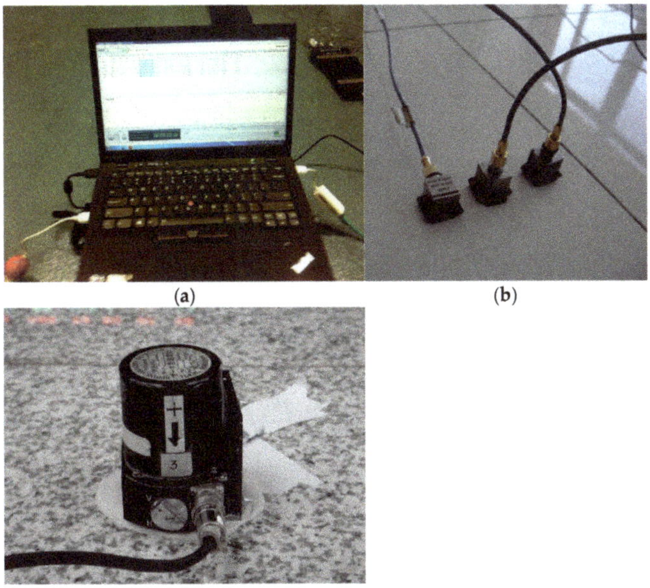
(a) (b) (c)

Figure 4. Data acquisition system and test instruments. (**a**) Bruel & Kjaer DAQ system, (**b**) 4507B accelerometers, and (**c**) VSE-15-D1 accelerometer.

3.1. Modal Testing of BU Stadium

Both ambient excitation and heel impact methods were employed for modal testing of BU stadium. For ambient vibration testing, modal properties of the floor were obtained using the data-driven SSI method [33,34]. The identified first three bending natural frequencies of the floor at Phase 1 stage are 5.00 Hz, 9.21 Hz, and 9.42 Hz, respectively, and they decreased to 4.61 Hz, 8.71 Hz, and 8.92 Hz accordingly after decoration. It should be noted that the mode shapes of the floor did not change after decoration. The 2nd and 3rd natural frequencies are close, owing to the similar length and width of the floor (45 m by 42 m).

In the following, results of the impacting method are presented to check the reliability of identified modal properties using the data-driven SSI method. The testing program was conducted in BU Stadium at different construction phases. TPs were installed at quarter points and midpoints in a grid, as shown in Figure 5. Figure 6 shows an example of the measured vertical acceleration response of the floor under a single-person (71.3 kg) excitation. As shown in the figure, the floor acceleration significantly increased once the heels touched the floor, then it gradually attenuated in a free decay manner to zero due to the presence of structural damping.

Figure 7a,b illustrate the Fourier spectrum of vertical accelerations at TP 3 (see Figure 5) under the same person's excitation before and after decoration, respectively. The first vertical bending frequency decreased from 5.00 Hz to 4.61 Hz, and the second bending frequency decreased from 9.21 Hz to 8.71 Hz after structural decoration. The structural damping increased since the spectrum became wider and flatter after structural decoration. It should be noted that the presence of a person's excitation on the floor would affect measured dynamic properties, known as human–structure interaction [35]; however, it should be noted that this effect could be ignored since the mass ratio of the single person to the test floor is relatively small (0.00018) [35]. Since TP 3 was placed at a quarter point along the length of the floor (see Figure 5), only two peaks (corresponding to the first symmetric bending mode and the second anti-symmetric bending mode along the slab length) were observed in Figure 7. After a careful examination of spectra at all TPs, we found that the natural frequencies identified using both methods agreed well with each other.

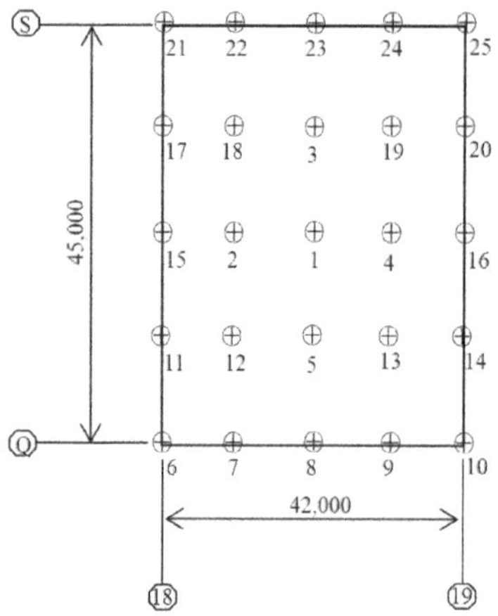

Figure 5. Plan view of slab B4 and sensor arrangements. All dimensions in mm.

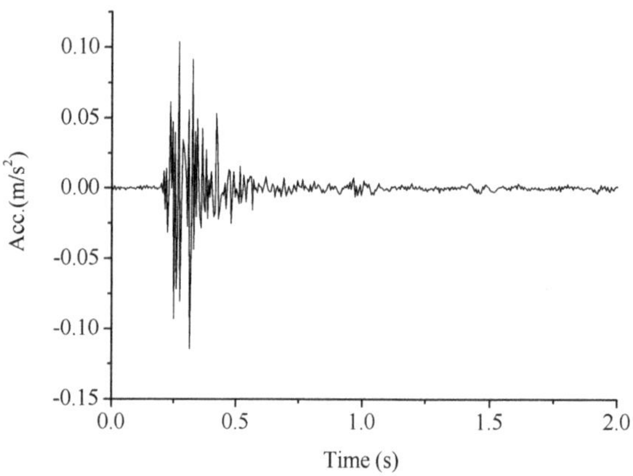

Figure 6. Measured floor responses with single-person impacting.

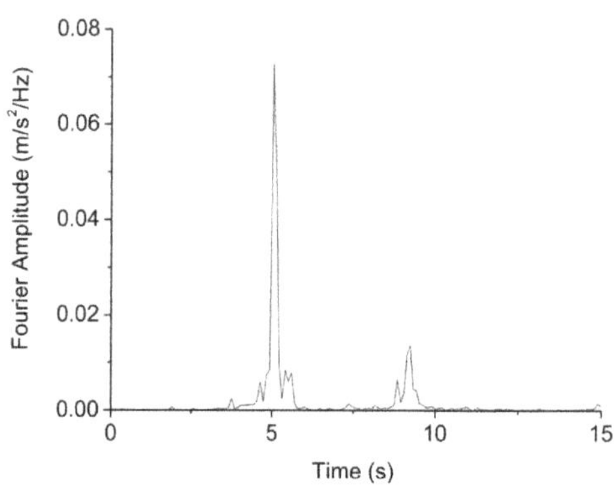

(a)

Figure 7. *Cont.*

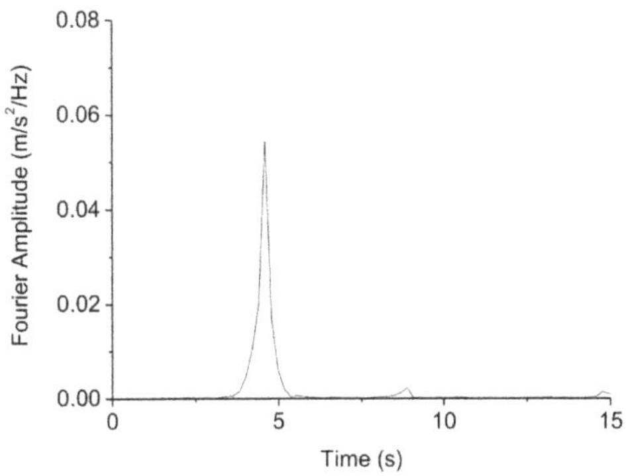

(b)

Figure 7. Measured floor responses with single-person impacting. (**a**) Time history; (**b**) frequency sepectrum.

3.2. Modal Testing of Dongfeng Bridge

The AV testing method was utilized for the L7 bridge owing to its flexibility and large size. A sufficiently dense grid of TPs was employed for L7 to identify the first few modes of interest. The sensor deployment of the L7 bridge is shown in Figure 8. Sensors installed on the bridge deck were arranged in three lines along the longitudinal direction of the bridge considering the large width of the deck, i.e., two lines on both sides of the bridge and the middle line was placed 0.6 m away from the central line, which was occupied by the central separation barrier. The testing campaign was divided into several setups due to limited accelerometers and coverage of wire length. Three reference TPs (marked as triangles in Figure 8) were placed on the bridge deck, i.e., one on each span, and the position of the reference TPs was carefully checked, such that the targeted vibration modes were observable. Since modal testing is not the main focus of this study, the details of test setups will not be shown here. Interested readers are referred to [36]. A time series of 60 min was recorded during individual trials in data acquisition.

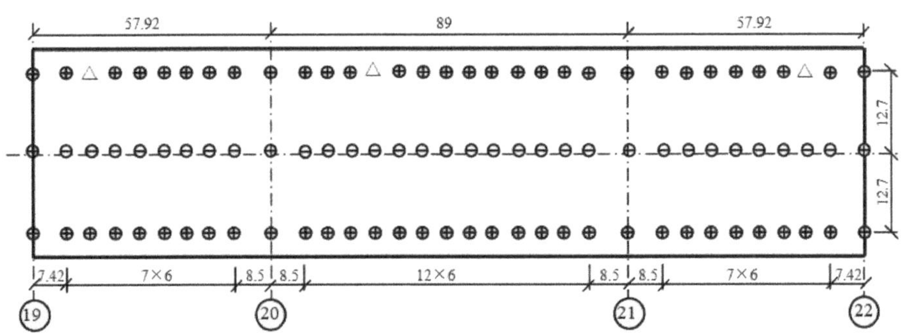

Figure 8. Test grid and sensors deployment of L7 bridge. Note: 1. All dimensions are in m; TP: test point; 2. △ Fixed reference TP, in three orthogonal directions; ⊕ mobile TP, in vertical and transverse directions; ⊖ mobile TP, in vertical direction.

Modal parameters of the L7 bridge were identified based on the reference-based data-driven SSI method as well as the peak-picking (PP) method [23]. The averaged normalized power spectrum density (ANPSD) [37] was employed in PP to increase the accuracy of identified modal frequencies. Figure 9 gives an example of a stabilization diagram at the reference point based on a data-driven SSI method combined with ANPSD. The red star (☆) represents stable points for frequency, the green quincunx (*) denotes stable points for both frequency and damping ratio, and the blue circle (○) denotes stable points for frequency, damping ratio, and mode shape. Obviously, the mode frequencies of the L7 bridge are closely spaced from 1 to 5 Hz. Table 2 compares the natural frequencies and damping ratios identified using the PP method and SSI method for the Phase 3 bridge, and the corresponding mode shapes are illustrated in Figure 10. It is observed that both methods can effectively identify the first few modes of interest. The identified frequencies by both methods agree well with each other, and the modal assurance criterion (MAC) values of the first 12 modes (except for mode 2, see Table 2) are close to the unit, offering promising results for the dynamic properties of the L7 bridge. The identified modes for transverse bending and torsion modes (see Figure 10b,d,f,g,j) are not as smooth as bending modes, partly because the excitation energy in longitudinal and transverse directions is much smaller than in vertical components (the amplitudes in longitudinal and transverse directions are about 25% of vertical one) under ambient vibration.

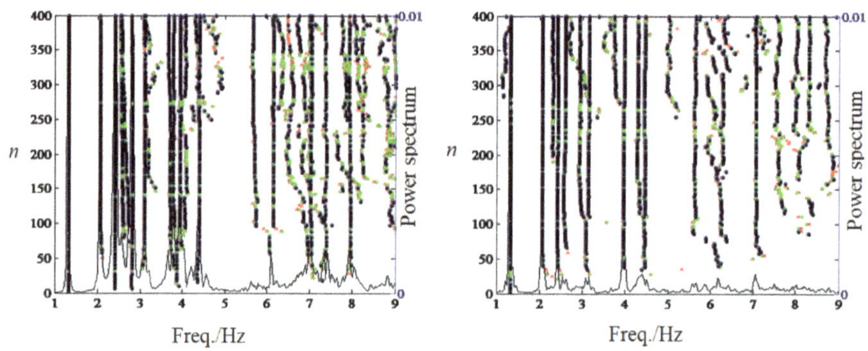

(a) (b)

Figure 9. Stabilization diagram at reference point: (**a**) vertical direction; (**b**) transverse direction. Note: ☆ represents stable points for frequency, * denotes stable points for frequency and damping ratio, ○ denotes stable points for frequency, damping ratio, and mode shape.

Table 2. Comparison of identified mode results using the PP method and SSI method for the Phase 3 bridge.

Mode No.	PP	SSI-DATA		Frequency Difference/%	MAC Value	Mode Description
	Frequency/Hz	Frequency/Hz	Damping Ratio/%			
1	1.318	1.320	0.60	−0.160	0.999	1st VB
2	NA	1.705	2.02	NA	NA	1st longitudina
3	2.053	2.064	1.78	−0.568	0.985	1st TB+ torsio
4	2.402	2.409	1.08	−0.299	0.994	2nd VB
5	2.563	2.582	2.90	−0.726	0.943	2nd TB+ torsio
6	2.789	2.794	1.42	−0.175	0.971	3rd VB
7	3.136	3.151	1.99	−0.465	0.993	1st torsion
8	3.656	3.683	1.20	−0.744	0.941	3rd TB+ torsio
9	3.775	3.787	1.53	−0.323	0.978	4th TB+ torsio
10	4.403	4.415	2.66	−0.281	0.969	4th VB
11	4.879	4.901	1.55	−0.463	0.942	5th TB+ torsio
12	5.657	5.706	1.41	−0.843	0.939	2nd torsion

Note: MAC-modal assurance criterion; NA-not available; VB-vertical bending; TB-transverse bending.

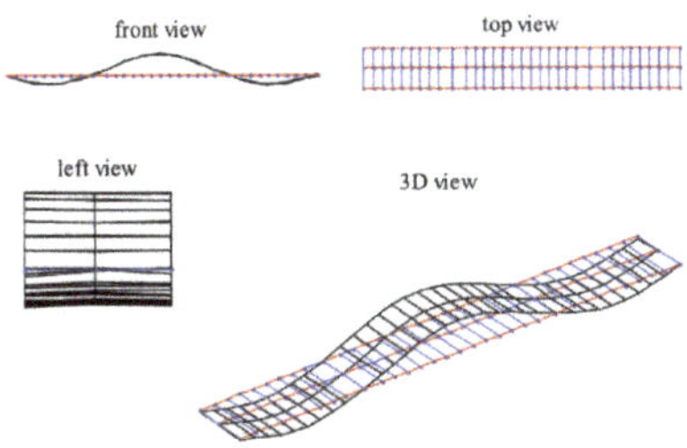

(**a**) first vertical bending (1.320 Hz)

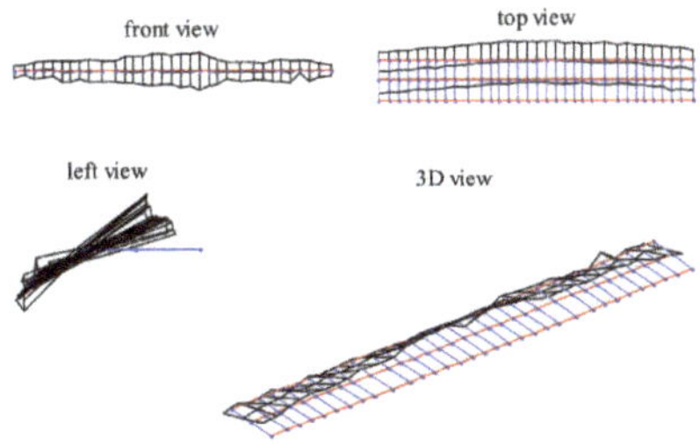

(**b**) 1st transverse bending+ torsion (2.064 Hz)

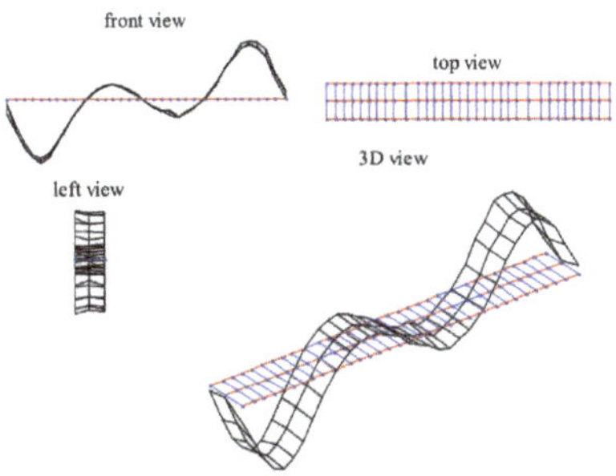

(**c**) 2nd vertical bending (2.409 Hz)

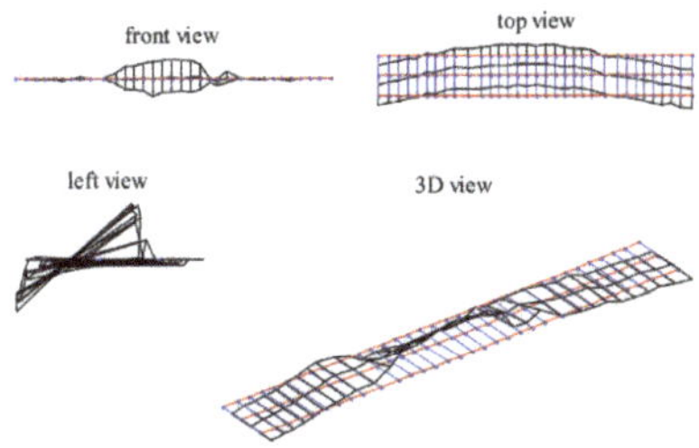

(**d**) 2nd transverse bending+ torsion (2.582 Hz)

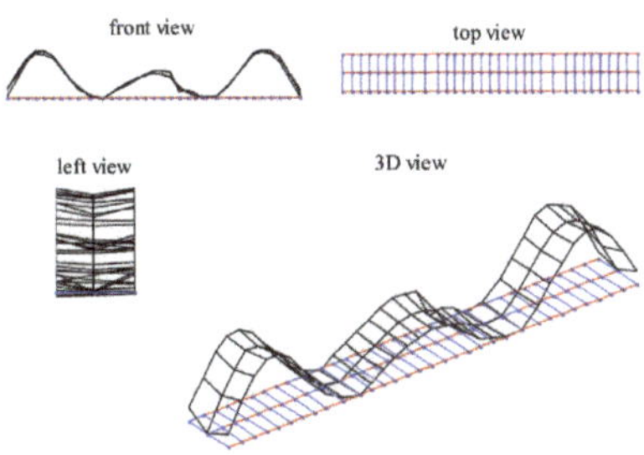

(**e**) 3rd vertical bending (2.794 Hz)

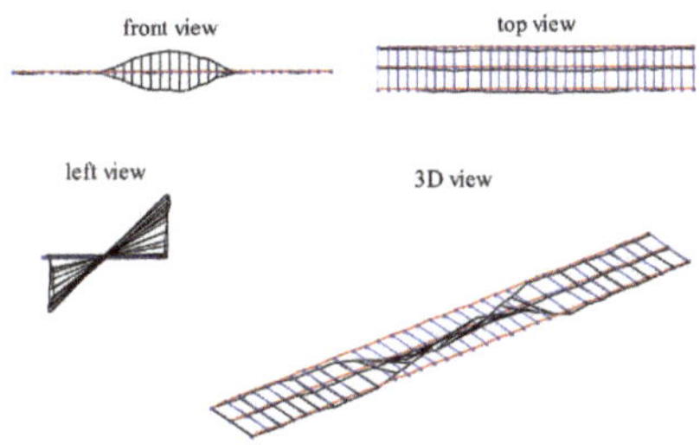

(**f**) 1st torsion (3.151 Hz)

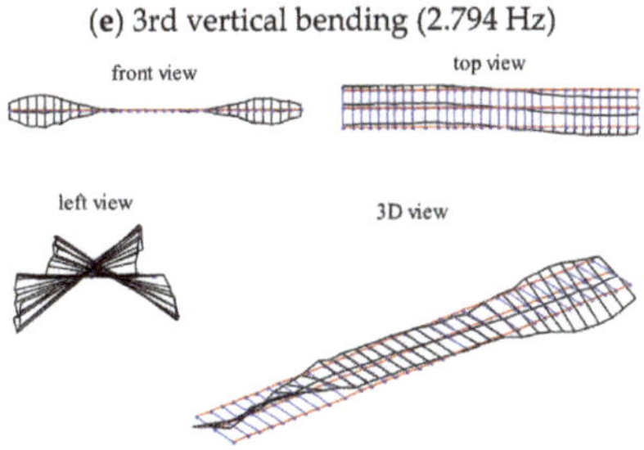

(**g**) 3rd transverse bending+ torsion (3.683 Hz)

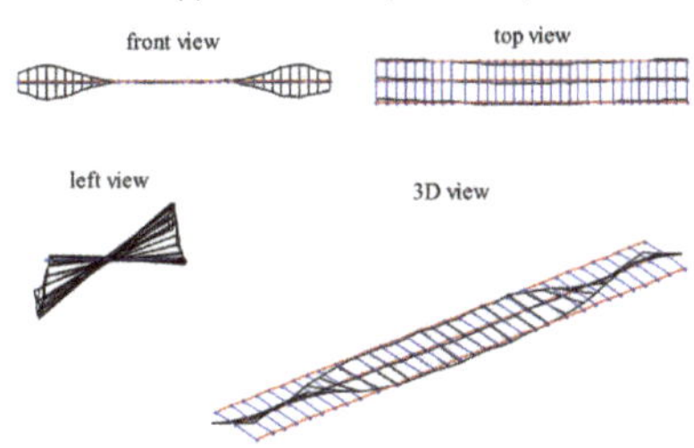

(**h**) 4th transverse bending+ torsion (3.787 Hz)

Figure 10. *Cont.*

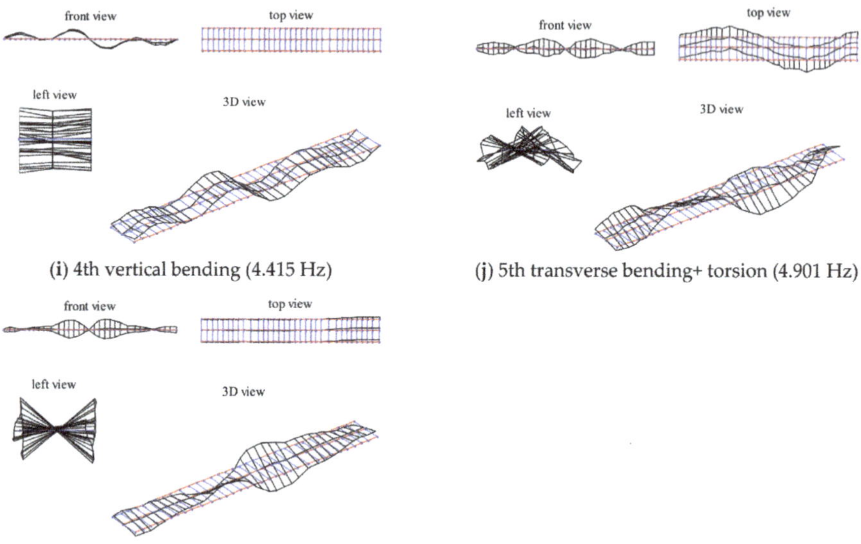

(**i**) 4th vertical bending (4.415 Hz)

(**j**) 5th transverse bending+ torsion (4.901 Hz)

(**k**) 2rd torsion (5.706 Hz)

Figure 10. The first few mode shapes of the L7 bridge based on the SSI method.

Based on these two methods, Table 3 summarizes the first 12 modes of the L7 bridge under different construction phases. As shown in Table 3, the bridge has low natural frequencies and small damping ratios, and the first few modes are closely spaced. The natural frequencies of the L7 bridge decreased after each stage of decoration. Specifically, they decreased by 15% to 21% in Phase 3 compared with Phase 1, whereas the damping ratios dramatically increased as non-structural components were included. It should be noted that introducing the pavements at Phase 2 and Phase 3 did not alter the mode shapes of the L7 bridge.

Table 3. Identified dynamic properties of the L7 bridge under different construction phases.

Mode No.	Phase 1		Phase 2		Phase 3		Mode Description
	Frequency /Hz	Damping ratio/%	Frequency /Hz	Damping Ratio/%	Frequency /Hz	Damping Ratio/%	
1	1.57	0.21	1.42	0.35	1.32	0.60	1st VB
2	2.07	0.68	1.90	1.91	1.71	2.02	1st longitudinal
3	2.38	0.62	2.24	1.29	2.06	1.78	1st TB+ torsion
4	2.76	0.48	2.58	0.69	2.41	1.08	2nd VB
5	3.16	1.12	2.84	2.20	2.58	2.90	2nd TB+ torsion
6	3.31	0.35	2.99	0.89	2.79	1.42	3rd VB
7	3.80	0.78	3.33	1.51	3.15	1.99	1st torsion
8	4.18	0.64	3.65	1.71	3.68	1.20	3rd TB+ torsion
9	4.52	0.52	3.96	0.80	3.79	1.53	4th TB+ torsion
10	5.49	0.67	4.85	1.47	4.42	2.66	4th VB
11	6.22	0.73	5.57	1.07	4.90	1.55	5th TB+ torsion
12	7.23	0.85	6.44	1.03	5.71	1.41	2nd torsion

Note: VB-vertical bending; TB-transverse bending.

4. Modeling of Non-Structural Elements

The aim of considering the contribution of NSEs is to let the dynamic properties of the numerical model approach the actual in-service structure, especially for the first few modes of interest. This is of great importance for serviceability-based performance evaluation and vibration mitigation, if necessary, of large-span flexible structures. The dynamic characteristics of a structure mainly depend on the distribution of its mass and stiffness, and connections and/or joints linking individual components would affect the damping of the system. In Section 3, experiments demonstrate that NSEs change the natural frequencies of a structure and increase structural damping. Note that the damping properties and mechanisms in structures are complex and would be affected by many factors [32,37], and the most effective and convincing method to obtain structural damping is through measurement or experimental tests. Hence, in the present study, we only focus on the contribution of NSEs on structural natural frequencies and mode shapes.

4.1. Main Types of NSEs in Large-Span Structures

For large-span structures investigated in this study, NSEs are mainly categorized into the following three types:

(1) slab/deck pavement;

(2) infill walls or crash barriers;

(3) joints/connections, such as temperature/settlement joints of large-span floors or bridge joints connecting two girders. Joints/connections, generally treated as free boundaries in structural design, are, in fact, filled with flexible materials (i.e., asphalt) or set as flexible connections to facilitate the normal use of structures. Since structures are usually in a linear elastic stage under in-service conditions, these joints/connections would behave much closer to rigid or semi-rigid connections rather than free boundary conditions (BCs).

The first two types of NSEs would alter the distribution of mass and stiffness of the structure, and the third one mainly affects structural stiffness through BCs, leading to the alteration in structural dynamic properties.

It should be noted that the contribution of additional mass, such as the presence of workers, construction machines, or preloading (construction materials stacked on floors), will not be included here since it will disappear soon after completion of construction.

4.2. Basic Principle and Modeling Methods of NSEs

The basic principle of appropriate modeling of NSEs is the equivalence of stiffness and mass as the actual structure, especially for the first few modes of interest.

For type (1), two methods could be employed here based on the equivalence of mass and stiffness as the actual structure. The first method is that the composite section of pavement and structural slab could be represented using a new section only if it has similar mass and moment of inertia as the original composite section such that one can update the parameters of structural slabs (i.e., thickness and material properties like density and elastic modulus) to consider the contribution of pavement without introducing new shell elements. The other method is using multi-layer shell elements with different thicknesses and material properties in the numerical model to consider the contribution of individual pavement. Obviously, the first method is an indirect method, and the latter is a direct method. It should be mentioned that these two methods are equivalent. These two methods will be introduced in the following section.

For type (2), the stiffness of infill walls or crash barriers should also be considered rather than additional mass since infill walls/crash barriers would not detach from the primary structure and would behave as a whole with the main structure. Both the in-plane and out-of-plane stiffness of the infill wall or crash barrier should be considered. For infill walls (i.e., masonry walls or lightweight partition walls), the equivalent pin-jointed strut model [2] (see Figure 11) is always employed to consider the stiffness of the structure. The in-plane and out-of-plane stiffness could be determined using the linear segment of the experimental results of a load–displacement curve under horizontal load, which are of increasing interest in the area of earthquake engineering, namely, to evaluate the effect of infill walls on seismic performance of main structures [5–10,38–40]. For a crash barrier that is made of reinforced concrete, one can treat it as a shear wall and use the elastic modulus and shear modulus to directly determine the in-plane and out-of-plane stiffness.

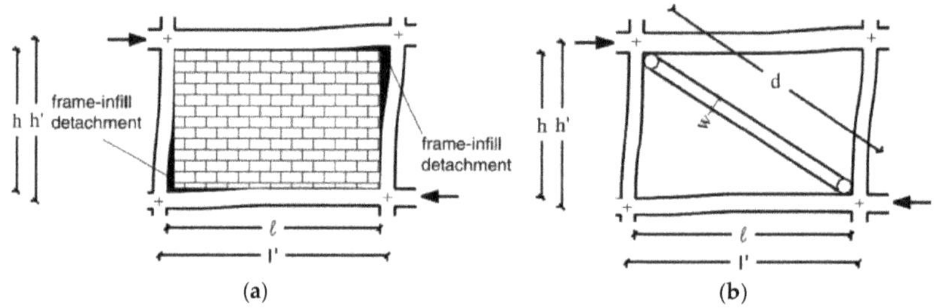

Figure 11. Infill walls under horizontal load: (**a**) actual system, (**b**) simplified model.

For type (3), the stiffness provided by NSEs in connections should be appropriately considered in the prediction model. The contribution of stiffness by NSEs could be modeled using additional linear spring elements, as shown in Figure 12. However, it remains challenging to accurately determine the stiffness of springs owing to the complex characteristics of joints and connections and the difficulty in quantifying the strains in BCs. In the current study, the trial-and-error method is suggested in determining the stiffness of springs.

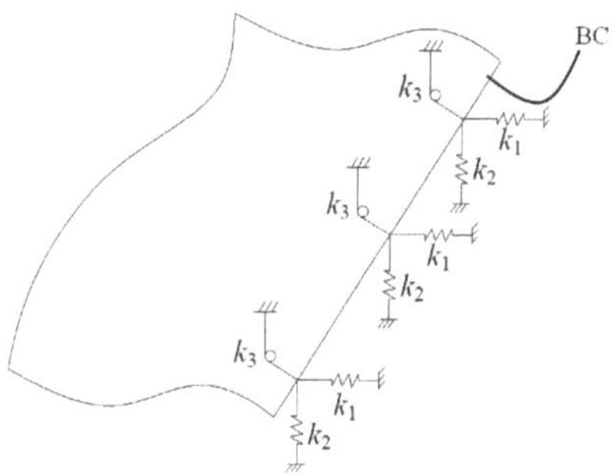

Figure 12. Schematic of boundary conditions to be processed.

It should be mentioned that although it is challenging to accurately determine the stiffness of the springs in joints/connections, one can roughly determine it by comparing the correlation of the first few local modes between predicted results against measured ones using the $Aver_{mac}$ indicator by the trial-and-error method. $Aver_{mac}$ is defined in the following form as:

$$Aver_{mac} = \min(MAC(1), \ldots, MAC(n)) \tag{1}$$

in which n is the number of modes of interest and MAC (modal assurance criterion) is the correlation between measured the predicated mode shapes, defined as follows [41–43]:

$$MAC(i) = \frac{\left[(\phi_i^A)^T \phi_i^B\right]^2}{\left[(\phi_i^A)^T \phi_i^A\right]\left[(\phi_i^B)^T \phi_i^B\right]} \tag{2}$$

where ϕ_i^A is the simulated ith modal shape vector by finite element method, and ϕ_i^B is the ith identified modal shape from measurement.

MAC value represents the correlation between two modal vectors, and a perfect correlation is achieved for $MAC = 1$ and no correlation when $MAC = 0$. In practice, it is generally accepted that the two vectors have acceptable correlations when $MAC > 0.80$. Thus, the $Aver_{mac}$ considers the overall correlation between the first few modes of interest. The larger the $Aver_{mac}$ is, the closer the predicted mode shapes to the measured ones. By setting the proper range of stiffness of springs and using $Aver_{mac}$ as an indicator, the stiffness of springs could be determined.

5. Validation of the Proposed Method

The performance of numerical models developed considering the effect of NSEs using the above-mentioned method will be presented in this section.

5.1. BU Stadium

An original FE model of BU stadium at Phase 1 stage (named M1) was set up using ANSYS software. The details of this model can be found in Ref. [44]. Figure 13 shows the first three bending modes of the large-span floor obtained from the modal analysis of the FE model. The natural frequencies for the second and third modes are close to each other owing to the similar dimensions (45 m by 42 m).

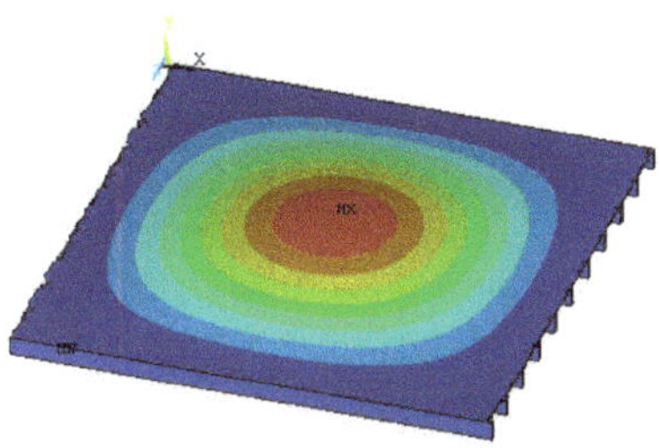

(**a**)1st vertical mode, $f_{v,1} = 4.39$ Hz

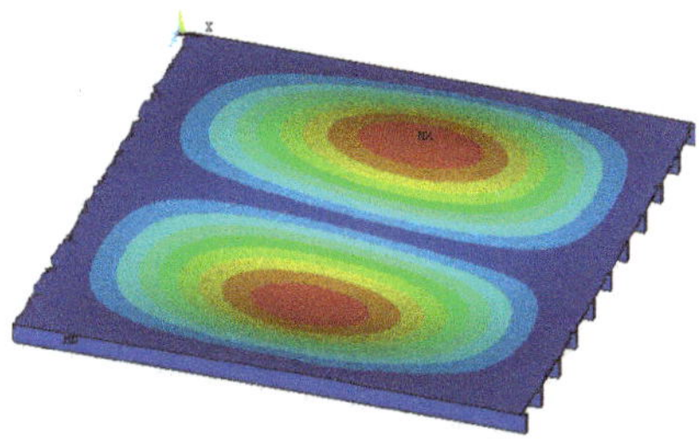

(**b**) 2nd vertical mode, $f_{v,2} = 8.41$ Hz

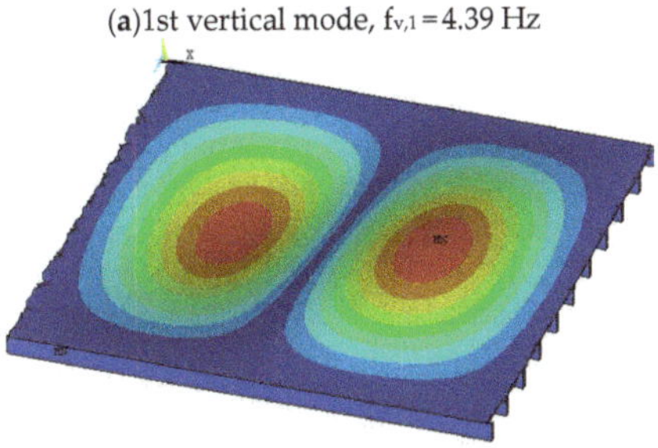

(**c**) 3rd vertical mode, $f_{v,3} = 8.59$ Hz

Figure 13. First three bending modes of large-span floors in BU stadium (Phase 2) by the FE method.

For BU Stadium, NSEs are mainly the pavement (ground leveling layer, self-leveling mortar, and wooden floor) on the structural concrete slab. The contribution of the pavement was achieved by modifying the parameters of the original concrete slabs according to the principle of equivalent stiffness and mass compared with the actual structure. The new model developed here was named M2. It is worth noting that only the mass of the wooden floor was included in M2, and the stiffness was ignored. As previously observed in a field measurement [45], the effective stiffness provided by the wooden floor to the overall structure is small due to the sliding between the wooden floor and underneath slabs. Modal properties, specifically the natural frequencies and modal shapes of M2, were obtained by modal analysis using ANSYS software. The mode shapes for M2 are similar to M1; namely, the mode shapes did not change after structural decoration.

Table 4 evaluates the performance of the M1 and M2 models on the prediction of the first three bending modes of the floor. As shown in Table 4, both M1 and M2 have satisfactory predictions for the first three modes. The natural frequencies of the slab slightly decreased after decoration owing to the more pronounced increase in structural mass compared with stiffness. The predicted natural frequencies of the M2 model are slightly smaller than the measured ones, partly because of neglecting the stiffness of wooden floors.

Table 4. Comparison of measured and calculated first three bending frequencies of large-span floors in BU stadium.

Mode No.	Calculated [Hz] M1	M2	Measured [Hz] T1	T2	Variance/% (M1–T1)/T1	(M2–T2)/T2
1	4.82	4.39	5.00	4.61	−3.55	−4.68
2	8.70	8.41	9.21	8.71	−5.58	−3.49
3	8.80	8.59	9.42	8.92	−6.59	−3.72

Note: T1: test results for structure before decoration; T2: test results for structure after decoration.

5.2. Dongfeng Bridge

The numerical model of the L7 bridge under different construction phases was developed using ANSYS software according to the construction drawings. The steel-box girder was modeled using shell-181 elements to represent the top, bottom, and web plate. Beam-188 elements with self-defined sections were employed to model U-shape, T-shape, and rectangle stiffeners in the girder. The piers were also modeled using beam-188 elements with self-defined sections, and the effect of bearing platforms and piles was not included in the model. Deck pavements were considered using multi-layer shell elements (shell 181) by attributing each layer with different thicknesses and material properties according to the actual construction method. The separating and crash barriers were also considered in the model. The spherical bearing was modeled using linear springs, and stiffness was determined based on the experimental results of the mechanical properties test (provided by the bearing supplier). The contribution of stiffness provided by bridge joints on both edges of the L7 bridge was considered using one-dimensional (longitudinal) spring elements, i.e., only the longitudinal constraint effect was considered. The stiffness of the spring was determined by comparing the first few modes of interest between simulated results against measured ones using the $Aver_{MAC}$ value [see Equation (1)] by the trial-and-error method. In the current study, only the first four modes were considered. A threshold value of $Aver_{mac} \geq 0.9$ is suggested after careful examination of all involved modes. The involved calculation parameters are listed as follows: elastic modulus of steel and concrete are 2.06×10^{11} and 3.25×10^{10} N/m^2, respectively; density for steel, concrete, and asphalt are 7850, 2500, and 2300 kg/m^3, respectively; the stiffness of springs at temperature joints along the longitudinal direction of the bridge is 2.01×10^6 N/m. Figure 14a,b illustrate the cross section of the bridge deck and the overall FE model of the L7 Bridge (Phase 3), respectively.

(a)

(b)

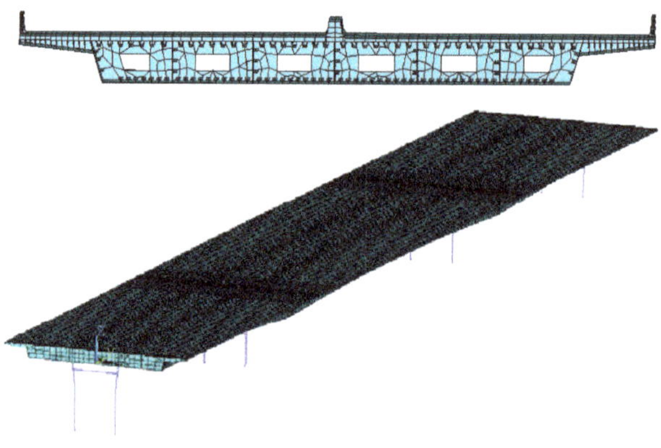

Figure 14. (**a**) cross section and (**b**) FE model of the L7 bridge.

Modal properties of the bridge were obtained using modal analysis using the ANSYS software.

Taking Phase 3 as an example, modal analysis was performed on the initially established finite element model of the L7 bridge, and the simulated and measured frequencies are presented in Table 5. Figure 15 depicts the first 12 mode shapes of the L7 bridge (Phase 3) obtained from the initial FE model.

Table 5. Comparison of measured and simulated first 12 modes of L7 bridge under Phase 3.

Mode No.	f_m	f_s	Variance (%)	Mode Description
1	1.32	1.36	3.58	1st VB
2	1.71	1.72	1.04	1st longitudinal
3	2.06	1.97	−4.50	1st TB+ torsion
4	2.41	2.44	1.18	2nd VB
5	2.58	2.65	2.63	2nd TB+ torsion
6	2.79	2.88	3.12	3rd VB
7	3.15	3.24	2.96	1st torsion
8	3.68	3.75	1.80	3rd TB+ torsion
9	3.79	3.88	2.64	4th TB+ torsion
10	4.42	4.60	4.21	4th VB
11	4.9	4.99	1.72	5th TB+ torsion
12	5.71	5.99	4.98	2nd torsion

Note: f_m—measured natural frequencies; f_s—simulated natural frequencies.

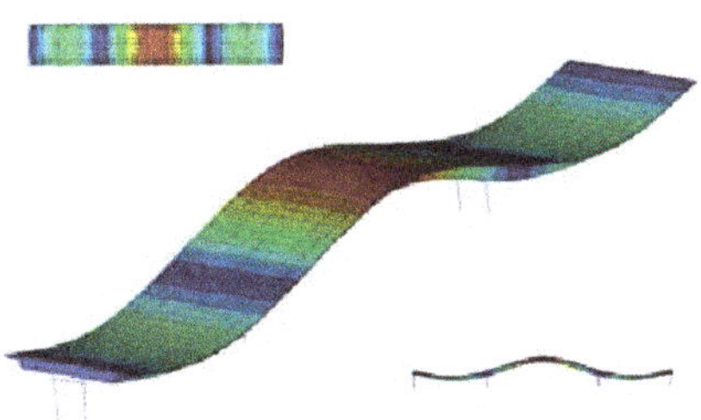

(**a**) 1st vertical bending (1.36 Hz)

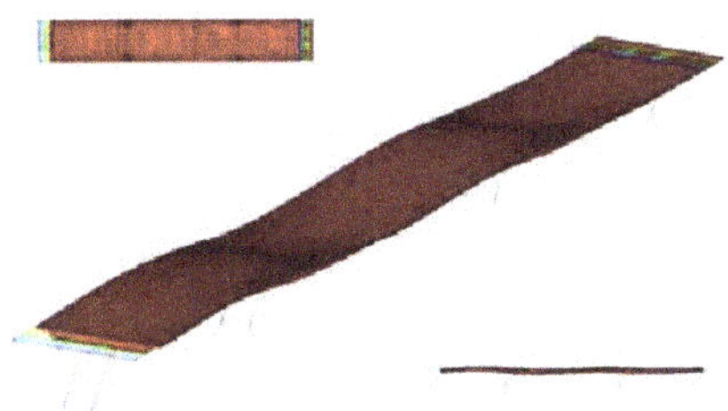

(**b**) 1st longitudinal (1.72 Hz)

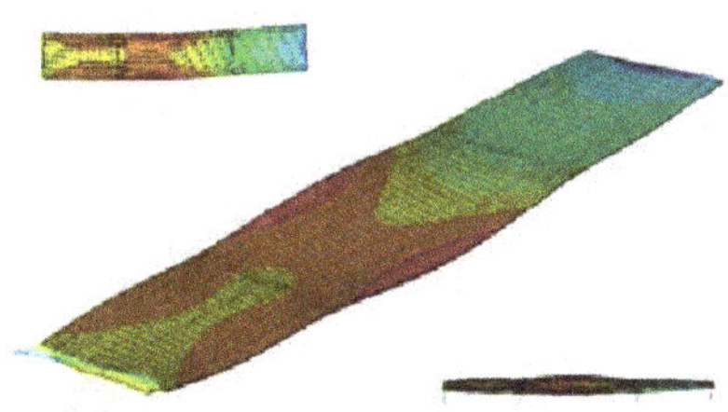

(**c**) 1st transverse bending+ torsion (1.97 Hz)

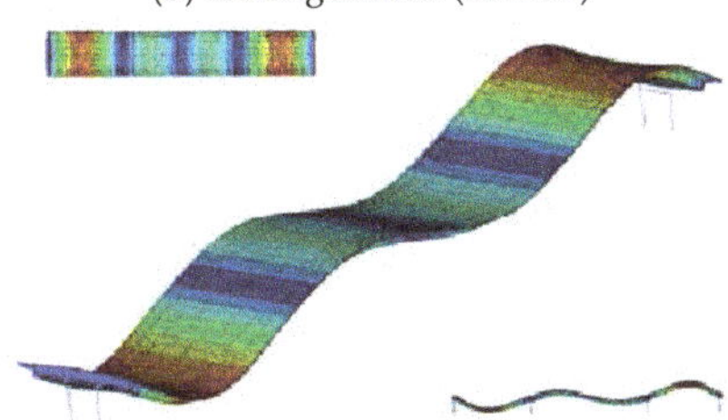

(**d**) 2nd vertical bending (2.44 Hz)

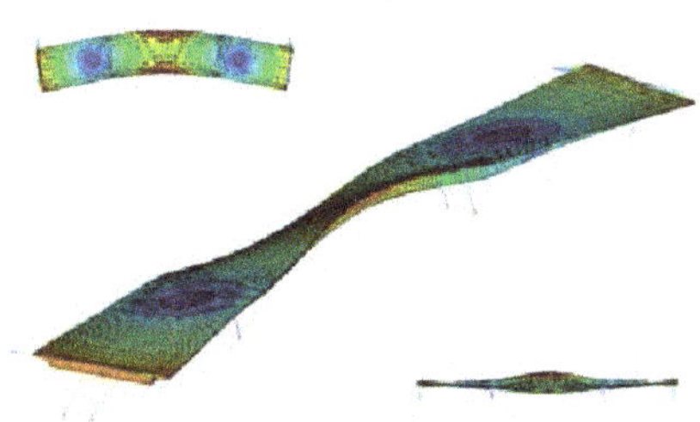

(**e**) 2nd transverse bending+ torsion (2.65 Hz)

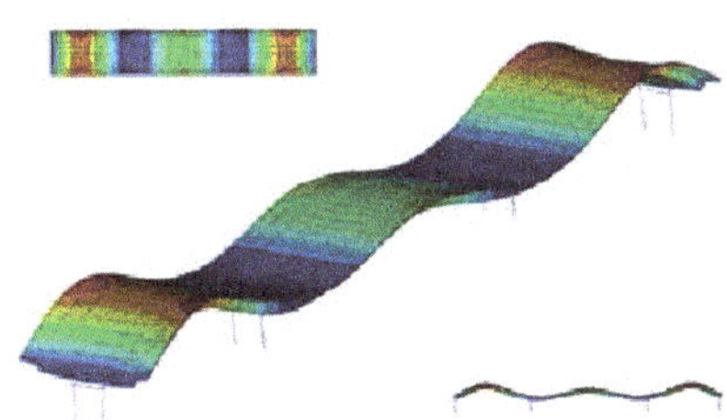

(**f**) 3rd vertical bending (2.88 Hz)

Figure 15. *Cont.*

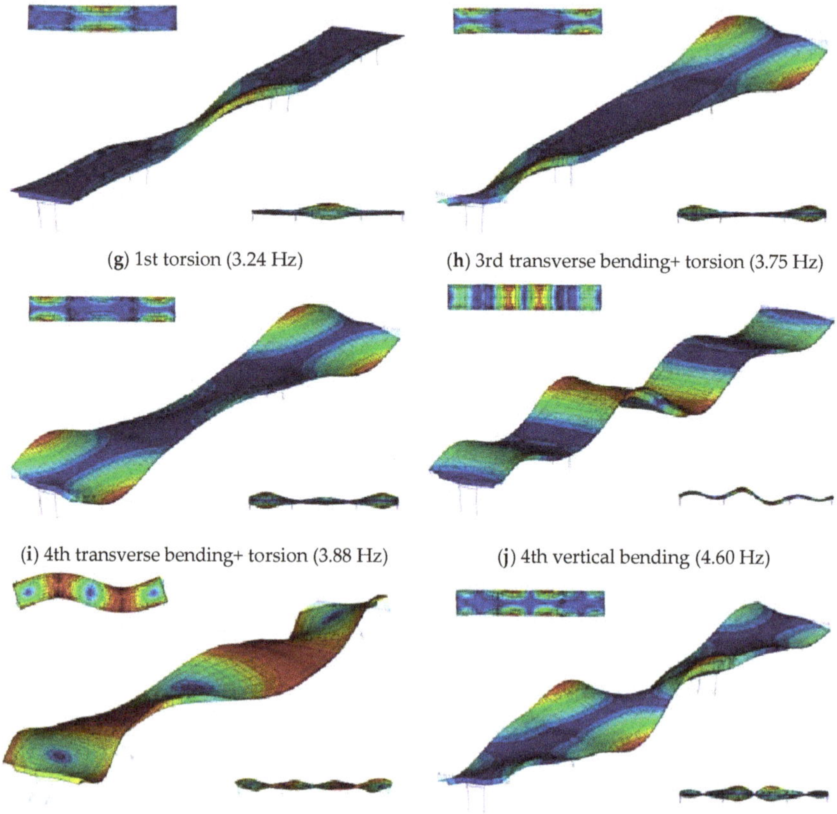

(**g**) 1st torsion (3.24 Hz) (**h**) 3rd transverse bending+ torsion (3.75 Hz)

(**i**) 4th transverse bending+ torsion (3.88 Hz) (**j**) 4th vertical bending (4.60 Hz)

(**k**) 5th transverse bending+ torsion (4.99 Hz) (**l**) 2rd torsion (5.99 Hz)

Figure 15. First 12 mode shapes of the L7 bridge (Phase 3) based on the initial FE model.

It is evident the predicted mode shapes are close to the ones identified in modal testing (shown in Figure 10). The relative errors between the frequencies obtained from finite element analysis and those measured ones are all within 5%, indicating a strong correlation between the simulated modal characteristics and the measurements.

To improve the results, the model updating procedures are further employed. Due to various assumptions, idealization, discretization, and parameterizations that are introduced in numerical modelling, the obtained numerical model may not always reflect the actual structural behavior. Currently, various methods for bridge finite element model calibration have been proposed, including sensitivity-based, maximum likelihood, non-probabilistic, probabilistic, response surface, and regularization methods. The sensitivity-based method treats the error between the theoretical and measured data as the objective function and modifies the sensitive parameter values to make the objective function optimal. This method requires iterative solutions during the process and needs to solve the sensitivity matrix of the system, which is of high cost for large structures. The response surface calibration method obtains the sample values by experimental design and obtains the explicit expression of the objective function and calibration parameter through regression analysis of the sample values. It approximates the complex implicit function relationship between the objective function and the calibration parameter using the explicit expression, which greatly improves the calibration efficiency. Suzana et al. [46] comprehensively reviewed the finite element model updating methods for structural applications. The whole process of model updating is described step by step: selection of updating parameters (design variables), definition of the model updating problem, and solution using different FEMU methods. Davide Raviolo et al. [47] systematically compared the optimization algorithms for finite element models, updating them on numerical and experimental benchmarks. Recently, they proposed a Bayesian sampling optimisation strategy for finite element model updating [48].

This paper uses the response surface method to calibrate the finite element model based on the modal test results of the bridge corresponding to each construction phase.

Taking Phase 3 as an example, five parameters are identified as correction factors based on sensitivity analysis and practical engineering insights: the elastic modulus of the steel plate (E_s), the mass density of the steel plate (ρ_s), the mass density of asphalt pavement (ρ_a), the elastic modulus of bridge piers (E_c), the longitudinal spring stiffness of expansion joints (k). Initially, value ranges for each parameter are established based on empirical data, followed by normalization of these parameters. The normalized levels for these parameters are presented in Table 6.

Table 6. Calibration items and level values.

Calibration Items	Level −1	Level 0	Level 1	Δ_i
Elastic modulus of steel plate E_s (10^{11} N/m^2)	1.85	2.06	2.26	0.21
Mass density of the steel plate ρ_s (kg/m^3)	7065	7850	8635	785
Mass density of asphalt pavement ρ_a (kg/m^3)	1840	2300	2760	460
Elastic modulus of bridge piers E_c (10^{10} N/m^2)	2.93	3.25	3.58	0.33
Longitudinal stiffness of expansion joints k (10^6 N/m)	1.8	2.0	2.2	0.2

As shown in Table 6, the density of asphalt concrete exhibits significant variation, which may be attributed to the influence of the mixture proportions on its density.

Firstly, the frequency residual is enlarged by 100 times as the objective function to avoid the rounding error. Subsequently, sensitivity analyses are conducted using the central difference method with regard to the five correction parameters. A center composite design method is used to design 27 orthogonal experiments for response surface equation fitting, and the F test is used to detect the significance of each coefficient. Insignificant correction coefficients and their combinations are eliminated, and after elimination, all the response surface equations have only one significant coefficient. The precision of the equations is verified using the R^2 test and the relative root mean square error (RMSE) value, and finally, the genetic algorithm in the MATLAB optimization toolbox is used to optimize the objective function and obtain the corrected values of the five parameters, as shown in Table 7.

Table 7. Selected parameters before and after updating.

Calibration Items	Initial Value	After Updating	Variance
ρ_s (kg/m^3)	7850	8493.8	8.2%
E_s (Pa)	2.06×10^{11}	1.90×10^{11}	−7.7%
E_c (Pa)	3.25×10^{10}	3.24×10^{10}	−0.03%
ρ_a (kg/m^3)	2300	1840.1	−19.9%
k (N/m)	2×10^6	1.91×10^6	−4.3%

The modal frequencies of the finite element model were recalculated using the updated parameters, and the results were compared with the measured values, as presented in Table 8.

From Table 8, it can be seen that, except for the 3rd and 4th-order frequencies, the errors of the bridge's higher-order frequencies after correction are significantly reduced and are all within 3%. The reasons for the increased errors of the 3rd and 4th-order frequencies are as follows: (1) The optimal solution of the multi-objective function optimization is the overall optimal solution, which may result in some objectives achieving the optimal solution while others deviate further from the optimal solution; (2) Sensitivity analysis can only guarantee effectiveness within the initial value neighborhood, and cannot objectively evaluate the global sensitivity of parameters. The correction parameters obtained through sensitivity analysis may not be the parameters with the largest errors, which may result in suboptimal correction effects for some objectives. The 3rd-order frequency error slightly increased, while the other higher-order frequency errors are extremely small (within 3% for the 3rd-order, within 1% for the 8th-order), indicating that the precision of the finite element model of the bridge has been significantly improved after correction.

Table 8. The relative error before and after updating.

Mode No.	f_m	f_s'	Initial Error (%)	Error After Updating (%)
1	1.32	1.31	3.58	−0.11
2	1.71	1.70	1.04	0.10
3	2.06	1.96	−4.50	−5.05
4	2.41	2.36	1.18	−2.06
5	2.58	2.61	2.63	1.16
6	2.79	2.79	3.12	−0.27
7	3.15	3.14	2.96	−0.06
8	3.68	3.67	1.80	−0.27
9	3.79	3.80	2.64	0.27
10	4.42	4.44	4.21	0.53
11	4.9	4.86	1.72	−0.90
12	5.71	5.84	4.98	2.40

Note: f_m—measured natural frequencies; f_s'—simulated natural frequencies after model updating.

Using similar methods, Table 9 shows the comparison of all simulated and measured dynamic properties of the L7 bridge at three different construction phases. The measured and simulated natural frequencies, as well as the mode shape correlation index, namely *MAC* values, are presented in the table. It is noted that the modal shapes have not changed for Phase 2 and Phase 3 compared with the Phase 1 structure.

Table 9. Comparison of measured and simulated first 12 modes of L7 bridge under different construction phases.

Mode No.	Phase 1				Phase 2				Phase 3				Mode Description
	f_m /Hz	f_s /Hz	Variance /%	*MAC*	f_m /Hz	f_s /Hz	Variance /%	*MAC*	f_m /Hz	f_s /Hz	Variance /%	*MAC*	
1	1.57	1.55	−1.05	0.999	1.42	1.44	1.59	0.999	1.32	1.31	−0.11	0.999	1st VB
2	2.07	1.99	−3.90	-	1.90	1.86	−2.22	-	1.71	1.70	0.10	-	1st longitudinal
3	2.38	2.45	3.10	0.964	2.24	2.29	2.01	0.965	2.06	1.96	−5.05	0.958	1st TB+ torsion
4	2.76	2.69	−2.45	0.991	2.58	2.53	−2.07	0.989	2.41	2.36	−2.06	0.983	2nd VB
5	3.16	3.34	5.60	0.877	2.84	2.91	2.46	0.867	2.58	2.61	1.16	0.848	2nd TB+ torsion
6	3.31	3.23	−2.35	0.995	2.99	2.91	−2.67	0.992	2.79	2.79	−0.27	0.987	3rd VB
7	3.8	3.95	3.90	0.931	3.33	3.18	−4.53	0.922	3.15	3.14	−0.06	0.921	1st torsion
8	4.18	4.07	−2.70	0.956	3.65	3.84	5.13	0.956	3.68	3.67	−0.27	0.943	3rd TB+ torsion
9	4.52	4.64	2.60	0.947	3.96	4.06	2.40	0.942	3.79	3.80	0.27	0.931	4th TB+ torsion
10	5.49	5.33	−2.85	0.979	4.85	5.06	4.41	0.975	4.42	4.44	0.53	0.965	4th VB
11	6.22	6.45	3.65	0.866	5.57	5.75	3.21	0.865	4.9	4.86	−0.90	0.849	5th TB+ torsion
12	7.23	7.53	4.10	0.943	6.44	6.64	3.09	0.943	5.71	5.84	2.40	0.939	2nd torsion

Note: f_m—measured natural frequencies; f_s—simulated natural frequencies. The 1st longitudinal mode shape is not available, and *MAC* for this mode is not listed in the table.

As shown in Table 9, the proposed method can effectively predict the first 12 modes of the L7 bridge at different construction phases. The largest discrepancies in natural frequencies for Phase 1–3 are 5.6%, 5.13%, and −5.05%, respectively. The *MAC* values for the first four vertical bending modes are over 0.96, indicating the predicted modes are well correlated with the measured ones. The predicted modes for transverse bending and torsion modes are not as perfect as vertical bending modes, partly because the excitation energy in longitudinal and transverse directions is much smaller than the vertical component (the amplitude is about 25% of the vertical one) under ambient vibration environment such that the identified modes are not as smooth (see Figure 10b,d,f,g,j), but they are generally acceptable since the minimum *MAC* value exceeds 0.84.

The natural frequencies of the L7 Bridge at Phase 3 decreased by 15–35% compared with Phase 1. In particular, the first vertical bending natural frequency decreased by 18.56% after deck pavement, implying the remarkable impact of NSEs on structural dynamic properties. Although the modes of the L7 bridge are closely spaced, the proposed method shows promising capability of predicting the first few modes of interest.

In a word, using the proposed method, one can accurately model the contribution of NSEs on structural dynamic properties.

6. Conclusions

The contribution of NSEs to the dynamic properties of large-span structures is systematically investigated in this study through field measurement, numerical simulation, and experimental techniques. Full-scale structures, including an indoor stadium and an elevated highway bridge, were measured at different construction phases to obtain the dynamic properties of interest. Results show that NSEs have a remarkable impact on serviceability-based dynamic properties of the target structure. In particular, it would alter natural frequencies, increase the structural damping, and sometimes even change the order of mode shapes of the structure. The evident difference in dynamic properties of structures under different construction phases reveals that the traditional load capacity models are incapable of reflecting the actual dynamic characteristics of in-service structures since they ignore the contribution of NSEs.

Another important conclusion is that different models should be employed for different calculation purposes. The main difference between the capacity-based model (i.e., for seismic performance assessment) and the serviceability-based model lies in the treatment of NSEs. Both models should consider the contribution of NSEs; however, since NSEs are in different states under these two apparently different conditions, different modeling methods should be employed to appropriately take into account the effect of NSEs.

A general modeling framework incorporating the contribution of slab/deck pavement, infill walls (or crash barriers), and joints/connections for large-span structures is developed to quantitatively consider the effect of NSEs based on the principle of equivalence of stiffness and mass on the actual structure. The effectiveness of the method is validated by field measurement results. It should be noted that the proposed method is somewhat empirical, especially in processing the boundaries (i.e., determining the stiffness of springs on BCs), since one needs to check the first few modes of interest by the trial-and-error method based on measured modal properties of the as-built structure. For structures to be built, it may show weakness since the actual dynamic properties (especially the mode shapes) are not available in advance. However, compared with the traditional load-bearing capacity model, the model developed here, taking into account the contribution of NSEs, would be much closer to the actual dynamic properties of the target structure. For structures having similarities with the test structures, the parameters obtained in this study could be utilized as useful references, which may further improve the accuracy of the predictions.

The contribution of this study is as follows: first of all, it is demonstrated that the traditional capacity-based models are incapable of reflecting the actual dynamic characteristics of in-service structures since they ignore the effect of NSEs. Secondly, this study provides a general framework for modeling the contribution of NSEs for serviceability-based modeling, which could serve as the first step toward a better understanding of the characteristics of NSEs for structures in operational stages. Thirdly, this study provides a detailed dataset of dynamic properties of various full-scale structures, giving light to further investigate the mechanism of interaction between main structure and NSEs. Still, much work needs to be conducted in the future, such as theoretical modeling of NSEs and quantitatively determining the potential influencing parameters.

Author Contributions: Conceptualization, methodology, writing—original draft preparation, W.H.; formal analysis, investigation, writing—review and editing, J.C. (Jialiang Chen); software, validation, writing—review and editing, C.S., S.H., J.C. (Junjie Chen) and Z.W.; resources, data curation, funding acquisition, W.H. All authors have read and agreed to the published version of the manuscript.

Funding: This research was funded by the project of the Ministry of Housing and Urban-Rural Development of the People's Republic of China (grant number 2022-K-153) and Zhejiang Engineering Center of Road and Bridge Intelligent Operation and Maintenance Technology (grant number 202404G).

Data Availability Statement: Data from the corresponding authors during this study are available upon request.

Conflicts of Interest: Author Jialiang Chen was employed by the company Beijing Oriental Yuhong Waterproof Technology Co., Ltd. Authors Jialiang Chen, Sen Hou and Junjie Chen was employed by the company Beiijao Zane Rail Technology (Beijing) Co., Ltd. Author Zhe Wang was employed by the company Beijing Metro Construction Administration Co., Ltd. The remaining authors declare that the research was conducted in the absence of any commercial or financial relationships that could be construed as a potential conflict of interest.

References

1. Devin, A.; Fanning, P.J. Non-structural elements and the dynamic response of buildings: A review. *Eng. Struct.* **2019**, *187*, 242–2 [CrossRef]
2. Cavaleri, L.; Papia, M.; Macaluso, G.; Di Trapani, F.; Colajanni, P. Definition of diagonal Poisson's ratio and elastic modulus f infill masonry walls. *Mater. Struct.* **2014**, *47*, 239–262. [CrossRef]
3. Klingner, R.E.; Bertero, V.V. Earthquake resistance of infilled frames. *J. Struct. Eng. ASCE* **1978**, *104*, 973–989. [CrossRef]
4. Chrysostomou, C.Z.; Asteris, P.G. On the in-plane properties and capacities of infilled frames. *Eng. Struct.* **2012**, *41*, 385–4 [CrossRef]
5. Telue, Y.; Mahendran, M. Behavior and design of cold formed steel wall frames lined with plasterboard on both sides. *Eng. Stru* **2004**, *26*, 567–579. [CrossRef]
6. Telue, Y.; Mahendran, M. Behavior of cold-formed steel wall frames lined with plasterboard. *J. Constr. Steel Res.* **2001**, *57*, 435–4 [CrossRef]
7. Gaiotti, R.; Smith, B. Stiffening of moment resisting frame by precast concrete cladding. *PCI J.* **1992**, *37*, 80–92. [CrossRef]
8. Madan, A.; Reinhorn, A.M.; Mander, J.B.; Valles, R.E. Modeling of masonry infill panels for structural analysis. *J. Struct. E* **1997**, *123*, 1295–1302. [CrossRef]
9. Saifullah, I.; Gad, E.; Shahi, R.; Wilson, J.; Lam, N.; Watson, K. Behavior of plasterboard-lined steel-framed ceiling diaphrag *Thin-Walled Struct.* **2019**, *141*, 1–14. [CrossRef]
10. Brandolese, S.; Fiorin, L.; Scotta, R. Seismic demand and capacity assessment of suspended ceiling systems. *Eng. Struct.* **2019**, *1* 219–237. [CrossRef]
11. Reynolds, P. The effects of raised access flooring on the vibrational performance of long-span concrete floors. Ph.D. Dissertati University of Sheffield, Sheffield, UK, 2000.
12. Miskovic, Z.; Pavic, A.; Reynolds, P. Effects of full-height nonstructural partitions on modal properties of two nominally identi building floors. *Can. J. Civ. Eng.* **2009**, *36*, 1121–1132. [CrossRef]
13. Su, R.K.L.; Chandler, A.M.; Sheikh, M.N.; Lam, N.T.K. Influence of non-structural components on lateral stiffness of tall buildin *Struct. Des. Tall Spec. Build.* **2005**, *14*, 143–164. [CrossRef]
14. Petrovic, S.; Pavic, A. Effects of non-structural partitions on vibration performance of floor structures: A literature Revi In Proceedings of the International Conference on Structural Dynamics (EURODYN 2011), Leuven, Belguim, 4–6 July 2011.
15. Devin, A.; Fanning, P.J. The evolving dynamic response of a four storey reinforced concrete structure during construction. *Sh Vib.* **2012**, *19*, 1051–1059. [CrossRef]
16. Falati, S. The Contribution of Non-Structural Components to the Overall Dynamic Behaviour of Concrete Floor Slabs. Ph.D. The New College, Oxford, UK, 1999.
17. Li, B.; Hutchinson, G.L.; Duffield, C.F. The influence of non-structural components on tall building stiffness. *Struct. Des. Tall S Build.* **2011**, *20*, 853–870. [CrossRef]
18. Craig, J.; Leistikow, R.; Fennell, C. Experimental studies of the performance of precast cladding connections. In Proceedings the Ninth World Conference on Earthquake Engineering, Tokyo, Japan, 2–9 August 1988; pp. 201–206.
19. Devin, A.; Fanning, P.J.; Middleton, C.J.; Pavic, A. Structural dynamic parameter identification and the effect of test techniqu In *Topics in Dynamics of Civil Structures, Volume 4: Proceedings of the 31st IMAC, A Conference on Structural Dynamics*; Springer: N York, NY, USA, 2013; Volume 2013, pp. 171–181.
20. Wyatt, T.A. *Design Guide on the Vibration of Floors*; Steel Construction Institute: London, UK, 1989.
21. Kim, Y.M.; You, K.P. Dynamic responses of a tapered tall building to wind loads. *J. Wind Eng. Ind. Aerodyn.* **2002**, *90*, 1771–17 [CrossRef]
22. Devin, A.; Fanning, P.J. Impact of nonstructural components on modal response and structural damping. *Top. Dyn. Civ. Str* **2012**, *1*, 415–421.

23. Jiménez-Alonso, J.F.; Pérez-Aracil, J.; Hernández Díaz, A.M.; Sáez, A. Effect of Vinyl Flooring on the Modal Properties of a Steel Footbridge. *Appl. Sci.* **2019**, *9*, 1374. [CrossRef]
24. He, W.; Xie, W. Characterization of stationary and walking people on vertical dynamic properties of a lively lightweight bridge. *Struct. Control Health Monit.* **2018**, *25*, e2123. [CrossRef]
25. Pavic, A.; Miskovic, Z.; Reynolds, P. Modal Testing and Finite Element model updating of a lively open-plan composite building floor. *J. Struct. Eng.* **2007**, *133*, 550–558. [CrossRef]
26. Ewins, D.J. *Modal Testing: Theory and Practice*, 2nd ed.; Research Studies Press Ltd.: Baldock, Hertfordshire, UK, 2000.
27. Zou, C.; Li, X.; He, C.; Zhou, S. An efficient method for estimating building dynamic response due to train operations in tunnel considering transmission path from source to receiver. *Comput. Struct.* **2024**, *305*, 107555. [CrossRef]
28. He, W.; Liu, J.; Song, S.; Liu, P. A non-contact vehicle weighing approach based on bridge weigh-in-motion framework and computer vision techniques. *Measurement* **2024**, *225*, 113994. [CrossRef]
29. Qiu, Y.; Zou, C.; Hu, J.; Chen, J. Prediction and mitigation of building vibrations caused by train operations on concrete floors. *Appl. Acoust.* **2024**, *219*, 109941. [CrossRef]
30. Li, X.; Chen, Y.; Zou, C.; Chen, Y. Train-induced vibration mitigation based on foundation improvement. *J. Build. Eng.* **2023**, *76*, 107106. [CrossRef]
31. He, W.; He, K.; Cui, H.; Wang, G. Using a rhythmic human shaker to identify modal properties of a stationary human body on a footbridge. *J. Sound Vib.* **2022**, *540*, 117309. [CrossRef]
32. He, W.; Cui, H.; Yang, S. Estimation of structural dynamic response induced by individual sit-to-stand loading using acceleration response spectrum approach. *Mech. Syst. Signal Process.* **2024**, *211*, 111233. [CrossRef]
33. Peeters, B.; De Roeck, G. Reference-based stochastic subspace identification for output-only modal analysis. *Mech. Syst. Signal Process.* **1999**, *13*, 855–878. [CrossRef]
34. Reynders, E.; De Roeck, G. Reference-based combined deterministic-stochastic subspace identification for experimental and operational modal analysis. *Mech. Syst. Signal Process.* **2008**, *22*, 617–637. [CrossRef]
35. Zivanovic, S.; Pavic, A.; Reynolds, P. Vibration serviceability of footbridges under human-induced excitation: A literature review. *J. Sound Vib.* **2005**, *279*, 1–74. [CrossRef]
36. Zhang, C. Research on Modal Parameters Identification and Riding Comfort of the Elevated Bridge. Master's Thesis, Wuhan University of Technology, Wuhan, China, 2016. (In Chinese).
37. Minette, R.S.; SilvaNeto, S.F.; Vaz, L.A. Experimental modal analysis of electrical submersible pumps. *Ocean Eng.* **2016**, *124*, 168–179. [CrossRef]
38. Avic, O. Amplitude-dependent damping in vibration serviceability: Case of a laboratory footbridge. *J. Archit. Eng.* **2016**, *22*, 04016005.
39. Mazza, F. In-plane-out-of-plane non-linear model of masonry infills in the seismic analysis of r.c.-framed buildings. *Earthq. Eng. Struct. Dyn.* **2019**, *48*, 432–453. [CrossRef]
40. Di Nino, S.; Luongo, A. A simple homogenized orthotropic model for in-plane analysis of regular masonry walls. *Int. J. Solids Struct.* **2019**, *167*, 156–169. [CrossRef]
41. Aras, F. Laboratory tests and vibration surveys for the mechanical properties of infill walls. *J. Perform. Constr. Facil.* **2018**, *32*, 04017117. [CrossRef]
42. Shankar, S.; Harmesh, K. Structural Dynamic Model Updating Techniques: A State of the Art Review. *Arch. Comput. Methods Eng.* **2016**, *23*, 515–533.
43. Ren, W.X.; Jaishi, B. Use of modal flexibility and normalized modal difference for vibration mode shape expansion. In Proceedings of the 2nd International Conference on Structural Condition Assessment, Monitoring and Improvement (SCAMI-2), Changsha, China, 19–21 November 2007.
44. Hong, W. Study on Floor Vibration Comfort of a Stadium. Dissertation, Wuhan University of Technology, Wuhan, China, 2009. (In Chinese).
45. He, W.; Xie, W. Study on sophisticated calculation model of large-span railway station structures based on vibration serviceability evaluation. *China Civ. Eng. J.* **2014**, *47*, 13–23. (In Chinese)
46. Ereiz, S.; Duvnjak, I.; Jiménez-Alonso, J.F. Review of finite element model updating methods for structural applications. *Structures* **2022**, *41*, 684–723. [CrossRef]
47. Raviolo, D.; Civera, M.; Zanotti Fragonara, L. A Bayesian sampling optimisation strategy for finite element model updating. *J. Civ. Struct. Health Monit.* **2024**. [CrossRef]
48. Raviolo, D.; Civera, M.; Fragonara, L.Z. A comparative analysis of optimization algorithms for finite element model updating on numerical and experimental benchmarks. *Buildings* **2023**, *13*, 3010. [CrossRef]

buildings

MDPI

Article

Field Measurement and Numerical Simulation of Sound Insulation Characteristics of Buildings Along Elevated Urban Rail Lines

Teng Wang [1,2], Wei He [3,*], Dongyang Wang [1,2] and Jialiang Chen [1,2,4]

1 Beijing Oriental Yuhong Waterproof Technology Co., Ltd., Beijing 101111, China; wangteng01@yuhong.com.cn (T.W.); wangdy@yuhong.com.cn (D.W.); chen_galeon@tsinghua.edu.cn (J.C.)
2 Beiijao Zane Rail Technology (Beijing) Co., Ltd., Beijing 101111, China
3 Faculty of Engineering, China University of Geosciences (Wuhan), Wuhan 430074, China
4 School of Civil Engineering, Tsinghua University, Beijing 100084, China
* Correspondence: hewei2018@cug.edu.cn

Abstract: The noise pollution issue along elevated urban rail transit lines has attracted considerable attention; however, research on the sound insulation characteristics of buildings during operational train noise remains limited. In this study, a field testing campaign was conducted in a building featuring facades constructed from aerated concrete blocks embedded with single-layer glass sliding windows along an elevated rail transit line. The sound insulation characteristics of the building were obtained based on measured data. By neglecting the influence of lateral sound transmission, a window-wall assembly model was proposed for simulating the building's sound insulation capability. To ensure the accuracy of the prediction results, initial simulations were first performed on individual elements and compared against the experimental sound insulation curves to identify key parameters. Subsequently, simulations of the window-wall assembly model were carried out and compared with test data to determine the acoustic boundary treatment. The results showed that the amount of sound insulation measured in the buildings ranged from 11.16 dB (A) to 19.85 dB (A) during train passage, exhibiting a notable dip in sound insulation within the frequency range of 80 Hz to 100 Hz. Conversely, the predicted sound insulation of 26.38 dB was much higher, also showing a significant dip at around 100 Hz. The simulated values aligned closely with the measured values below 500 Hz; however, discrepancies were observed in mid-to-high frequency ranges above this threshold. To this end, further simulations were performed on a window-wall assembly model incorporating air gaps to take into account air leakage effects. The results indicated that the presence of air gaps considerably diminished high-frequency sound insulation; when accounting for window air leakage effects, the simulated results approached closer alignment with the measured ones.

Keywords: elevated urban rail transit; sound insulation characteristics; wall-windows assembly model; field measurement; finite element analysis

1. Introduction

By the end of 2023, a total of 59 cities in China had operational urban rail transit lines, boasting an aggregate operating mileage exceeding 11,000 km [1]. While urban rail transit significantly enhances travel convenience for the public, it also generates notable environmental noise and vibration issues. These problems are particularly pronounced for elevated lines situated near residential areas, where frequent noise and vibration complaints have become a major concern [2–9].

1.1. Sources and Characteristics of Noise from Elevated Urban Rail Lines

The noise generated by elevated urban rail lines during the operation period can be classified into several categories according to the noise source [5,6], including wheel-rail noise, aerodynamic noise, traction system noise, and structural secondary noise. Wheel-rail noise refers to the noise generated by the contact between the wheels of the train and the tracks during the train's movement, which mainly includes rolling noise, impact noise, and curve whistle noise, of which rolling noise accounts for the largest proportion; aerodynamic noise is the noise generated by the train's high-speed operation when interacting with the air; traction system noise refers to the noise generated by the train's power source and cooling module during normal operation; structural secondary noise refers to the noise generated by the vibration of the elevated bridge and the surrounding structures of the platform when the train passes by.

In addition to the classification based on the noise source, the elevated line train noise can be classified into airborne noise and structural secondary sound according to the way it is transmitted into the building interior [3]. Airborne noise is emitted by the source and then transmitted through the air and the building's sound insulation to the indoor environment, while secondary structural noise is generated by the building's own vibration. Both of them interact with each other and jointly affect the sound environment quality in the building.

1.2. Research Progress on Sound Insulation of Buildings

Current research on noise and vibration along elevated urban rail transit lines mainly focuses on noise characteristics and sound field distribution laws [10,11], as well as comprehensive noise and vibration reduction measures (including structural acoustic optimization [12,13], track damping [14,15], sound barriers [16,17] or sound barriers on the trackside [18]), with little attention paid to the sound insulation performance of buildings. In fact, the airborne noise emitted by the elevated urban rail transit line can be transmitted through the building's exterior walls to the interior, and the exterior walls can effectively reduce airborne noise and serve as sound barriers. In recent years, considerable research has been conducted on enhancing sound insulation performance within the area of building acoustics:

(1) Sound insulation performance of individual building components, using either the finite element method or laboratory testing of specific wall structures and window assemblies [19–23]. For example, Granzotto et al. [19] systematically studied the influence of glass frame, glass thickness, and the presence of PVB on the sound insulation performance of windows using the sound laboratory testing protocol; Wu et al. [20] studied the sound insulation characteristics of several lightweight partition walls by employing both laboratory testing and finite element methods; Zhu et al. [21] investigated the influence of glass thickness and air layer thickness on the sound insulation performance of double-glazed windows using COMSOL finite element software. Liu et al. [22] tested the sound insulation performance and mechanical properties of medium density fiberboard (MDF) and rubber multilayer panels. The weighted sound reduction was 28.0 dB for 6 mm MDF, while it was 37.4 dB for 6 mm wood composite damping material, an increase of 25.1%. Wang et al. [23] investigated the influence factors on the impact sound insulation performance of floor lining type, thickness, mass density and dynamic stiffness of elastic material, and the quality and thickness of floating plates.

(2) Sound insulation performance of composite components, with particular focus on the sound insulation capabilities of composite wall structures and composite window structures [24–30]. These studies are crucial for improving acoustic comfort in various environments like residential buildings, commercial spaces, and industrial facilities. For instance, Liu [24] explored the sound insulation performance of composite panel walls by employing finite element simulation techniques. This approach allowed for a detailed analysis of how different materials and structural configurations influence sound transmission loss. Zhao [12] proposed a theoretical calculation formula that quantifies the sound insulation properties of these composite systems. His research emphasizes the importance of understanding material properties such as density, thickness, and damping characteristics in predicting overall acoustic performance. Furthermore, ongoing studies

are exploring innovative materials and design strategies aimed at enhancing sound isolation while maintaining other essential building attributes like thermal efficiency and structural integrity.

In general, current research on the sound insulation performance of buildings primarily focuses on the sound insulation characteristics of individual or composite components. However, there remains a scarcity of studies addressing the sound insulation performance of actual buildings, particularly those involving field measurements. Furthermore, existing experiments at the component level have predominantly utilized idealized artificial sound sources (such as white noise). This leads to an equal sound intensity source for all concerned frequencies. However, it has been demonstrated that the sound insulation performance of buildings is closely related to the characteristics of operational sound sources, which means that using idealized artificial sound sources may not reflect the real sound insulation characteristics of buildings. Hence, it is crucial to consider this effect under actual conditions.

1.3. Structure of This Paper

This study conducted noise field tests in both indoor and outdoor areas of buildings situated near an elevated urban rail line to obtain the sound insulation properties. Utilizing the measured data, a window-wall assembly model was proposed to simulate and analyze sound insulation performance. The findings from this research can give insights for implementing effective noise control measures in buildings located along elevated urban rail lines.

The remainder of this paper is structured as follows. Section 2 describes the test site, test setup and results of the field measurement. A numerical window-wall assembly model is proposed to simulate the acoustic insulation performance in Section 3. Section 4 further discusses the impact of windows and walls on the building's sound insulation performance. The main conclusions and future work are presented in the closing section.

2. Field Measurement

2.1. Description of Test Site

A test campaign was conducted on an elevated urban rail transit line in Wuhan, China. The bridge employs a single-box, single-chamber concrete box girder with a span of 25 m and a bridge pier height of H = 12 m. The subway train was grouped using four B-type vehicles with an operating speed of 50 km/h. The test building was a 23-storey frame-shear wall structure with a total height of h = 70. The horizontal distance between the building and the elevated line was L = 25 m.

2.2. Test Setup

The sensors employed are Brüel & Kjær microphones from Denmark, paired with a DH9581 dynamic data collector. In addition to the collector and microphones, the setup includes the necessary adapters, external data transmission cables, and sensor brackets. Sound measurement locations have been strategically established both indoors and outdoors on the fifth floor of the structure. According to Chinese Guidelines (GB/T 19889.5-2006 Acoustics—Measurement of sound insulation in building elements—Part 5: Field measurements of airborne sound insulation of façade elements and facades) [31], indoor and outdoor microphone measurement points are positioned 1.5 m from the wall and at an elevation of 1.2 m above ground level. The sampling frequency of the microphones is set as 10,000 Hz. The configuration of these measurement points is illustrated in Figure 1. The exterior wall of the test room comprises aerated concrete blocks with embedded single-glazed sliding windows. Detailed parameters regarding the building facade and windows are presented in Table 1.

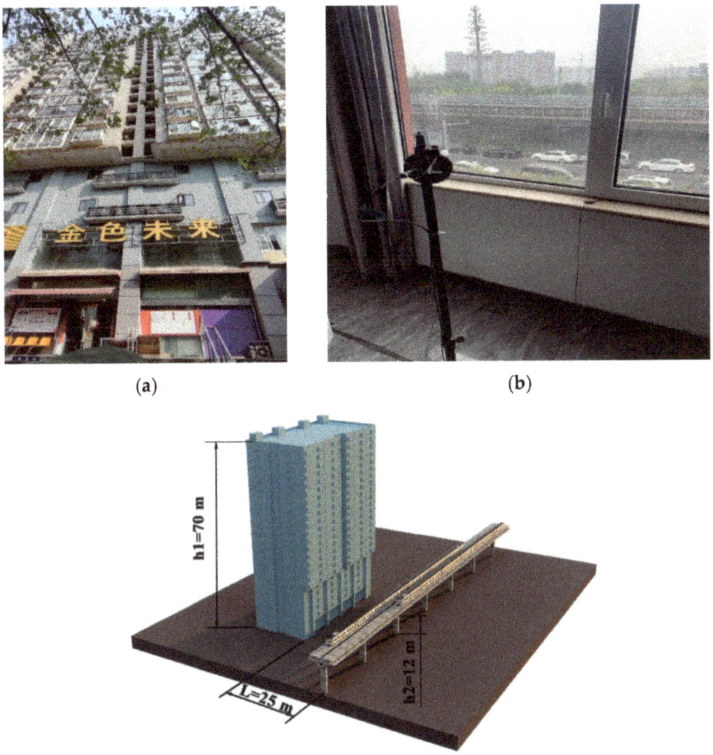

Figure 1. Test setup: (**a**) real picture of the test building; (**b**) arrangement of microphone indoor (**c**) side view.

Table 1. Parameters of building facade and windows.

Type of Wall	Type of Window	Wall Size	Window Size
Aerated concrete block	Single-glazed sliding windows	5000 × 3000 × 200	2700 × 1500 (Single glazing 5 mm thick)

2.3. Test Results

Long-term continuous monitoring was conducted to capture the complete process of sound pressure variation during train passage. The representative time series of sound pressure recorded by indoor and outdoor microphones during the train passage are shown in Figure 2.

As the train passes, there is a significant increase in sound pressure, and a distinct spindle-shaped waveform can be observed in the time history curve, with outdoor sound pressure notably exceeding indoor sound pressure. The sound pressure time history is further analyzed in the 1/3 octave frequency band and processed using the A-weighted network. In accordance with the testing specifications, the analysis frequency band covers 18 center frequencies ranging from 100 Hz to 5000 Hz [31]. Figure 3 illustrates the sound pressure levels in the one-third octave frequency band for both outdoor and indoor test points of the building during multiple train passages (exceeding 20 trains). The test results are presented as envelope curves, along with their corresponding mean values.

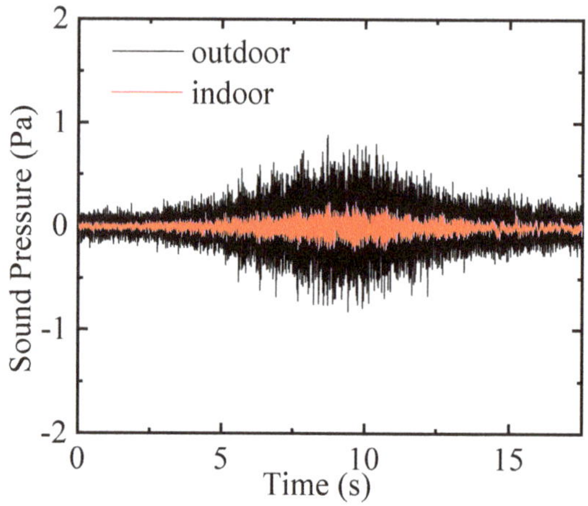

Figure 2. Time series of indoor and outdoor sound pressure under train passage.

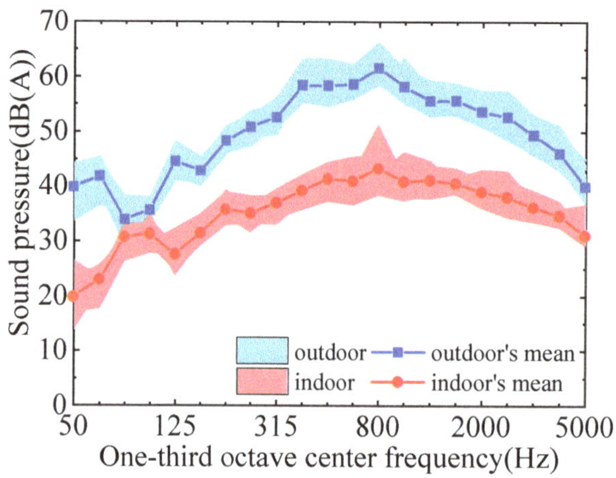

Figure 3. Indoor and outdoor A-weighted sound pressure level.

Utilizing the measured average sound pressure level curves for both indoor and outdoor environments, the difference between these two can be calculated to determine the sound insulation (or sound transmission loss) of the building, as illustrated in Figure 4. Analysis of the sound insulation curve reveals that the sound insulation of the building is not uniformly distributed across the frequency spectrum. The sound insulation ranges from 3 to 22 dB for frequencies from 50 Hz to 5000 Hz, exhibiting a peak in the low-frequency band below 60 Hz, with minimal sound insulation capacity observed between 80 Hz and 100 Hz. In contrast, there is a gradual increase in sound insulation capacity from 125 Hz to 800 Hz, followed by a gradual decrease from 1000 Hz to 5000 Hz.

Further calculations were conducted to determine the overall sound insulation R value of the building. The sound insulation values corresponding to all train pass-by events are presented in Figure 5. The measured sound insulation range for the test building is between 11.16 dB (A) and 19.85 dB (A), with an average value of 16.4 dB (A).

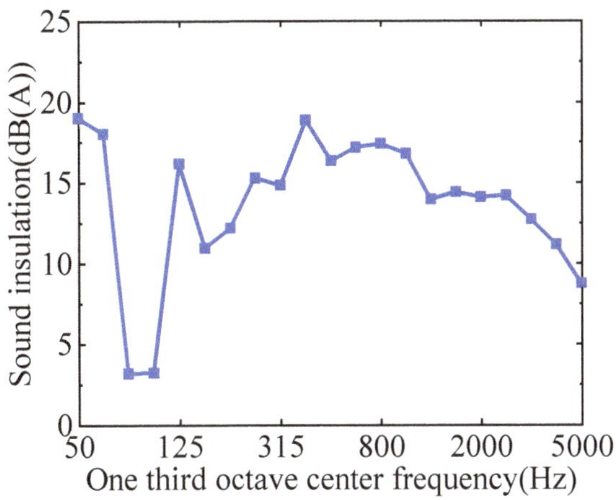

Figure 4. Sound insulation curves for window-wall combination components.

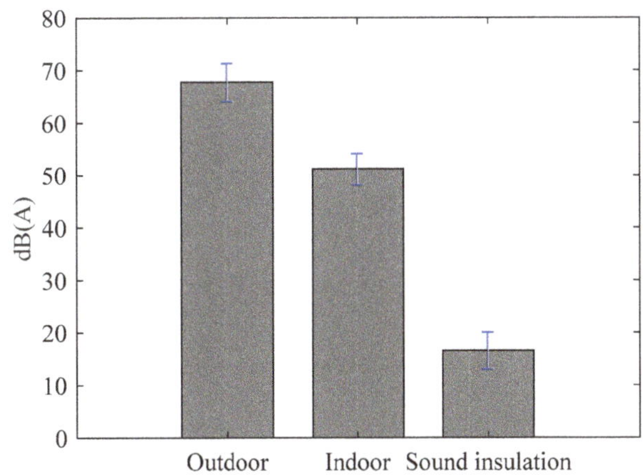

Figure 5. Sound insulation for window-wall assembly component.

3. Numerical Simulation of Acoustic Insulation Performance Based on the Window-Wall Assembly Model

In this section, a numerical model is presented to give an alternative tool to predict the acoustic insulation performance of buildings along elevated urban rail transit.

3.1. Principles and Assumptions of the Numerical Model

The simplified numerical model is established based on the following assumptions:

(1) The low-frequency secondary structure-borne noise is not considered. The noise generated by a train passing buildings along the railway line encompasses not only wheel-rail noise, aerodynamic noise, and other airborne radiated sounds, but also induces vibrations in the building structure that result in low-frequency secondary structure-borne noise. Given that the frequency range of secondary structural noise is relatively narrow (20 Hz to 200 Hz), this study disregards its influence and focuses solely on the sound insulation performance of buildings against airborne radiated noise.

(2) Only the direct sound transmission of the building facade is considered and the lateral sound transmission loss from adjacent walls is ignored. Figure 6 illustrates a schematic view of the transmission paths for outdoor noise entering indoor spaces. D_d denotes the direct sound transmission path, while F_f and D_f indicate lateral sound

transmission paths. In comparison to lateral sound transmission, direct sound transmission predominates; thus, the acoustic insulation performance of the exterior wall directly affects the overall sound insulation performance of the building.

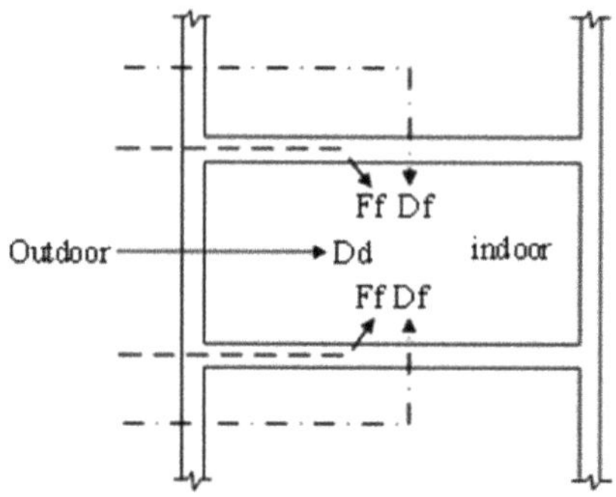

Figure 6. Sound transmission pathways.

In light of these facts, the influence of the lateral sound transmission effects of the building is ignored, and a finite element model of the exterior wall (window-wall assembly components) has been developed. Simulations have been performed to evaluate the acoustic insulation performance of the building.

To ensure the accuracy of the simulation results, initial simulations were first performed on individual elements (wall and window) and compared against the experimental sound insulation curves to identify key parameters. Based on this, a window-wall assembly model is further established. Subsequently, simulations for the window-wall assembly model were carried out and compared with test data to determine the acoustic boundary treatment. The main idea of the simulation flowchart is illustrated in Figure 7.

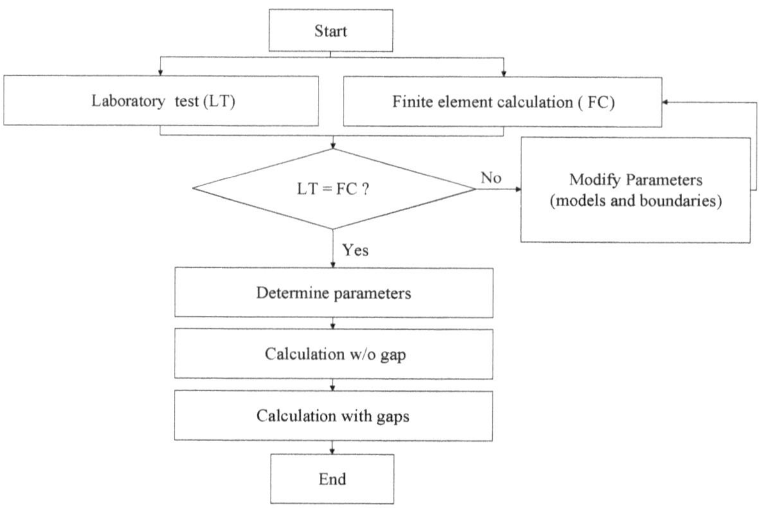

Figure 7. Procedures of the proposed method.

The finite element acoustic calculations are conducted using the LMS Virtual.Lab platform. The acoustic module within the LMS Virtual.Lab software is a specialized CAE tool designed for finite element analysis in the field of acoustics, incorporating methods such as FEM and BEM, among other acoustic analysis techniques.

3.2. Parameters for Acoustic Calculations

According to the laboratory sound insulation test specifications for individual elements [32], the planar dimensions of the window component are 1500 mm × 1250 mm or proportionally enlarged according to this ratio; the planar dimensions of the wall component should be no less than 10 m^2, and its short side should be no less than 2.3 m. Based on these dimensions, a finite element model of a concrete wall embedded with a single-layer glass window is established, as shown in Figure 8.

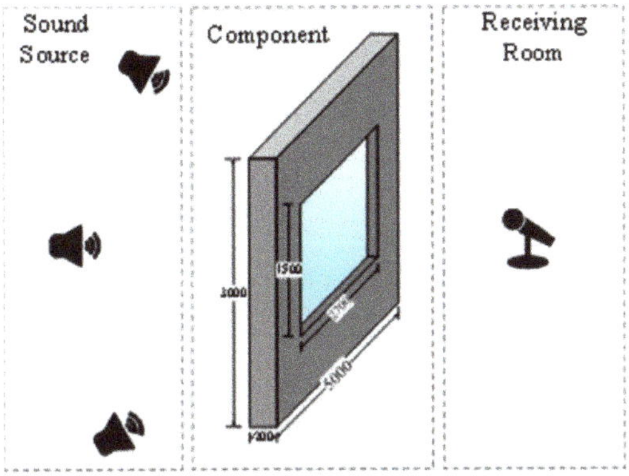

Figure 8. Calculation model of individual component for sound insulation.

The upper limit frequency f_{max} in acoustic calculations is related to the grid size L as described by Equation (1) [33]:

$$f_{\max} \leq \frac{c}{6L} \tag{1}$$

where c denotes the speed of sound propagation in the medium—340 m/s for air.

According to Equation (1), the grid size of the model has been determined as 10 mm to obtain the maximum analysis frequency of 5000 Hz. The specific parameters for the single-layer glass window and concrete wall are presented in Table 2.

Table 2. Model Parameters.

Material	Size (mm)	Density (kg/m^3)	Elastic Modulus (Pa)	Poisson's Ratio
Single-pane glass window	1500 × 1250 × 3	2500	7.2×10^{10}	0.2
Concrete wall	5000 × 5000 × 200	2500	3×10^{10}	0.2

Utilizing the parameters presented in Table 2, a structural model and an acoustic grid were developed, with an acoustic envelope surface incorporated through acoustical preprocessing. The acoustic radiation boundary conditions were established by applying the AML attribute layer to the left and right sides of the acoustic surfaces, designating one side as anechoic chamber and the other as a reverberation chamber (see Figure 8). Following the configuration of the acoustic boundary conditions, structural boundary conditions were also defined. The lateral faces of the structural elements were designated as constrained element groups under simply supported constraints. The contact interfaces between the acoustic grid and structural grid were configured as coupling faces, with a tolerance for these coupling faces at 10 mm.

First, sound–structure coupling calculations were conducted on the individual window and wall element to obtain its sound insulation curves. These calculated curves were then compared against those obtained from lab tests in a reverberation-absorption chamber. Figures 9 and 10 give comparative results for the calculated and experimental values of sound insulation performance for the single-layer glass window and concrete wall components, respectively.

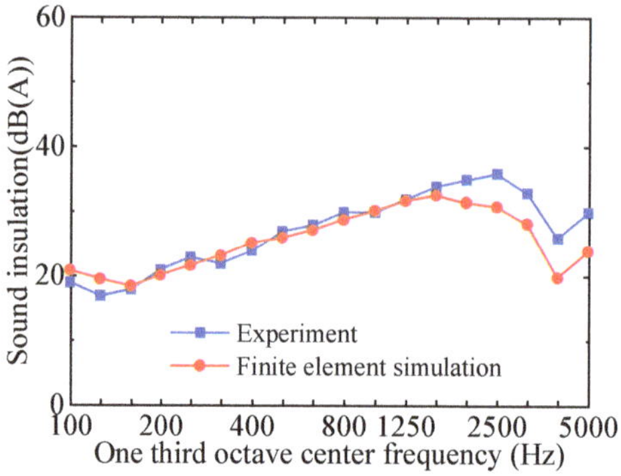

Figure 9. Comparison of crossover frequency sound insulation of single pane glass.

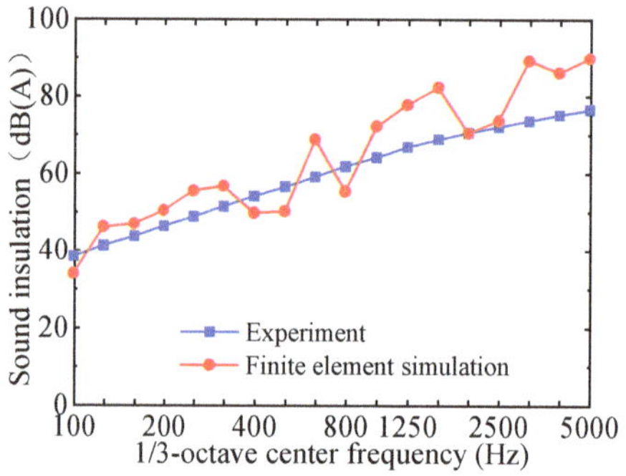

Figure 10. Comparison of crossover frequency sound insulation of concrete.

As shown in Figure 9, for the single-layer glass window component, the trend of the calculated sound insulation curve aligns closely with the experimental ones. Specifically, the sound insulation calculation results at frequencies below 1600 Hz closely match with the experimental data, while discrepancies of approximately 10% are observed in the range of 2000 Hz to 5000 Hz. For the concrete wall component, although the trend in sound insulation calculations is consistent with the measured values, some frequency points display larger errors primarily attributable to the size effects. Overall, both types of components demonstrate commendable consistency between the calculated and measured sound insulation results. This demonstrated that the acoustic parameters employed in finite element analysis are reasonable and suitable for subsequent investigations into structural sound insulation for window-wall assembly components.

3.3. Sound Insulation Analysis Using the Window-Wall Assembly Model

Utilizing the window and wall dimensions of the test site, a finite element model of the window-wall assembly components was developed, with acoustic parameters configured in accordance with the results presented in Section 3.2. The finite element model is illustrated in Figure 11. The parameters for the model are presented in Table 3.

Figure 11. Finite element model of window-wall assembly components.

Table 3. Model Parameters of the window-wall assembly model.

Materials	Size of Components (mm)	Densities (kg/m^3)	Elastic Modulus (Pa)	Poisson's Ratio
Single-pane glass window	2700 × 1500 × 5	2500	7.2× 10^{10}	0.2
Aerated concrete wall	5000 × 3000 × 200	1300	4.5× 10^9	0.2

In contrast to the modeling of individual components, the assembling of the wall and window results in a recessed wall surface due to discrepancies in thickness dimensions. Consequently, it is necessary for the air layer to be protruding to ensure the accuracy of sound–vibration coupling calculations.

To improve computational efficiency, two grid sizes for the air layer—20 mm and 10 mm—were employed. According to the relationship between analytical frequency and grid size (as described in Equation (1)), the 20 mm grid supports analysis frequencies up to 2500 Hz, while the 10 mm grid supports frequencies up to 5000 Hz. Therefore, a 20 mm grid is utilized for frequency ranges from 50 Hz to 2500 Hz, whereas a 10 mm grid is applied for frequencies ranging from 2500–5000 Hz.

The sound–vibration coupling simulations were conducted on the window-wall assembly model, with results compared against measured values as illustrated in Figure 12.

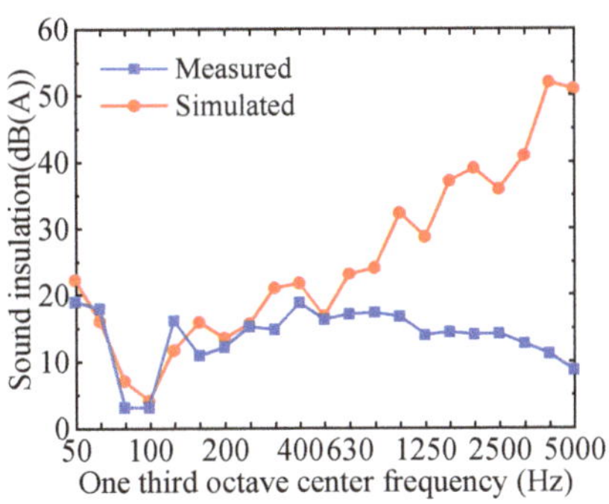

Figure 12. Predicted sound insulation results using window-wall assembly model against measured ones.

As illustrated in Figure 12, the simulated results exhibit minimal error within the 50 Hz to 500 Hz frequency band, and the trend of the sound insulation curve aligns closely with the measured values. Conversely, in the 1000 Hz to 5000 Hz frequency band, simulated values are significantly higher than their measured counterparts; that is, the theoretical sound insulation performance is much better than the measured results. A potential explanation for this discrepancy is that instrument cables traversed through the window during testing, preventing a proper seal and resulting in sound leakage. The subsequent discussion will further explore how sound leakage affects sound insulation performance.

3.4. Sound Insulation Analysis Using the Window-Wall Assembly Model with Air Gaps

Considering the effect of sound leakage, the window gap was determined based on site conditions and positioned centrally with a width of 10 mm. The finite element model of the window-wall assembly component with an air gap is illustrated in Figure 13.

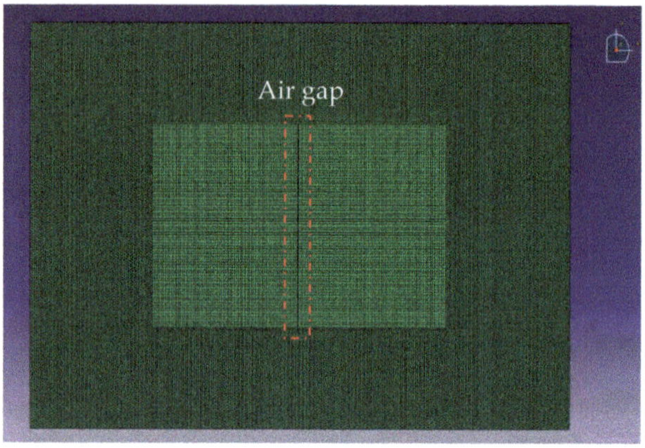

Figure 13. Finite element model of the window-wall assembly component with air gap.

In terms of coupling surfaces, the window-wall assembly component with a gap exhibits a distinct configuration compared to its gapless counterpart. A 10 mm gap has been artificially introduced at the interface where it contacts the member's gap.

Figure 14 shows a comparison between the calculated and measured sound insulation performance of the window-wall assembly model with a gap. Compared to the gapless condition, the sound insulation performance in the frequency band above 630 Hz decreases significantly, and it is more consistent with the measured results. The sound insulation valley in the frequency band of 80 Hz to 100 Hz disappears, and the overall trend of the curve becomes smoother. The overall calculated sound insulation performance is still greater than the measured value, mainly because the lateral sound transmission effect is ignored in the theoretical model.

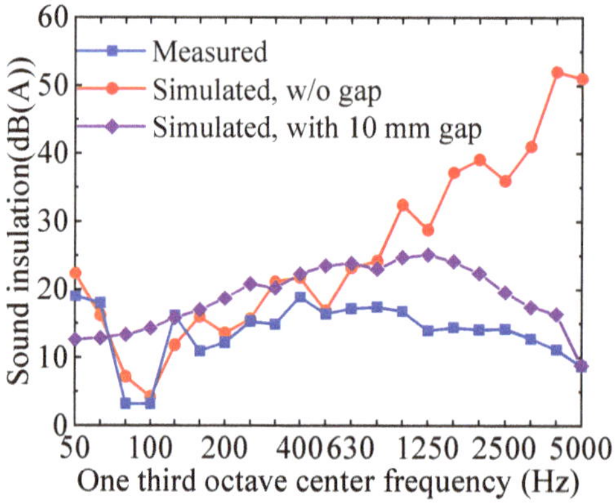

Figure 14. Calculation results of sound insulation of window-wall assembly model with air gap.

To investigate the effect of gap width on the sound insulation performance, four additional gap sizes—20 mm, 30 mm, 40 mm, and 50 mm—were considered for comparative analysis, with the results presented in Figures 15 and 16.

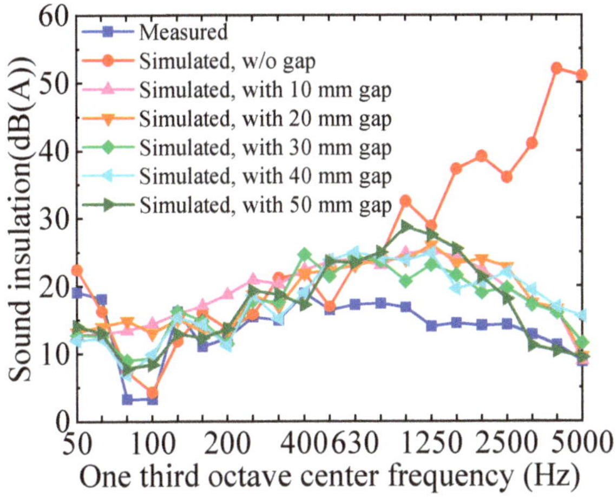

Figure 15. Calculated results for different gap widths.

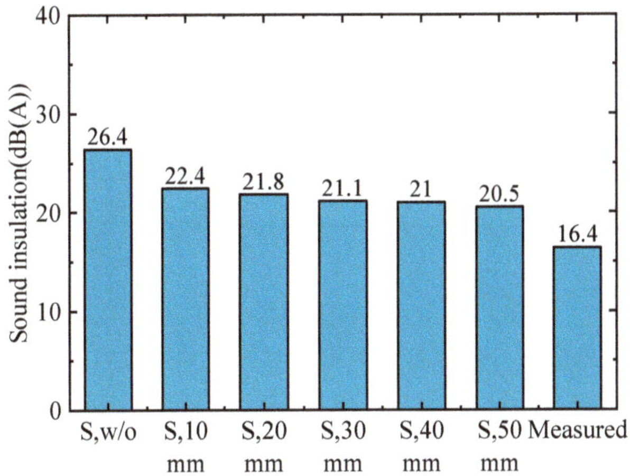

Figure 16. Sound insulation with different air gap widths. Note: S,w/o, simulated w/o gap; S,10 mm, simulated with 10 mm gap; S,20 mm, simulated with 20 mm gap; S,30 mm, simulated with 30 mm gap; S,40 mm, simulated with 40 mm gap; S,50 mm, simulated with 50 mm gap.

The sound insulation amount generally decreases with the increase in gap width; however, the rate of change varies. The most pronounced reduction in sound insulation occurs when transitioning from no gap to a 10 mm gap, resulting in a decrease of 4 dB (A), whereas at a 50 mm gap, the reduction is approximately 6 dB (A). Notably, there is no obvious pattern regarding sound insulation across different gap widths. Furthermore, the calculated sound insulation values for various gap widths remain higher than the measured ones, probably due to the omission of lateral sound transmission effects in the theoretical model.

4. Discussion

Section 3 introduces the sound insulation calculation method for the window-wall assembly components and verifies the feasibility of obtaining building sound insulation characteristics by using a simplified window-wall assembly model. This section further discusses two main factors, namely windows and walls, on the sound insulation properties of buildings.

A benchmark window-wall assembly model, shown in Figure 17, features a wall measuring 3600 × 3600 × 240 mm and a window measuring 1500 × 1800 × 5 mm.

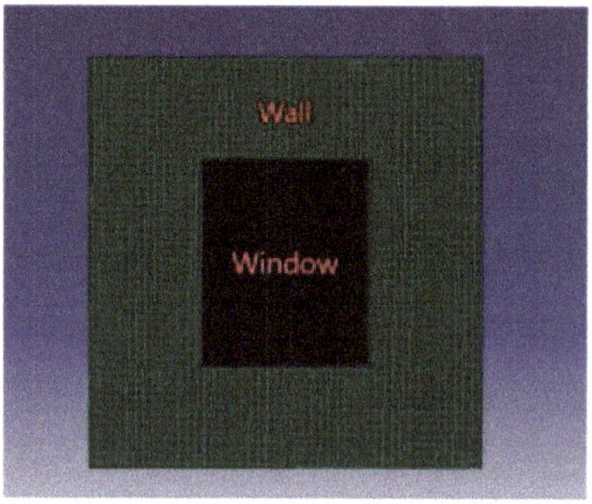

Figure 17. Schematic diagram of the computational model.

4.1. Effect of Windows on Sound Insulation Capability

Four types of windows, including single-pane glass, double-pane glass, PVB laminated glass (with different thicknesses), and double-layer composite structure glass windows, are considered to assess the influence of window type on sound insulation performance. The walls are constructed using aerated concrete blocks. The geometrical parameters and material properties of these four types of glass windows are listed in Table 4.

Table 4. Parameters for windows.

Type of Windows	Thickness (mm)	Density (kg/m^3)	Elastic Modulus (Pa)	Poisson's Ratio	Damping Ratio
Single-pane glass	5	2500	7.2×10^{10}	0.2	0.05
Double-pane glass	5 + 6air + 5				
PVB Laminated glass	5 + 0.38PVB + 5				
	5 + 0.76PVB + 5				
Double-layer composite structure	5 + 0.38PVB + 6air + 5				

Sound insulation calculations were conducted for the aforementioned five cases, with the results presented in Figure 18.

It is observed that the sound insulation amount of the window-wall assembly component is approximately 26 dB for the single-layer glass window. It improves by about 1.5 dB to 2 dB for the double-layer insulated glass window. The sound insulation property of the PVB laminated glass window surpasses that of the double-layer insulated glass window; specifically, the sound insulation of the 0.38PVB laminated glass window increases by approximately 6 dB, while that of the 0.76PVB laminated glass window rises by around 8.5 dB. Notably, the double-layer composite structure glass window exhibits the highest sound insulation improvement of 13 dB.

The sound insulation curves in the frequency domain of all the considered types of windows show a consistent trend across various conditions in the mid-frequency range (200 Hz to 3150 Hz), with all cases exhibiting a trough in sound insulation at 3150 Hz followed by an upward trend beyond this frequency. Compared to the single-layer glass window, the other three types of windows increase the overall sound insulation capability across all frequency points, particularly against low-frequency noise interference. Furthermore, employing an air layer alongside PVB lamination significantly improves acoustic isolation capabilities, most notably at a low-frequency attenuation near 100 Hz within single-layer configurations where a trough exists in sound transmission loss data. The double-layer composite structure window has the best sound insulation capability, followed in descending order by the PVB laminated window and the double-layer insulated glass window, and the single-layer glass window gives the poorest sound insulation capability.

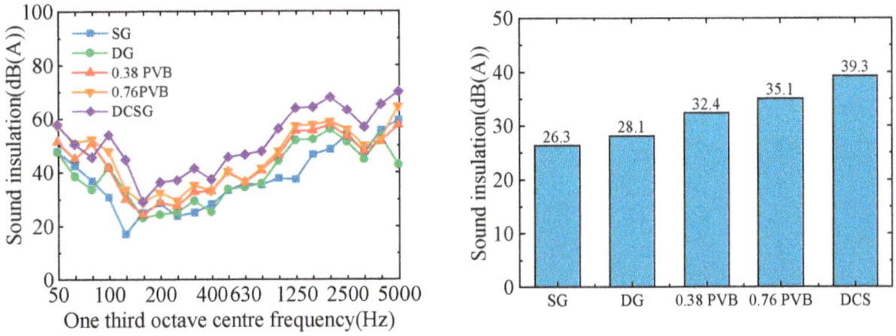

Figure 18. Sound insulation amounts for different window types. Note: SG, single-layer glass; DG, double-layer glass; 0.38 PVB, 0.38 PVB laminated glass; 0.76 PVB, 0.76 PVB laminated glass; DCSG, double-layer composite structure glass.

4.2. *Effect of Wall on Sound Insulation Capability*

Five wall thicknesses—100 mm, 150 mm, 200 mm, 250 mm, and 300 mm, are considered to assess the influence of wall thickness on sound insulation performance. These specific thicknesses are based on common construction practices in engineering applications. A single-pane glass window is adopted for the analysis.

Four types of wall materials are considered, including common concrete, brick walls, lightweight concrete, and foam concrete. They are all common construction materials of walls widely employed in application. The parameters of the different walls are given in Table 5.

Table 5. Parameters for different walls.

Material	Density (kg/m^3)	Elastic Modulus (Pa)	Poisson's Ratio
Concrete wall	2500	3×10^{10}	0.2
Brick wall	2100	1.22×10^{10}	0.2
Lightweight concrete wall	1950	2.1×10^{10}	0.2
Foam concrete wall	1300	4.5×10^{9}	0.2

Figure 19a–d illustrates the simulation results of sound insulation under different wall materials with various wall thicknesses.

The results presented in Figure 19a–d are further fitted in plots, as shown in Figure 20, and the sound insulation capability R (unit in dB) of the four types of wall materials could be represented in terms of wall thickness t (unit in mm), as expressed in Equation (2).

$$R = \begin{cases} 22.64\left(e^{0.00042t} - e^{-0.031t}\right), \text{ Odinary concrete} \\ 21.23\left(e^{0.00047t} - e^{-0.043t}\right), \text{ Brick masonry} \\ 18.81\left(e^{0.00045t} - e^{-0.054t}\right), \text{ Aerated concrete} \\ 19.67\left(e^{0.00026t} - e^{-0.030t}\right), \text{ Foam concrete} \end{cases} \quad (2)$$

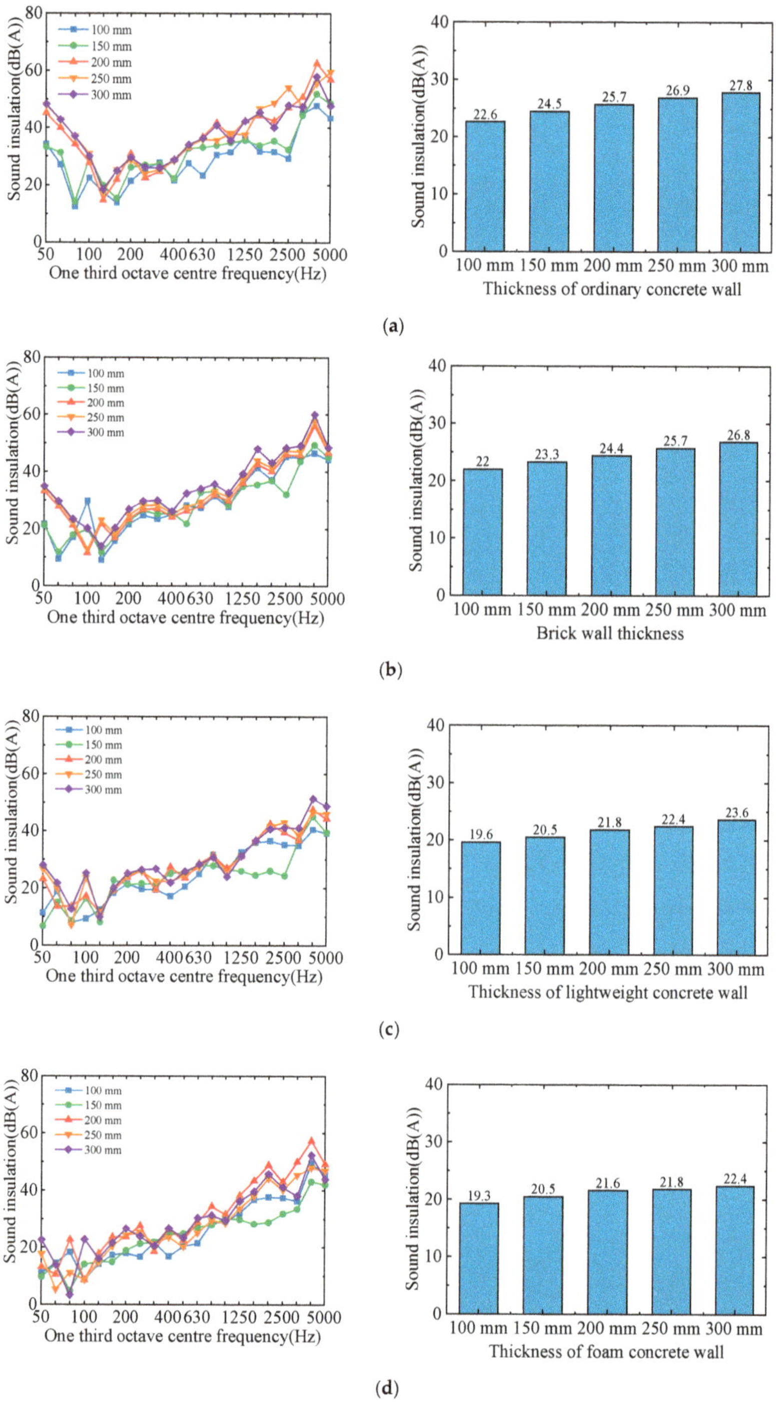

Figure 19. Sound insulation capability for different wall materials and thicknesses: (**a**) common concrete walls; (**b**) brick walls; (**c**) lightweight concrete walls; (**d**) foam concrete walls.

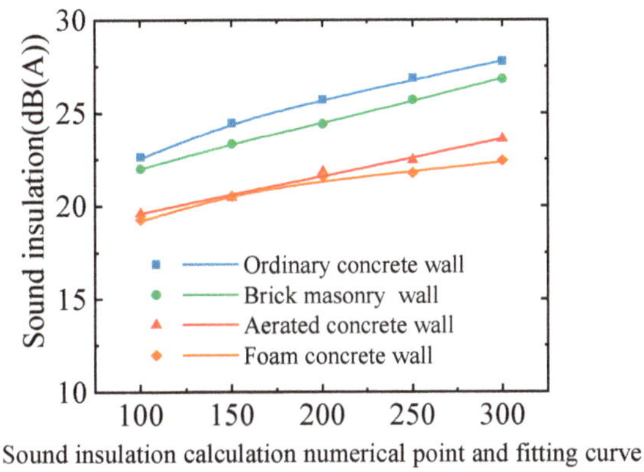

Figure 20. Relationship between sound insulation and wall thickness.

The sound insulation analysis in the frequency domain indicates that an increase in mass density and elastic modulus can significantly improve the low-frequency sound insulation performance of building components. Specifically, for varying wall thicknesses, it has been observed that the sound insulation valley for components with thicknesses of 100 mm and 150 mm occurs at frequencies below 100 Hz. In contrast, thicker components with 200 mm, 250 mm, and 300 mm thickness exhibit sound insulation valleys at approximately 125 Hz.

In terms of high-frequency performance, all four types of wall material components demonstrate notable sound insulation peaks around frequencies of 2500 Hz and 3150 Hz. These phenomena are closely linked to both the physical dimensions and modal characteristics inherent to each component's design. Furthermore, within the mid-frequency band (ranging from 800 Hz to 2500 Hz), it is evident that both the 100 mm- and 150 mm-thick components display significantly lower levels of sound insulation compared to their thicker counterparts—specifically those measuring either 200 mm or greater.

The analysis of the overall sound insulation values reveals that, when the thickness of the component is increased from 100 mm to 150 mm, the sound insulation value increases by approximately 1 to 2 dB; an increase from 150 mm to 200 mm results in a gain of about 1 to 1.5 dB; transitioning from 200 mm to 250 mm yields an increase of roughly 0.5 to 1 dB; and finally, increasing the thickness from 250 mm to 300 mm leads to an enhancement of about 0.7 to 1.1 dB. Notably, when the component's thickness is doubled from 100 mm to 200 mm, one would expect a corresponding increase in sound insulation value between approximately 2 and 4 dB across various materials.

Furthermore, reductions in the mass density and elastic modulus of wall materials contribute significantly to decreases in their sound insulation capabilities. For instance, brick walls exhibit lower sound insulation performance compared with ordinary concrete walls, decreasing by approximately 0.6 to 1.3 dB; lightweight concrete walls show a reduction of about 3 dB compared with ordinary concrete walls; while foam concrete walls demonstrate a decrease ranging between 3 and 4 dB compared with ordinary concrete walls.

Compared with the acoustic insulation capabilities of individual components, it is observed that windows are the most significant component that dominates the sound insulation capability of the window-wall assembly components.

4.3. Limitations of the Current Work

This study tries to provide a simple but efficient tool to take into account the sound insulation characteristics of buildings along elevated urban rail lines. A window-wall assembly model has been proposed based on several assumptions. Since the lateral sound transmission loss from adjacent walls is ignored, the model may give higher sound insulation values compared to the measured ones. The presence of an air gap was found to have an impact on the sound insulation performance of buildings. In this current study, a window-wall assembly model with an air gap was employed to consider its effect. The model can generally reveal the sound insulation characteristics of buildings.

Additionally, for high-rise buildings, the measured outdoor sound source at different floors varies along the height of building during train operation due to sound propagation. Since only limited test points were installed, the effect of the building's height on sound insulation characteristics is not considered.

This study may serve as the first step toward a better understanding of the sound insulation characteristics of buildings. More effort will be made in the future to explore more efficient tools to quantify this effect.

5. Conclusions

Based on the above analyses, the following conclusions could be drawn:

The measured overall sound insulation of buildings situated along the elevated urban rail transit lines was found to range from 11.16 dB (A) to 19.85 dB (A). The sound insulation curve showed a sound insulation valley at 80 Hz to 100 Hz, an upward trend from 100 Hz to 800 Hz, and a downward trend from 1000 Hz to 5000 Hz. The established window-wall assembly model can effectively reveal the sound insulation characteristics of the test building. The air gap in windows should be taken into account in the sound insulation analysis since it has a notable impact on the high-frequency sound insulation capability. Windows are the most significant component that dominates the sound insulation capability of the window-wall assembly components.

In this paper, a window-wall assembly model was proposed for simulating building sound insulation by neglecting the influence of lateral sound transmission from adjacent walls, resulting in a slightly higher calculated sound insulation value than the measured one. This issue will be taken into account in future studies. In addition, sound insulation analysis for different wall types and window types is also a worthwhile research topic. The research findings can provide a reference for noise control in buildings along the elevated urban rail transit system.

Author Contributions: Conceptualization, methodology, writing—original draft preparation, W.H.; formal analysis, investigation, writing—review and editing, T.W.; validation, writing—review and editing, D.W. and J.C. All authors have read and agreed to the published version of the manuscript.

Funding: This research was funded by the project of the Ministry of Housing and Urban-Rural Development of the People's Republic of China (grant number 2022-K-153).

Data Availability Statement: The original contributions presented in the study are included in the article, further inquiries can be directed to the corresponding author.

Conflicts of Interest: Authors Teng Wang, Dongyang Wang and Jialiang Chen were employed by the companies Beijing Oriental Yuhong Waterproof Technology Co., Ltd. and Beijiao Zane Rail Technology (Beijing) Co., Ltd. The remaining authors declare that the research was conducted in the absence of any commercial or financial relationships that could be construed as a potential conflict of interest.

References

1. Zhou, L.; Yu, Q.; Zeng, X. Analysis of current situation of urban rail transit noise and related countermeasures. *China Enviro Prot. Ind.* **2023**, *4*, 53–59. (In Chinese)
2. Welch, D.; Shepherd, D.; Dirks, K.N.; Reddy, R. Health effects of transport noise. *Transp. Rev.* **2023**, *43*, 1190–1210. [CrossRef]
3. Tetiranont, S.; Sadakorn, W.; Rugkhapan, N.; Prasittisopin, L. Enhancing Sustainable Railway Station Design in Tropical Climate Insights from Thailand's Architectural Theses and Case Studies. *Buildings* **2024**, *14*, 829. [CrossRef]
4. Gangwar, M.; Sharma, S.M. Evaluating choice of traction option for a sustainable Indian Railways. *Transp. Res. Part D Trans Environ.* **2014**, *33*, 135–145. [CrossRef]
5. He, W.; He, K.; Zou, C.; Yu, Y. Experimental noise and vibration characteristics of elevated urban rail transit considering the effe of track structures and noise barriers. *Environ. Sci. Pollut. Res.* **2021**, *28*, 45903–45919. [CrossRef] [PubMed]

Qiu, Y.; Zou, C.; Hu, J.; Chen, J. Prediction and mitigation of building vibrations caused by train operations on concrete floors. *Appl. Acoust.* **2024**, *219*, 109941. [CrossRef]
He, W.; Liu, J.; Song, S.; Liu, P. A non-contact vehicle weighing approach based on bridge weigh-in-motion framework and computer vision techniques. *Measurement* **2024**, *225*, 113994. [CrossRef]
Zou, C.; Li, X.; He, C.; Zhou, S. An efficient method for estimating building dynamic response due to train operations in tunnel considering transmission path from source to receiver. *Comput. Struct.* **2024**, *305*, 107555. [CrossRef]
Hu, J.; Zou, C.; Liu, Q.; Li, X.; Tao, Z. Floor vibration predictions based on train-track-building coupling model. *J. Build. Eng.* **2024**, *89*, 109340. [CrossRef]
Yu, L.; Bu, Z.; Li, L.; Zhu, X.; Lei, Z.; Luo, Y. Characteristics analysis of urban rail viaduct structure noise considering Doppler effect. *Urban Mass Transit* **2020**, *23*, 74–79+85. (In Chinese)
He, W.; Zou, C.; Pang, Y.; Wang, X. Environmental noise and vibration characteristics of rubber-spring floating slab track. *Environ. Sci. Pollut. Res.* **2021**, *28*, 13671–13689. [CrossRef]
Zhao, Y.; Li, X.; Lv, Q.; Jiao, H.; Xiao, X.; Jin, X. Measuring, modelling and optimising an embedded rail track. *Appl. Acoust.* **2017**, *116*, 70–81. [CrossRef]
Li, X.; Chen, Y.; Zou, C.; Chen, Y. Train-induced vibration mitigation based on foundation improvement. *J. Build. Eng.* **2023**, *76*, 107106. [CrossRef]
He, W.; He, K.; Yao, C.; Liu, P.; Zou, C. Comparison of damping performance of an aluminum bridge via material damping, support damping and external damping methods. *Structures* **2022**, *45*, 1139–1155. [CrossRef]
Li, X.; Liang, L.; Zhao, Q. Influence of track structure type on noise radiated from an elevated box-girder. *China Civ. Eng. J.* **2018**, *51*, 78–87. (In Chinese)
Song, L.; Li, X.; Zhang, L.; Liu, Q.; Feng, Q.; Luo, Y. Characteristics and prediction of structure-borne noise from urban rail transit bridge-sound barrier system. *J. Traffic Transp. Eng.* **2021**, *21*, 193–202. (In Chinese)
Song, L.; Gao, K.; Liu, Q.; Liu, L.; Feng, Q. Study on the structure-borne noise of U-shaped girder bridges with fully-enclosed sound barriers. *Appl. Acoust.* **2023**, *211*, 109497. [CrossRef]
Frid, A. Skirts and barriers for reduction of wayside noise from railway vehicles—An experimental investigation with application to the BR185 locomotive. *J. Sound Vib.* **2003**, *267*, 709–719. [CrossRef]
Granzotto, N.; Bettarello, F.; Ferluga, A.; Marsich, L.; Schmid, C.; Fausti, P.; Caniato, M. Energy and acoustic performances of windows and their correlation. *Energy Build.* **2017**, *136*, 189–198. [CrossRef]
Wu, W.; He, W.; Yu, H. Study and analysis on sound insulation performance of several common kinds of light weight panel for building partition. *Sichuan Build. Sci.* **2021**, *47*, 83–89. (In Chinese)
Zhu, X.; Wang, L.; Wang, X.; Jiang, J. Simulation Study of Sound Insulation Performance of the Double-glazed Window. *Noise Vib. Control.* **2022**, *42*, 256–262. (In Chinese)
Liu, M.; Peng, L.; Fan, Z.; Wang, D. Sound Insulation and Mechanical Properties of wood damping composites. *Wood Res.* **2019**, *64*, 743–758.
Wang, J.; Du, B. Experiment on the optimization of sound insulation performance of residential floor structure. *Appl. Acoust.* **2021**, *174*, 107734. [CrossRef]
Liu, X. Acoustic Analysis of Windows and Walls Using LMS Virtual.Lab. Master's Thesis, Nanchang Hangkong University, Nanchang, China, 2016. (In Chinese).
Lakshmikandhan, K.N.; Harshavardhan, B.S.; Prabakar, J.; Saibabu, S. Investigation on Wall Panel Sandwiched with Lightweight Concrete. *IOP Conf. Ser. Mater. Sci. Eng.* **2017**, *225*, 012275. [CrossRef]
Arjunan, A.; Baroutaji, A.; Robinson, J.; Vance, A.; Arafat, A. Acoustic metamaterials for sound absorption and insulation in buildings. *Build. Environ.* **2024**, *251*, 111250. [CrossRef]
Osipov, A.; Mees, P.; Vermeir, G. Numerical simulation of airborne sound transmission at low frequencies: The influence of the room and the partition parameters. *Proc. Inter-Noise* **1997**, *2*, 759–762.
Maluski, S.P.S.; Gibbs, B.M. Application of a finite-element to low-frequency sound insulation in dwellings. *Acoust. Soc. Am.* **2000**, *108*, 1741–1751. [CrossRef] [PubMed]
Ljunggren, L. Airborne sound insulation of single walls at low frequencies: A discussion on the influence of boundary and mounting conditions. *Build. Acoust.* **2001**, *8*, 257–267. [CrossRef]
Tadeu, A.; António, J. Acoustic insulation of single panel walls provided by analytical expressions versus the mass law. *Sound Vib.* **2002**, *257*, 457–475. [CrossRef]
GB/T 19889.5-2006; Acoustics—Measurement of Sound Insulation in Building Elements Part 5: Field Measurements of Airborne Sound Insulation of Façade Elements and Facades. China Architecture & Building Press: Beijing, China, 2006.
GB/T 19889.3-2005; Acoustics—Measurement of Sound Insulation in Buildings and of Building Elements Part 3: Laboratory Measurements of Airborne Sound Insulation of Building Elements. China Architecture & Building Press: Beijing, China, 2006.
Ferziger, J.H.; Peric, M. *Computational Methods for Fluid Dynamics*; Springer: Berlin/Heidelberg, Germany, 2002.

buildings

MDPI

Article

Characteristics and Control of Subway Train-Induced Environmental Vibration: A Case Study

Lizhong Song [1,2], Xiang Xu [1,2], Quanmin Liu [1,2,*], Haiwen Zhang [1,2] and Yisheng Zhang [1,2]

1 State Key Laboratory of Performance Monitoring and Protecting of Rail Transit Infrastructure, East China Jiaotong University, Nanchang 330013, China; songlizhong@ecjtu.edu.cn (L.S.); 2023138086100094@ecjtu.edu.cn (X.X.); 2022138086100044@ecjtu.edu.cn (H.Z.); 2022138086100050@ecjtu.edu.cn (Y.Z.)

2 MOE Engineering Research Center of Railway Environmental Vibration and Noise, East China Jiaotong University, Nanchang 330013, China

* Correspondence: liuquanmin@ecjtu.edu.cn

Abstract: With the widespread construction of the subway in the Chinese mainland, the environmental vibration caused by subway operation has attracted increasing attention. Train-induced environmental vibrations can cause structural deformation, uneven settlement of line foundations, and tunnel leakage, affecting the structural safety of lines and foundations. This research focuses on a segment of the Nanchang Metro Line 3, which has been chosen as the subject of investigation. A numerical model was developed to analyze the subway train-induced environmental vibration, employing the finite element method (FEM). Utilizing a numerical model, an investigation was conducted to examine the impact of train speed on the subway train-induced environmental vibration, the train-induced environmental vibration transmission characteristics were analyzed, and the control effects of vibration reduction tracks on train-induced environmental vibration were discussed. Train-induced vibration tests were also conducted on Nanchang Metro Line 3 to verify the control effects of various vibration reduction tracks. The results indicate that the subway train-induced environmental vibration rises as the train speed goes up, and the vibration peaks always appear around 63 Hz. When the train speed doubles, the Z-vibration level increases from about 5.1 dB to 5.9 dB. Subway train-induced environmental vibration shows a fluctuating decreasing trend with increasing distance from the centerline of the tunnel. The Z-vibration level reaches its maximum 4 m away from the centerline of the tunnel. Compared with the embedded sleeper, the vibration-damping fastener exhibits a vibration reduction effect of about 9 dB to 18 dB, the rubber vibration-damping pad exhibits a better vibration reduction effect of about 16 dB to 24 dB, and the steel spring floating plate exhibits the best vibration-damping effect of about 18 dB to 28 dB. The calculated Z-vibration levels are basically consistent with the measured values, indicating the accuracy of the calculated results of the control effects of the vibration reduction tracks.

Keywords: subway train-induced vibrations; vibration characteristics; vibration control; numerical simulations; case study: Nanchang Metro Line 3

1. Introduction

With the widespread construction of the subway in the Chinese Mainland, the environmental vibration caused by subway operation has attracted increasing attention. During the operation of subway trains, the wheel–rail roughness can cause severe dynamic interaction between the track and the train, leading to vibration of the train systems and track structures. The vibration is transmitted to the area near the line through the track, which, in turn, causes vibration of the soil, ground, and buildings. Train-induced environmental vibrations can cause structural deformation, uneven settlement of line foundations, and tunnel leakage, which can affect the structural safety of lines and foundations. Therefore, train-induced environmental vibration is an unavoidable and important issue in subway operation and development.

With the successive appearance of many numerical computation methods, such as the infinite element method (IFEM), finite element method (FEM), and boundary element method (BEM), numerical calculation methods have been widely used for solving the dynamic problems of tunnel–soil systems. Due to its advantages of clear physical concepts, wide applicability, versatility, and ease of implementation, the FEM has been applied in a large number of theoretical studies on environmental vibration caused by subway train movements. Chua et al. [1,2] developed a two-dimensional (2D) finite element (FE) model to analyze the vibration response of tunnels, buildings, and soils under train loading. Jones et al. [3] developed a 2D coupled finite element–boundary element model (FE–BE model) to analyze the train-induced environmental vibration in a tunnel. Ma et al. [4] investigated the characteristics of vibration sources near subway tunnels and the propagation patterns of vibration waves in soil layers and ground surface by conducting on-site vibration measurements. Then, 2D numerical models under field conditions were established using an FE model to investigate how the parameters of the vibration isolation system of utilizing pile barriers, a widely used vibration mitigation measure in tunnels, affect its properties. Hussein et al. [5] analyzed the traffic-induced building vibration in a tunnel using the substructure method, combining a PiP model of a tunnel–soil system with a 2D frame beam FE model of the building. Hesami et al. [6] used 2D FE model analysis to simulate train movement and its impact on neighboring buildings. Lan et al. [7] combined Padé approximation-based preview repetitive control (POPRC) with equivalent input disturbances (EID) to improve track performance and obtained a simplified POPRC law with EID in the original system, illustrating the robustness and validity of the controller. When analyzed with a 2D FE model, the dynamic parameters of the system and the train loads need to be processed in an equivalent or simplified manner, which will inevitably lead to a large discrepancy between the calculation results and the actual situation. However, tunnel–soil systems are generally very complex, and it is difficult to accurately predict the train-induced environmental vibration using a 2D FE model. To perform quantitative analysis, it is more advantageous to perform calculations with a three–dimensional (3D) FE model. Zhang et al. [8] developed a 3D model of the interaction between the soil and the pile foundation for the purpose of exploring the structural and ground vibrations that are generated by trains operating on viaducts. Dai et al. [9] employed a semi-analytical approach to develop a 3D model of the train–rail–tunnel–ground system and proceeded to analyze the amplitude–frequency characteristics and the spread rule of the environmental vibration that was generated by train operations within underground high–speed railroads. Kouroussis et al. [10,11] established a time-domain track–train vertical coupled dynamics model and a track–soil–structure 3D model, and used joint simulation methodologies and step–by–step calculations to examine the vibration of the building and the ground brought on by passing of the train, respectively. Author He and his colleagues [12,13] developed a 3D analytical model. The purpose of this model was to predict the ground vibration of two fully spaced embedded parallel tunnels. They also analyzed the vibration fields generated at different frequencies and with different soil properties. Sun et al. [14] investigated the X–section cast-in-place concrete (XCC) pile raft (BTXPR for short) foundations buried in air-dry sand and saturated sandy subsoil, respectively. The vibration speed response was measured and analyzed at different cyclic loading frequencies. Measurement shows that the size and change of the vibration speed of the BTXPR foundation are closely related to the saturation of the bottom subsoil. Khajehdezfuly et al. [15] evaluated the vibration induced by the concurrent transit of rail and road vehicles, analyzed the effect of different parameters on the vibrations, and conducted comprehensive field tests. Ma et al. [16] established a 3D model of building–soil–tunnel–track system and analyzed the vibration effects of train passage on the ancient buildings of Xi'an Bell Tower by using the train loads calculated as the external load input. Zhai et al. [17,18] calculated the train dynamic load as an external load input applied to a 3D model of a soil–building system and investigated the vibration of soil, ground, and a subway superstructure caused by train operation. Fiala et al. [19] developed a 3D coupled FE–BE model for calculating the effect of high–speed train operation on the ground on the vibration of buildings along the route. Wu et al. [20] investigated the seismic isolation effect of RSM seismic isolation pads for different site classes. The main influencing factors included base pressure, seismic excitation intensity, and liner thickness. The study showed that the seismic isolation performance increased with seismic excitation intensity, base pressure, and liner thickness.

The noise which is produced by the running of a rail transit system is a prominent problem that bothers residents residing along the railway line [21]. An increasing number of acoustic metamaterials have been studied in recent years with the aim of diminishing the consequences of secondary noise alongside railroads. Talebitooti et al. [22] used the Biot theory to model porous elastic materials—which consider three types of wave propagation, including two compression terms and one shear term—to investigate the effect of embedded porous cores on structural sound transmission loss (STL). Acoustic metamaterials have also been used in noise reduction in rail transportation. Amadasi et al. [23] explored the reduction in train noise in urban areas by miniature sound barriers made of metamaterials and optimized their acoustic design. Zhang et al. [24] elaborated an active control system which can directly apply control torques or moments to the target structures with the properties of rotational motion or vibration and discussed its theory, modeling, comprehensive parameter analysis, and case studies for the flutter vibration control of bridges as well as its application prospects in various other occasions. Many researchers have undertaken investigations into the simulation and prediction of vibrations caused by subway operation. However, the subway train-induced environmental vibration characteristics are not yet clear, and the vibration problem has not been effectively controlled. Consequently, within this paper, a specific tunnel section along Nanchang Metro Line 3 has been designated as the research focus. Employing the FEM, a numerical model dedicated to simulating train-induced environmental vibration has been constructed. Based on the numerical model, the influence of train speed on train-induced environmental vibration was studied, the train-induced environmental vibration transmission characteristics were analyzed, and the control effects of vibration reduction tracks on train-induced environmental vibration were discussed. Vibration tests induced by trains were additionally conducted on Nanchang Metro Line 3 for the purpose of verifying the control effects of diverse vibration reduction tracks.

2. Numerical Model

2.1. Wheel–Rail Interaction Model

When a train runs on rails, it is affected by the roughness of the rails and the wheels, resulting in interaction forces between them, namely wheel–rail forces. Under the action of wheel–rail forces, the rail vibrates, and the vibration energy is transmitted to the tunnel and surrounding soil through the track, thereby causing vibration of the tunnel and surroundings soil. A wheel–rail interaction model was first established based on the dynamic flexibility method, as shown in Figure 1, with the basic equation shown below [25]:

$$\mathbf{F}(\omega) = \frac{\mathbf{R}}{\mathbf{Y}_c + \mathbf{Y}_w + \mathbf{Y}_r} \tag{1}$$

where $\mathbf{F}$ represents the wheel–rail forces; ω represents the circular frequency, unit: rad/s; $\mathbf{R}$ represents the wheel–rail roughness; and $\mathbf{Y}_c$, $\mathbf{Y}_w$, and $\mathbf{Y}_r$ denote the dynamic flexibility of the wheel–rail contact spring, the wheel, and the rail, respectively.

The wheel–rail roughness spectrum given by the ISO3095:2005 standard [26] was chosen as the excitation of the model of wheel–rail interaction as illustrated in Figure 2.

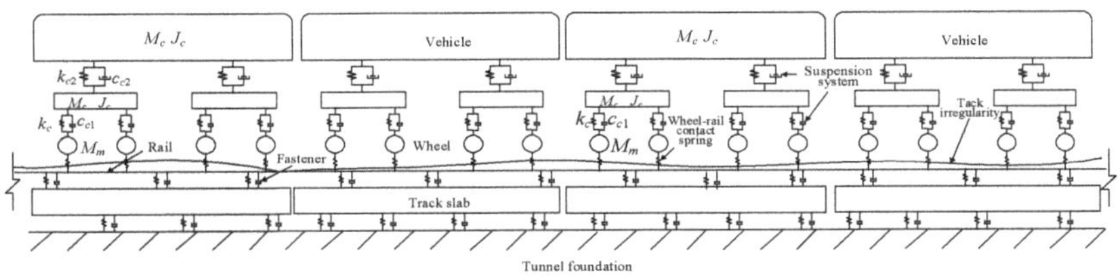

Figure 1. Schematic diagram of train–track coupling vibration analysis model.

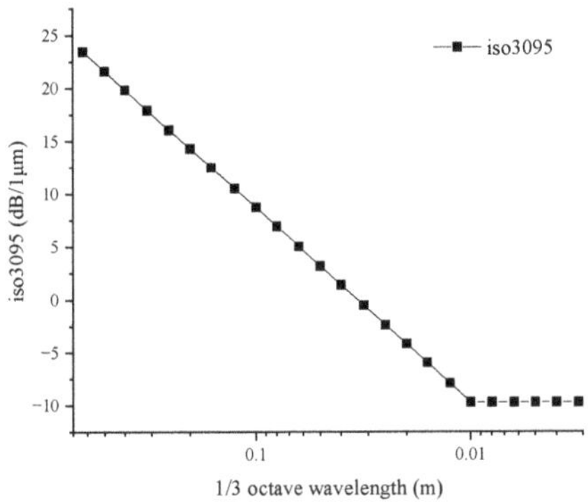

Figure 2. Rail roughness spectrum.

The wheel–rail forces on track types such as rubber vibration-damping pads, steel spring floating plates, vibration-damping fasteners, and embedded sleepers were calculated using the wheel–rail interaction model. Table 1 shows the parameters that are used for calculating the track structures.

Table 1. The parameters utilized for the calculation of the track structures.

Parameter	Unit	Value
Density of rail	kg/m^3	7800
Poisson's ratio of rail	—	0.3
Elastic modulus of rail	MPa	2.1×10^5
Fastener stiffness	MN/m	80
Vibration-damping fastener stiffness	MN/m	10
Fastener spacing	m	0.625
Fastener damping	kN·s/m	75
Rubber vibration-damping pad stiffness per unit area	N/m^3	0.025
Steel spring stiffness	MN/m	12
Steel spring spacing	m	1.2

2.2. Train-Induced Environmental Vibration Model

By using the FEM for structural vibration calculation, the responses of the node acceleration, velocity, and displacement of the structure over time subjected to moving loads can be obtained. The basic equation is as follows:

$$\mathbf{Kx} + \mathbf{C\dot{x}} + \mathbf{M\ddot{x}} = \mathbf{F} \tag{2}$$

where **K**, **C**, and **M** are the stiffness matrix, damping matrix, and structural mass matrix, respectively; **x**, $\dot{\mathbf{x}}$, and $\ddot{\mathbf{x}}$ are the system vibration displacement, velocity, and acceleration, respectively, and **F** represents the wheel–rail forces.

A representative section of Nanchang Metro Line 3 was selected as the research object. According to the details of the soil layer distribution in the design data, the longitudinal geological profile along the line, and the mechanical properties and physical of the foundation soil, the soil layers used for finite element calculation were simplified to six layers. Table 2 presents the calculation parameters.

Table 2. Soil layer calculation parameters.

Soil Layer	Thickness (m)	Density (kg/m^3)	Shear Wave Velocity (m/s)	Poisson's Ratio	Elastic Modulus (MPa)
Miscellaneous fill	3	1850	150.7	0.38	115.96
Silt clay	4	1950	189.9	0.38	194.09
Sandy soil	6	1980	201.5	0.35	217.06
Gravel sand and round gravel	12	2000	290.5	0.32	445.58
Moderately weathered mudstone–sandstone	15	2300	813.2	0.25	3802.44
Slightly weathered mudstone–sandstone	20	2500	1085.5	0.22	7187.69

As illustrated in Figure 3, a numerical model of train-induced environmental vibration was established using the FEM. The cross-section of the tunnel is circular, with CHN60 rails and sleepers embedded in an integral track bed, an outer diameter of 6 m, a lining thickness of 0.3 m, an inner diameter of 5.4 m, and a tunnel burial depth of 10 m. In the model, the rail was represented through the utilization of spatial beam elements; the fasteners were simulated using spring-damping elements; shell elements were used to model the tunnel lining; and the soil layers and foundation of the tunnel were simulated using solid elements. To reduce the number of elements and improve computational efficiency, a semi-model of the tunnel–soil structure was established based on symmetry. Symmetry constraints were applied on the symmetry plane, and the bottom of the model was fixed, while the front and right sides were equipped with equivalent viscoelastic boundaries. The parameters utilized for the calculations pertaining to the tunnel and track are presented in Table 3.

Table 3. Track and tunnel calculation parameters.

Component	Density (kg/m^3)	Elastic Modulus (MPa)	Poisson's Ratio
Rail	7800	2.1×10^5	0.3
Track bed	3000	3.25×10^4	0.2
Tunnel foundations	2500	3.15×10^4	0.2
Tunnel lining	2500	3.15×10^4	0.2

Table 4. The locations of the calculation points.

Point	Location
P1	At the center of the track bed
P2	On the tunnel wall 1.5 m above the rail surface
P3	On the ground directly above the centerline of the tunnel
P4	On the ground 2 m away from the centerline of the tunnel
P5	On the ground 4 m away from the centerline of the tunnel
P6	On the ground 8 m away from the centerline of the tunnel
P7	On the ground 16 m away from the centerline of the tunnel
P8	On the ground 32 m away from the centerline of the tunnel

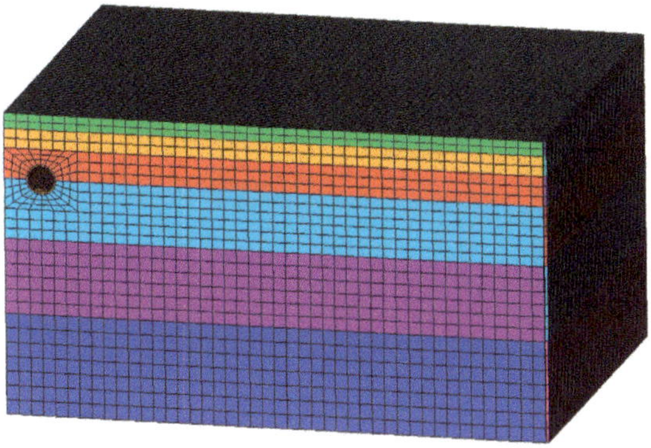

(a)

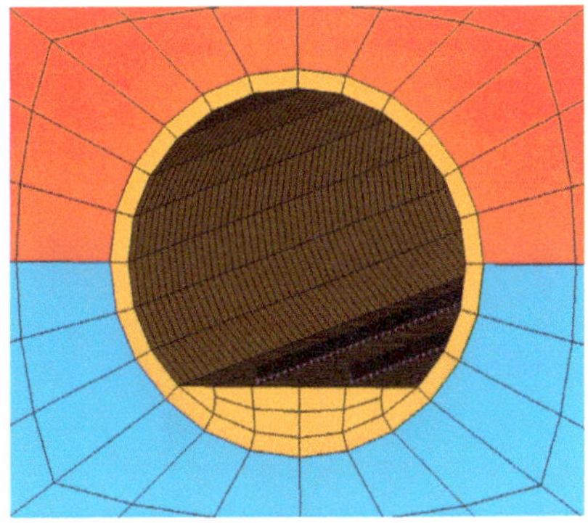

(b)

Figure 3. Numerical model of train-induced environmental vibration: (**a**) overall model diagram; (**b**) partial enlarged view of the subway tunnel.

Figure 4 presents a schematic diagram illustrating the calculation points, while Table 4 provides a detailed enumeration of their respective locations.

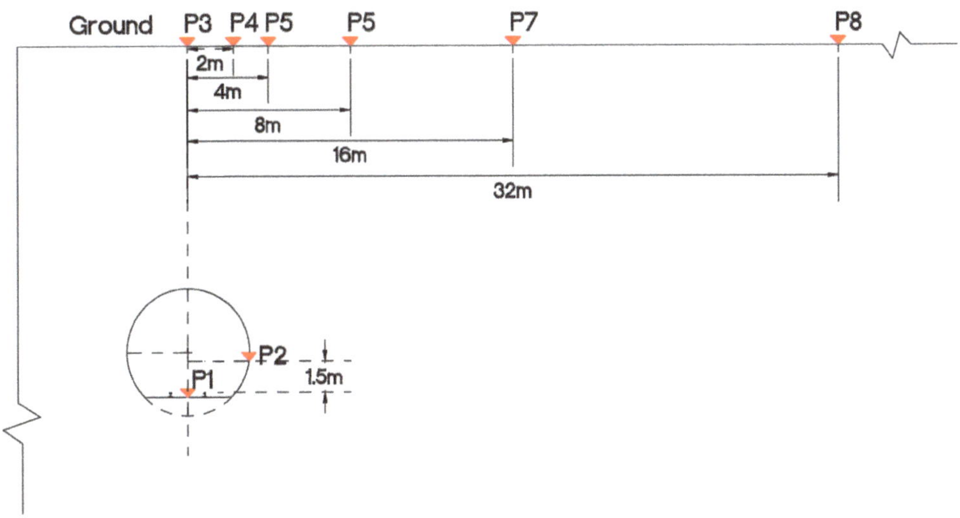

Figure 4. A schematic diagram of the calculation points.

3. Vibration Characteristics

The train-induced vibration responses passing through a subway tunnel with embedded sleepers at speeds of 60 km/h, 80 km/h, 100 km/h, 120 km/h, 140 km/h, and 160 km/h were computed using the aforementioned numerical model. Based on the calculated results, the influence of train speed on the subway train-induced environmental vibration characteristics was studied, and the vibration transmission characteristics were analyzed.

The formula for calculating the vibration acceleration level (*VAL*) can be written as follows:

$$VAL = 20\,\lg\left(\frac{a}{a_0}\right) \tag{3}$$

where a is the RMS value of vibration acceleration, a_0 is the reference acceleration, and $a_0 = 1 \times 10^{-6}$ m/s^2.

Z-vibration level is the *VAL* corrected by the Z-weighting factors according to ISO 2631-1:1997 [27]. Z-weighting factors at center frequencies of 1/3 octave band are listed in Table 5.

Table 5. Z-weighting factors at center frequencies of 1/3 octave band.

Center Frequency (Hz)	Z—Weighting Factor (dB)
1	−6
1.25	−5
1.6	−4
2	−3
2.5	−2
3.15	−1
4	0
5	0
6.3	0
8	0
10	−2
12.5	−4
16	−6
20	−8
25	−10
31.5	−12
40	−14
50	−16
63	−18
80	−20

3.1. *Influence of Train Speed*

Figure 5 presents the spectra of wheel–rail force amplitudes at various train speeds. It can be observed that the trends of wheel–rail force variation with changing frequency are almost the same at varying speeds of the train. The maximum values of the wheel–rail forces in the frequency domain are 4.60, 5.99, 7.37, 8.72, 10.05, and 11.37 kN for train speeds of 60, 80, 100, 120, 140, and 160 km/h, respectively, and they all occur at 65 Hz.

Figure 6 illustrates the Z-vibration level spectra of typical points resulting from the passage of the train through the subway tunnel at different speeds. The analysis indicates that the subway train-induced vibrations at various points increase in conjunction with an elevation in train speed, and the vibration peaks always appear around 63 Hz. This is because the peaks of wheel–rail forces always occur in the frequency band of 63 Hz.

To analyze further the influence that train speed has on the environmental vibration characteristics generated by subway trains, the linear fittings of the relationship between Z-vibration levels at different points and the base-10 logarithm of the train speed are shown in Figure 7.

According to the linear fittings in Figure 7, the Z-vibration levels at different points can be expressed as follows:

$$VLZ_{\mathrm{P}i}(v) = VLZ_{\mathrm{P}i}(v_0) + C_{v\mathrm{P}i}(i = 1,\ 2,\ \cdots,\ 8) \quad (4)$$

where $VLZ_{\mathrm{P}i}(v)$ and $VLZ_{\mathrm{P}i}(v_0)$ represent the Z-vibration levels of the point Pi at speeds of v and v_0; $C_{v\mathrm{P}i}$ denotes the speed correction; and $C_{v\mathrm{P}i}$ can be written as follows:

$$C_{v\mathrm{P}i} = S_{v\mathrm{P}i}\lg\frac{v}{v_0}(i = 1,\ 2,\ \cdots,\ 8) \quad (5)$$

where S_{vPi} is the slope of the linear fitting of the relationship between Z-vibration level at the point Pi and the base-10 logarithm of the train speed.

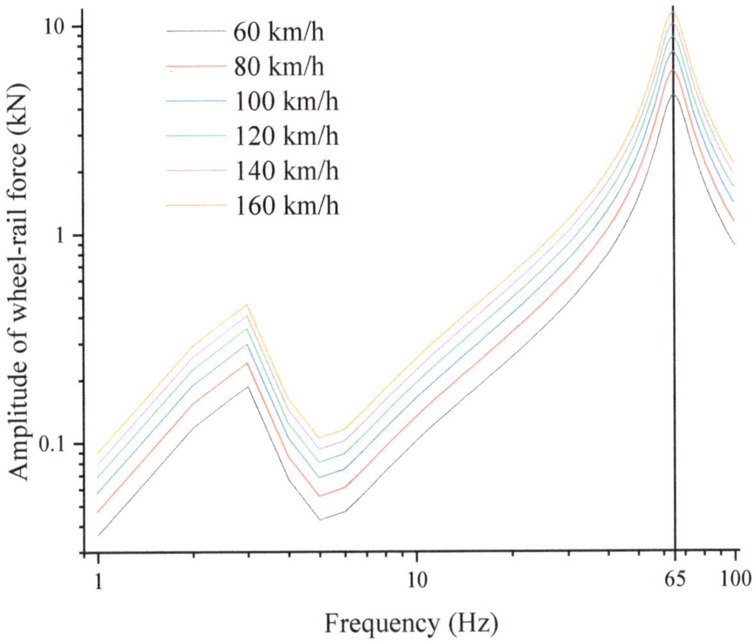

Figure 5. Spectra of wheel–rail force amplitudes at different speeds.

In accordance with the aforementioned formulas, the Z-vibration level at the point Pi increases by about $S_{vPi}\lg 2$ dB when the train speed doubles. Based on the linear fitting results in Figure 7, it can be concluded that when the train speed doubles, the Z-vibration level increases from about 5.1 dB to 5.9 dB.

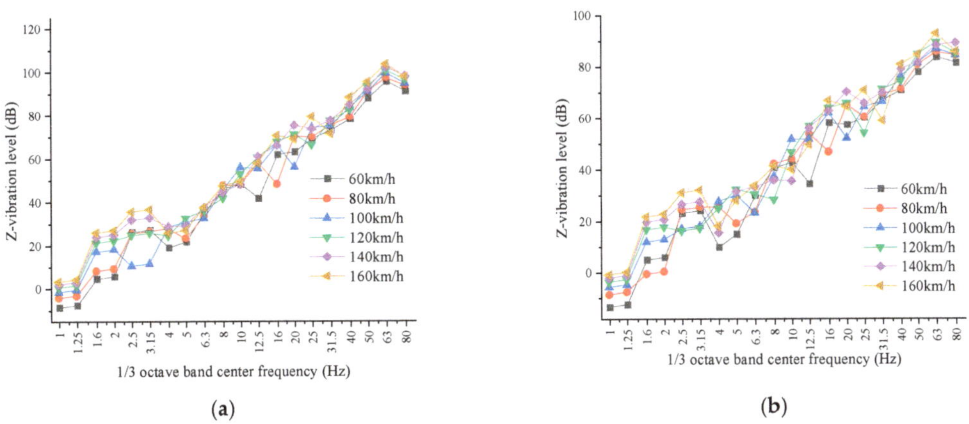

Figure 6. *Cont.*

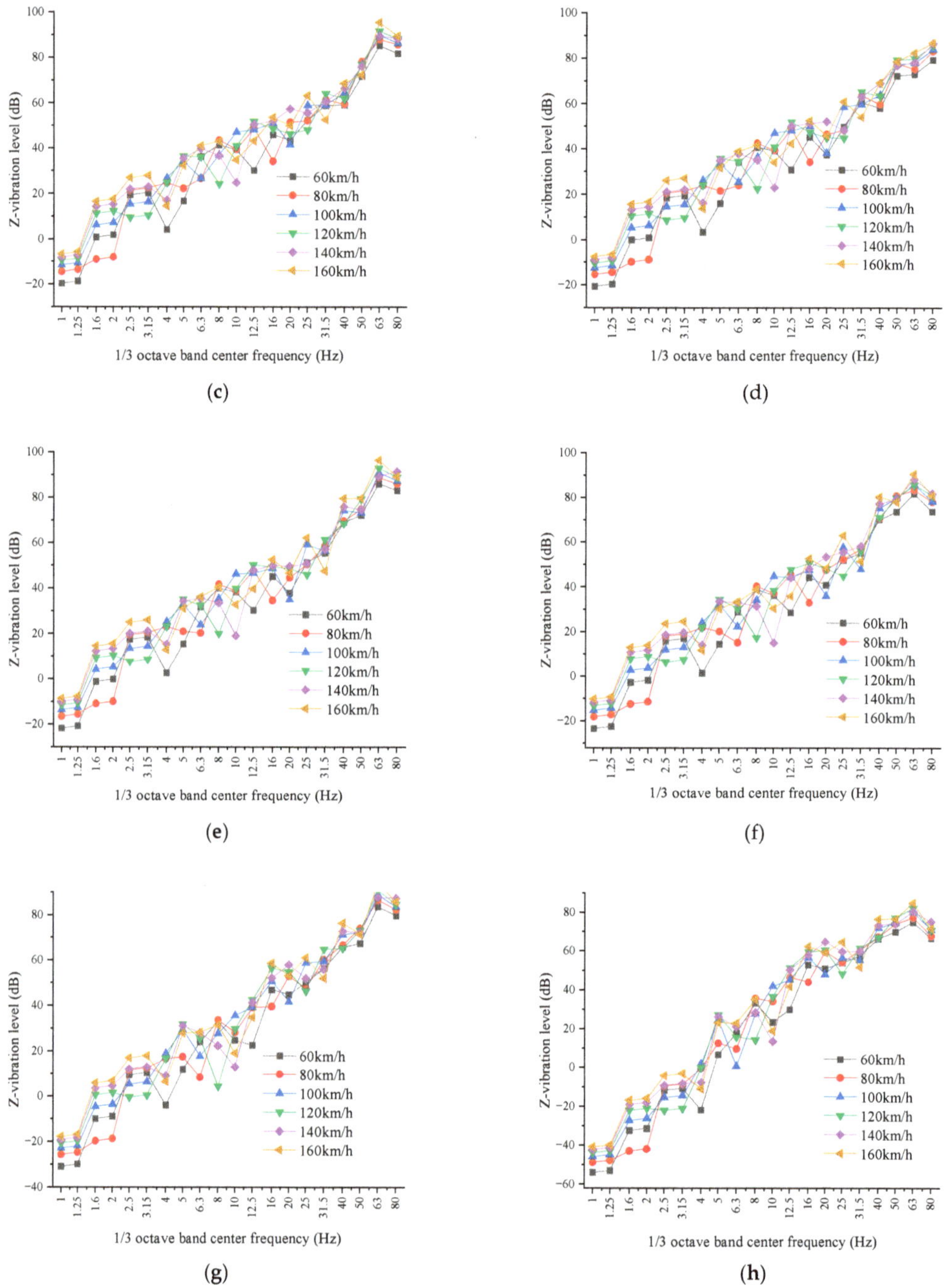

Figure 6. The Z-vibration-level spectra of typical points caused by the train passing through the subway tunnel at different speeds: (**a**) P1; (**b**) P2; (**c**) P3; (**d**) P4; (**e**) P5; (**f**) P6; (**g**) P7; (**h**) P8.

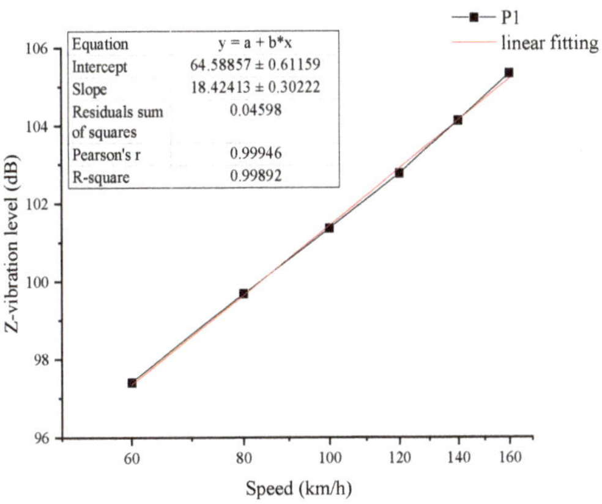

(a)

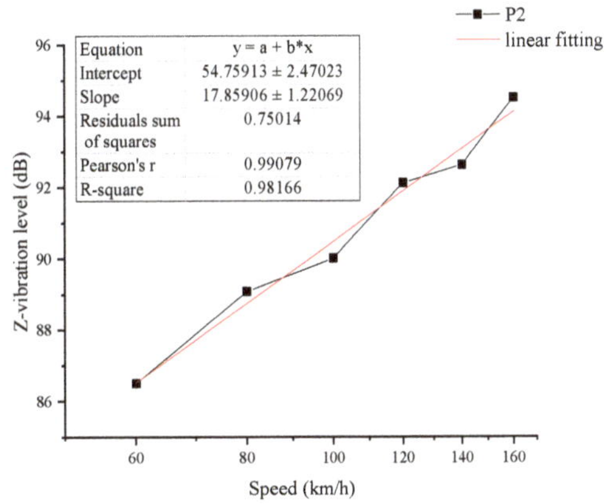

(b)

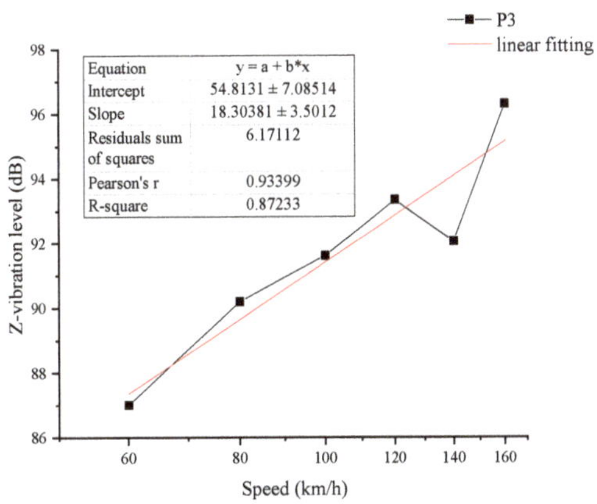

(c)

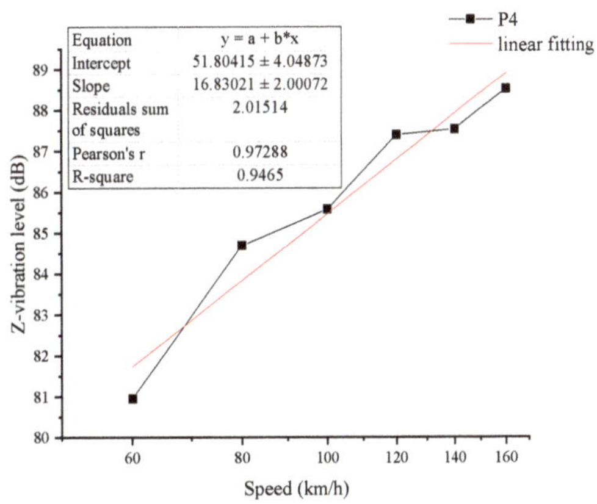

(d)

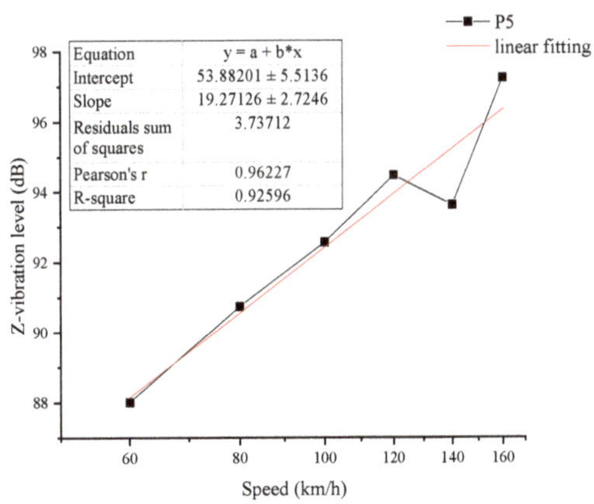

(e)

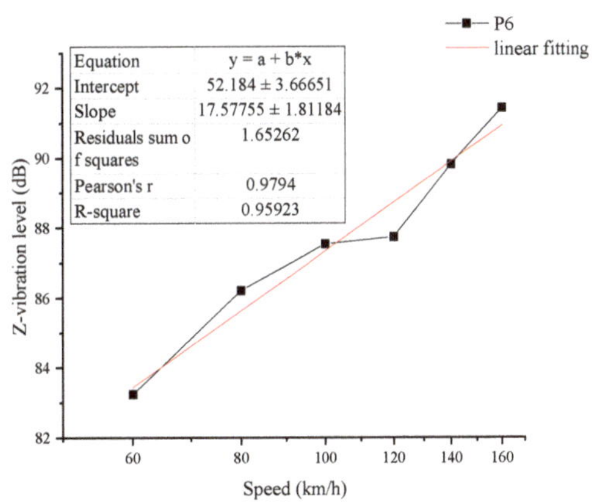

(f)

Figure 7. *Cont.*

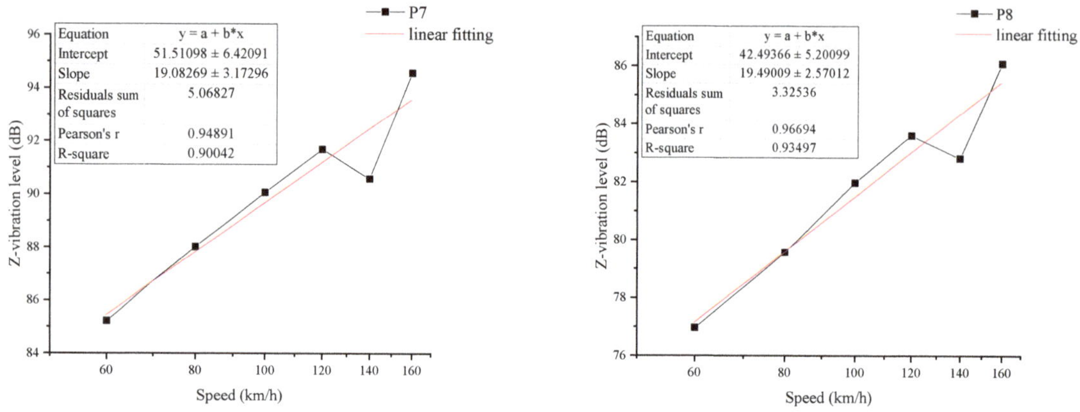

(**g**) (**h**)

Figure 7. The linear fittings of the relationship between Z-vibration levels and the base-10 logarithm of the train speed: (**a**) P1; (**b**) P2; (**c**) P3; (**d**) P4; (**e**) P5; (**f**) P6; (**g**) P7; (**h**) P8.

3.2. Vibration Transmission Characteristics

Figure 8 shows the Z-vibration level spectra at various points resulting from the train's passage in the subway tunnel at speeds of 60, 80, 100, 120, 140, and 160 km/h. At different train speeds, the patterns of Z-vibration level variation with frequency at various points are generally consistent, with vibration peaks occurring around 63 Hz. The maximum vibration occurs at P1 on the track bed, and the minimum vibration occurs at P8, which is located 32 m from the centerline of the tunnel.

Figure 9 shows the variation curves of subway train-induced ground vibration with changing distance at speeds of 60, 80, 100, 120, 140, and 160 km/h. The train-induced ground vibration passing through the subway tunnel at different speeds shows a fluctuating decreasing trend with increasing distance from the centerline of the tunnel. The Z-vibration level reaches its maximum 4 m away from the centerline of the tunnel and its minimum 32 m away from it. This phenomenon is associated with the propagation path of vibration waves through the foundation soil. The vibration waves induced by rail transportation mainly propagate in the form of longitudinal, transverse, and Rayleigh waves, in which Rayleigh waves are superimposed by transverse and longitudinal waves. The propagation trajectory is not a straight line but an ellipse, so the localized peaks on the surface are related to the superposition propagation effect of Rayleigh waves.

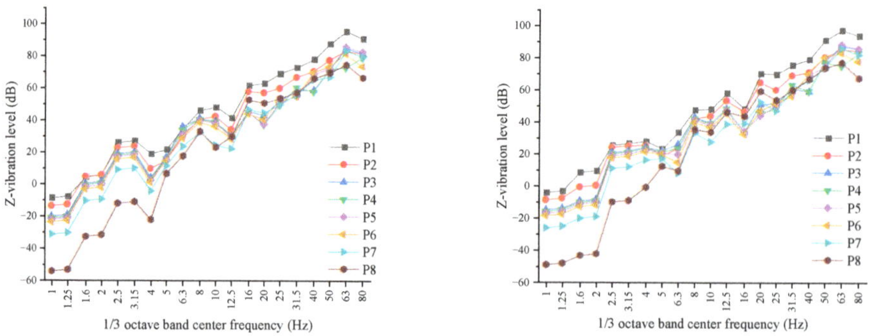

(**a**) (**b**)

Figure 8. *Cont.*

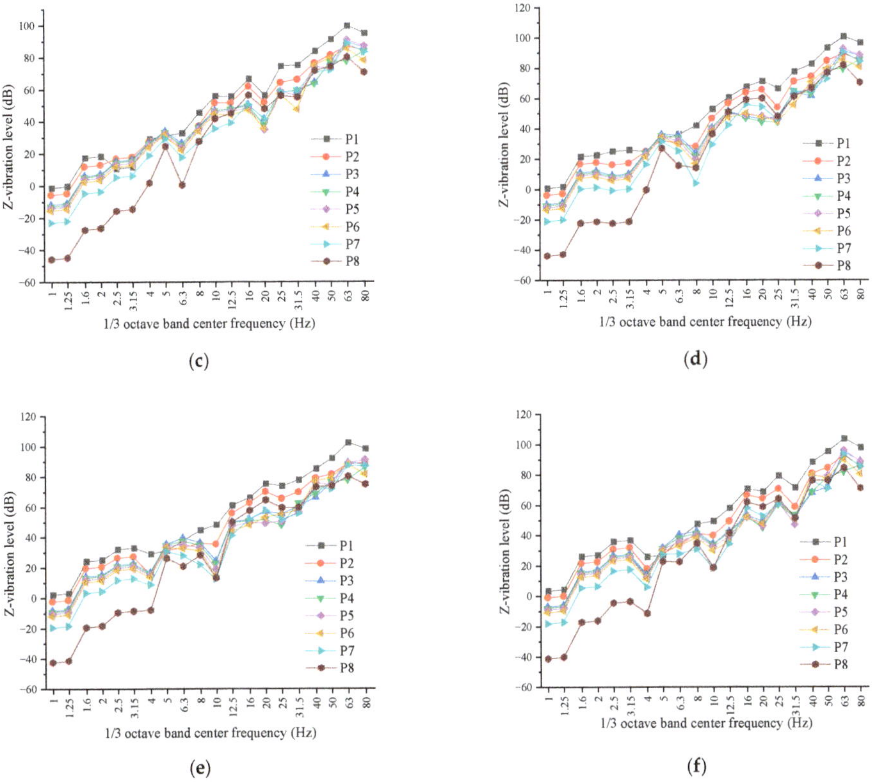

Figure 8. The Z-vibration level spectra at various points caused by the train passing through the subway tunnel at different speeds: (**a**) 60 km/h; (**b**) 80 km/h; (**c**) 100 km/h; (**d**) 120 km/h; (**e**) 140 km/h; (**f**) 160 km/h.

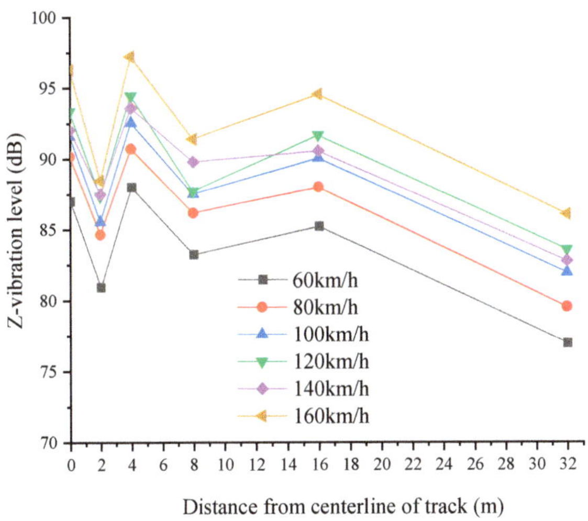

Figure 9. The variation curves of subway train-induced ground vibration with changing distance at speeds of 60, 80, 100, 120, 140, and 160 km/h.

4. Vibration Control

4.1. Numerical Analysis

Based on the numerical model, the ground vibrations that are produced when the train travels through a subway tunnel furnished with embedded sleepers, steel spring floating plates, vibration-damping fasteners, and rubber vibration-damping pads at a speed of 80 km/h were calculated. Based on the calculated results, research on vibration control measures for subway tunnels was carried out.

The Z-vibration level spectra at various points with different tracks are shown in Figure 12. Compared with the embedded sleepers, the steel spring floating plates, vibration-damping fasteners, and rubber vibration-damping pads all show good vibration reduction effects. Among the three types of vibration reduction tracks, the steel spring floating plates are the most effective when it comes to vibration reduction, followed by rubber vibration-damping pads, and the vibration-damping fasteners are the least effective.

In order to quantitatively analyze the vibration reduction effect of different vibration reduction tracks, Table 6 provides the Z-vibration levels at various points on different tracks, and Table 7 presents the effects of vibration reduction that different vibration reduction tracks have on various points.

Table 6. The Z-vibration levels at various points on different tracks (unit: dB).

Track Type	P2	P3	P4	P5	P6	P7	P8
Embedded sleeper	89.08	90.20	84.69	90.75	86.21	88.03	79.56
Vibration-damping fasteners	79.43	72.67	72.56	73.07	71.76	71.21	70.45
Rubber vibration-damping pads	71.09	66.45	63.03	66.49	62.96	63.82	63.94
Steel spring floating plates	69.45	65.03	63.29	64.66	61.66	60.42	61.71

Table 7. Vibration reduction effects of different vibration reduction tracks at various points (unit: dB).

Track Type	P2	P3	P4	P5	P6	P7	P8
Vibration-damping fasteners	9.65	17.54	12.13	17.67	14.45	16.82	9.11
Rubber vibration-damping pads	18.00	23.75	21.66	24.26	23.25	24.21	15.62
Steel spring floating plates	19.63	25.17	21.40	26.08	24.55	27.62	17.85

Compared with the embedded sleepers, the vibration-damping effect of vibration-damping fasteners is about 9 dB to 18 dB, that of rubber vibration-damping pads is about 16 dB to 24 dB, and that of steel spring floating plates is about 18 dB to 28 dB.

4.2. Test Verification

For the purpose of validating the computational outcomes regarding the vibration-damping effects of various vibration-damping tracks, sections with steel spring floating plates, rubber vibration-damping pads, and embedded sleepers were selected on Nanchang Metro Line 3 to conduct tests of train-induced vibration, as shown in Figure 10. Accelerometers were positioned on the tunnel wall 1.5 m above the rail surface, i.e., point P2, to record the vibration caused by the train passing at a speed of about 80 km/h.

(a) (b)

Figure 10. Test photos: (**a**) a test section; (**b**) an accelerometer installed on the tunnel wall.

Figure 11 compares the measured and calculated values of train-induced vibration at point P2. Obviously, the measured values of the Z-vibration levels essentially conform to the calculated values, indicating the accuracy of the calculated results of the vibration reduction tracks.

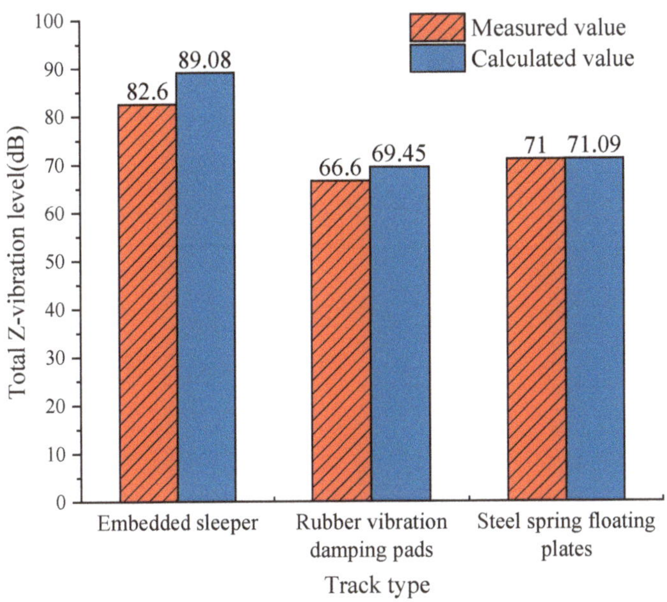

Figure 11. Comparison between the calculated and measured values of Z-vibration levels at P2.

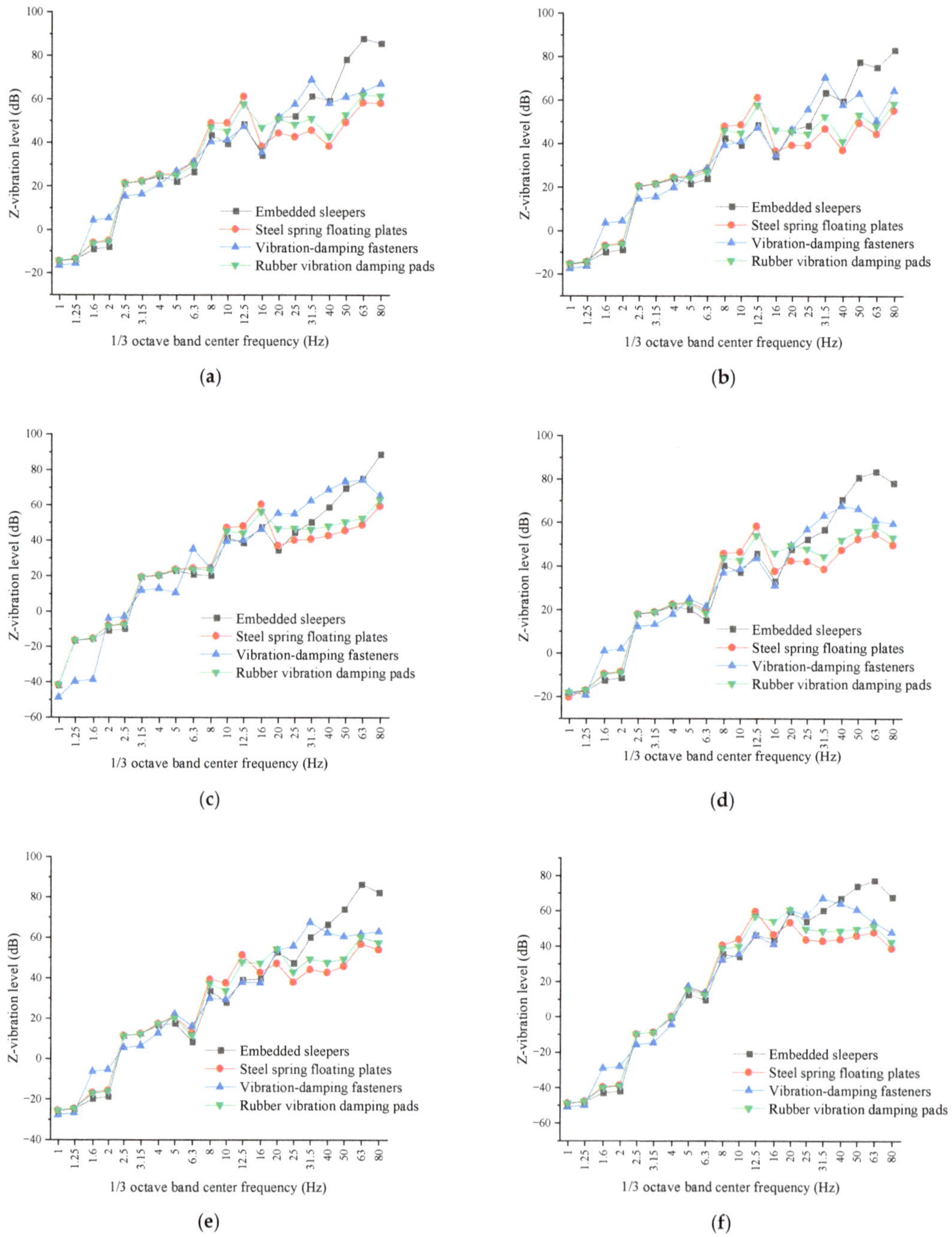

Figure 12. The Z-vibration level spectra at various points with different tracks: (**a**) P3; (**b**) P4; (**c**) P5; (**d**) P6; I P7; (**f**) P8.

4.3. Discussion

To further verify the feasibility of the vibration control measures, the environmental vibration limits specified in the Standard of vibration in urban area environment (GB 10070-1988) [28] for each type of urban area are listed in Table 8.

Table 8. Z-vibration level limits for various types of urban areas (unit: dB).

Applicable Area	Daytime	Night
Special residential areas	65	65
Residential areas and areas used for cultural and educational purposes	70	67
Areas used for mixed purposes and commercial centers	75	72
Industrial areas	75	72
Areas along both sides of the trunk line	75	72
Areas along both sides of the trunk rail line	80	80

By comparing the ground vibration on different vibration reduction tracks with the vibration limits in Table 8, it can be seen that, for tracks with embedded sleepers, only the vibration at point P8 meets the vibration limits for the area along both sides of the trunk rail line, while the vibrations at other points all exceed the limits. On tracks with vibration-damping fasteners, the vibrations at all points on the ground basically meet the vibration limits for areas used for mixed purposes and commercial centers, industrial areas, and areas along both sides of the trunk line. On tracks with rubber vibration-damping pads, the vibrations at points P3 and P5 meet the vibration limits for residential areas and areas used for cultural and educational purposes, while other points on the ground meet the vibration limits of special residential areas. On tracks with steel spring floating plates, the vibrations at all points on the ground basically meet the vibration limits of special residential areas.

5. Conclusions

In this study, a tunnel section of Nanchang Metro Line 3 was selected as the research object, and a numerical model of train-induced environmental vibration was established using the FEM. The numerical model was used to investigate the effect of train speed on train-induced environmental vibration, the train-induced environmental vibration transmission characteristics were analyzed, and the control effects of vibration reduction tracks on train-induced environmental vibration were discussed. Train-induced vibration tests were also conducted on Nanchang Metro Line 3 to verify the calculated effects of various vibration reduction tracks. The following are the primary conclusions:

(1) Subway train-induced environmental vibration enhances with an increase in train speed, and the vibration peaks always appear around 63 Hz. When the train speed doubles, the Z-vibration level increases from approximately 5.1 dB to 5.9 dB.

(2) Subway train-induced ground vibration exhibits a fluctuating reduction trend as distance from the centerline of the tunnel increases. The Z-vibration level reaches its maximum at 4 m away from the centerline of the tunnel.

(3) Compared with embedded sleepers, vibration-damping fasteners have a vibration reduction effect of about 9 dB to 18 dB, rubber vibration-damping pads have a better vibration reduction effect of about 16 dB to 24 dB, and steel spring floating plates have the best vibration reduction effect of about 18 dB to 28 dB.

(4) The calculated Z-vibration levels are basically consistent with the measured ones, indicating the accuracy of the calculated results of the control effects of the vibration reduction tracks.

This research has a certain reference value for subway vibration control, and research will be conducted on the active control of subway vibration in the future.

Author Contributions: Conceptualization, L.S. and Q.L.; methodology, L.S. and X.X.; software, X.X.; validation, H.Z. and Y.Z.; formal analysis, X.X.; data curation, X.X.; writing—original draft preparation, L.S.; writing—review and editing, Q.L.; project administration, Q.L.; funding acquisition, L.S. All authors have read and agreed to the published version of the manuscript.

Funding: This research was funded by the China Postdoctoral Science Foundation, grant numbers 2023T160214 and 2023M731077, and the Natural Science Foundation of Jiangxi Province, grant numbers 20224BAB214074 and 20232BAB204087.

Data Availability Statement: The data presented in this study are available on request from the corresponding author.

Conflicts of Interest: The authors declare no conflicts of interest.

References

1. Chua, K.H.; Balendera, T.; Lo, K.W. Ground-borne vibrations due to trains in tunnels. *Earthq. Eng. Struct. D* **1992**, *21*, 445–46 [CrossRef]
2. Chua, K.H.; Lo, K.W.; Balendra, T. Building response due to subway train traffic. *Int. J. Geotech. Eng.* **1995**, *121*, 747–75 [CrossRef]
3. Jones, C.J.C.; Thompson, D.J.; Petyt, M. A model for ground vibration from railway tunnels. *Proc. Inst. Civil. Eng.-Transp.* **200** *153*, 121–129. [CrossRef]
4. Ma, X.F.; Cao, M.Y.; Gu, X.Q.; Zhang, B.M.; Yang, Z.H.; Guan, P.F. Vibration-isolation performance of a pile barrier in an area soft soil in shanghai. *Shock Vib.* **2020**, *2020*, 8813476. [CrossRef]
5. Hussein, M.; Hunt, H.; Kuo, K.; Costa, P.A.; Barbosa, J. The use of sub-modeling technique to calculate vibration in buildin from underground railways. *Proc. Inst. Mech. Eng. F-J. Ral.* **2015**, *229*, 303–314. [CrossRef]
6. Hesami, S.; Ahmadi, S.; Ghalesari, A.T. Numerical modeling of train-induced vibration of nearby multi-story building: Ca study. *KSCE J. Civ. Eng.* **2016**, *20*, 1701–1713. [CrossRef]
7. Lan, Y.H.; Zhao, J.Y. Improving Track Performance by Combining Padé-Approximation-Based Preview Repetitive Control ar Equivalent-Input-Disturbance. *J. Electr. Eng. Technol.* **2024**, *19*, 3781–3794. [CrossRef]
8. Zhang, Z.J.; Li, X.Z.; Zhang, X.; Fan, J.; Xu, G.H. Semi-analytical simulation for ground-borne vibration caused by rail traffic c viaducts: Vibration-isolating effects of multi-layered elastic supports. *J. Sound Vib.* **2022**, *516*, 116540. [CrossRef]
9. Dai, F.; Yang, J.Z.; Bi, L.X.; Liu, Y.T. Analysis on characteristics of ground vibration induced by underground high-speed trai based on in-situ test. *J. Low Freq. Noise Vib. Act. Control* **2024**, *43*, 1097–1111. [CrossRef]
10. Kouroussis, G.; Connolly, D.; Alexandrou, G.; Vogiatzis, K. Railway ground vibrations induced by wheel and rail singular defec *Veh. Syst. Dyn.* **2015**, *53*, 1500–1519. [CrossRef]
11. Olivier, B.; Verlinden, O.; Kouroussis, G. A vehicle/track/soil model using co-simulation between multibody dynamics and fin element analysis. *Int. J. Rail Transp.* **2020**, *8*, 135–158. [CrossRef]
12. He, C.; Zhou, S.H.; Guo, P.J.; Di, H.G.; Zhang, X.H. Analytical model for vibration prediction of two parallel tunnels in a full-spa *J. Sound Vib.* **2018**, *423*, 306–321. [CrossRef]
13. He, C.; Zhou, S.H.; Di, H.G.; Guo, P.J.; Xiao, J.H. Analytical method for calculation of ground vibration from a tunnel embedd in a multi-layered half-space. *Comput. Geotech.* **2018**, *99*, 149–164. [CrossRef]
14. Sun, G.C.; Kong, G.Q.; Liu, H.L.; Amenuvor, A.C. Vibration velocity of X-section cast-in-place concrete (XCC) pile-raft foundati model for a ballastless track. *Can. Geotech. J.* **2017**, *54*, 1340–1345. [CrossRef]
15. Khajehdezfuly, A.; Shiraz, A.A.; Sadeghi, J. Assessment of vibrations caused by simultaneous passage of road and railw vehicles. *Appl. Acoust.* **2023**, *211*, 109510. [CrossRef]
16. Ma, M.; Li, W.; Qian, C.Y.; Deng, G.H.; Li, Y.D. Study of the train-induced vibration impact on a historic bell tower above t spatially overlapping metro lines. *Soil Dyn. Earthq. Eng.* **2016**, *81*, 58–74. [CrossRef]
17. Zhai, W.M.; He, Z.X.; Song, X.L. Prediction of high-speed train induced ground vibration based on train-track ground syste model. *Earthq. Eng. Eng. Vibr.* **2010**, *9*, 545–554. [CrossRef]
18. Yang, J.J.; Zhu, S.Y.; Zhai, W.M.; Kouroussis, G.; Wang, Y.; Wang, K.Y.; Lan, K.; Xu, F.Z. Prediction and mitigation of train-induc vibrations of large-scale building constructed on subway tunnel. *Sci. Total Environ.* **2019**, *668*, 485–499. [CrossRef]
19. Fiala, P.; Degrande, G.; Augusztinovicz, F. Numerical modeling of ground-borne noise and vibration in buildings due to surfa rail traffic. *J. Sound Vib.* **2007**, *301*, 718–738. [CrossRef]
20. Wu, M.T.; Tian, W.H.; He, J.; Liu, F.C.; Yang, J. Seismic isolation effect of rubber-sand mixture cushion under different site clas based on a simplified analysis model. *Soil Dyn. Earthq. Eng.* **2023**, *166*, 107738. [CrossRef]
21. Zou, C.; Wang, Y.M.; Tao, Z.Y. Train-induced building vibration and radiated noise by considering soil properties. *Sustainabil* **2020**, *12*, 937. [CrossRef]
22. Talebitooti, R.; Zarastvand, M.R. The effect of nature of porous material on diffuse field acoustic transmission of the sandwi aerospace composite doubly curved shell. *Aerosp. Sci. Technol.* **2018**, *78*, 157–170. [CrossRef]

23. Amadasi, G.; Iannace, G.; Bevilacqua, A. City-Train noise reduction in urban area by using acoustic mini-screens made of metamaterials. In Proceedings of the Inter-Noise 2023, Chiba, Tokyo, Japan, 20–23 August 2023.
24. Zhang, C.W. The active rotary inertia driver system for flutter vibration control of bridges and various promising applications. *Sci. China Technol. Sci.* **2023**, *66*, 390–405. [CrossRef]
25. Wu, T.; Thompson, D.J. Vibration analysis of railway track with multiple wheels on the rail. *J. Sound Vib.* **2001**, *239*, 69–97. [CrossRef]
26. *ISO3095:2005*; Railway Applications-Acoustics-Measurement of Noise Emitted by Railbound Vehicles. International Organization for Standardization: Geneva, Switzerland, 2005.
27. *ISO2631-1:1997*; Mechanical Vibration and Shock-Evaluation of Human Exposure to Whole-Body Vibration-Part 1: General Requirements. International Organization for Standardization: Geneva, Switzerland, 1997.
28. *GB 10070-1988*; Standard of Vibration in Urban Area Environment. Ministry of Ecology and Environment: Beijing, China, 1988.

buildings

MDPI

Article

Research on the Vibration Propagation Characteristics of Non-Uniform Speed Trains Entering and Leaving Stations Based on Field Measurements

Ying Shi [1], Na Cai [2] and Yekai Chen [3,*]

[1] Guangdong-Hongkong-Macao Joint Laboratory of Energy Saving and Intelligent Maintenance for Modern Transportations, Guangzhou Railway Polytechnic, Guangzhou 511300, China; shiying@gtxy.edu.cn
[2] Zhejiang Taicheng Environmental Technology Co., Ltd., Taizhou 318000, China; caina46@yeah.net
[3] School of Civil Engineering and Transportation, South China University of Technology, Guangzhou 510641, China
* Correspondence: cheops@scut.edu.cn

Abstract: Urban rail transit systems, while alleviating traffic congestion, generate environmental vibrations that impact adjacent structures and residents, particularly during train acceleration and deceleration near stations. Existing research predominantly focuses on constant-speed operations, leaving a gap in understanding vibration propagation during variable-speed phases. This study investigates vibration characteristics and propagation behaviors using field measurements from a subway station in Foshan, China. Wireless vibration sensors were deployed across nine measuring points at varying distances (15–35 m) from the subway station's external wall, capturing time-domain and frequency-domain data during train operations. The analysis incorporated China's JGJ/T 170-2009 standards, evaluating vibration acceleration levels (VAL) and 1/3 octave band spectra. Key findings revealed background vibrations (0–10 Hz) exhibited negligible interference, whereas vehicle-induced vibrations (40–60 Hz) demonstrated directional disparities: urban-bound trains produced higher accelerations (0.004–0.008 m/s^2 vertically) than suburban-bound ones (0.001–0.005 m/s^2) due to track damping measures and propagation distance. Vibration attenuation with distance was found to be non-linear, influenced by soil hardening and train speed. Vertical vibrations near the station (15 m) approached the 70 dB regulatory limit, emphasizing proximity risks. Doppler effects were observed during train acceleration/deceleration, though data limitations precluded precise quantification of speed impacts. This work supplements knowledge on non-uniform train-induced vibrations, offering insights for urban planning and mitigation strategies.

Keywords: vibration propagation behavior; urban rail transit; non-uniform speed; vibration attenuation; field measurement

1. Introduction

In conjunction with the sustained momentum of China's economic growth and urbanization, the urban rail transit system has experienced rapid development [1–3]. Characterized by substantial capacity and swift transit capabilities, this system serves as an efficacious solution to mitigate urban traffic congestion, thereby emerging as the preferred choice of urban transportation for the majority of the residents [4–6]. However, as cities expand vertically and horizontally, the environmental vibrations induced by subway operations have emerged as a critical concern. Particularly, when the subway trains enter and exit the station, vibrations are transmitted through the track bed and soil to adjacent commercial buildings and residential structures, resulting in floor vibrations [7,8]. These vibrations not only imperil the structural integrity of buildings but also disrupt the daily lives and work of nearby residents [9,10]. Moreover, such vibrations may precipitate damage to the buildings' decorative layers, such as the fracturing or detachment of paint. Prolonged exposure to vibrations can also exert a variety of adverse effects on human health, including disturbances to vision, movement, and concentration, culminating in diminished work efficiency, fatigue, and even safety incidents [11–13]. Consequently, an in-depth investigation into the environmental vibrations caused by subway operation is of paramount importance [14–16].

Existing research predominantly focuses on vibration characteristics generated by constant-speed train operations. Researchers engage in the development of vibrational models of trains and track systems to study the vibration responses of trains traversing diverse tunnel tracks [17–20], focusing on the variations in the intensity of vibrational sources, the design of trains, and the types of tracks. For instance, Huang et al. delved into the vibrational characteristics of trains navigating through cured tracks [12], whereas Zhang conducted an exploration into the deformation resistance and seismic mitigation properties of various ballast bed configurations under the influence of train-induced vibrations [21]. Additionally, researchers have constructed comprehensive dynamic interaction models integrating wheel-track-soil systems and soil-pile-building structures, thereby elucidating the impact resultant from wheel–rail collisions instigated by vehicles [22–24]. For instance, Tao et al. developed semi-analytical models to predict train-induced vibrations in multi-layered soils [23]. Utilizing these analytical model analyses and field measurement data [15,16,25], as well as the numerical models to forecast the vibrational impact of train operations on buildings [26–28], researchers have also proposed a range of vibrational mitigation strategies [10,27,29,30]. For instance, Jin et al. demonstrated the efficacy of resonant floating slabs in mitigating structural responses [29].

In conclusion, most existing studies on environmental vibrations caused by subway trains focus on the impacts of trains traveling at constant speeds. Notably, there are few studies pertaining to the effects of trains during variable speed operations, especially during the acceleration and deceleration phases. The vibration characteristics and propagation mechanisms generated during these acceleration and deceleration phases significantly impact environmental vibrations during the startup and braking phases of subway trains [31,32]. Taking an existing railway station and a proposed construction project under construction as the background, this study adopts the computational method prescribed by extant normative standards to scrutinize the vibrational propagation patterns of subway trains in both the time and frequency domains, utilizing data from the experiment results. Specifically, the study examines three key aspects: the orientation of the route toward and away from the station, the acceleration or deceleration of the train during entry or exit of the station, and the distances from the subway station's outer wall. By analyzing the vibration characteristics and propagation behaviors generated during the acceleration and deceleration processes of trains, the research fills a critical gap by providing an enhanced understanding of environmental vibrations caused by subway trains under non-uniform operational states, thereby furnishing a more robust scientific foundation for the investigation into the vibration characteristics resultant from subway operations.

Against the backdrop of an operational railway station and an ongoing construction project, this study employs the computational methodologies prescribed by current standards to investigate the vibration propagation rules of subway trains in both the time and frequency domains based on experimental results. The research examines three key aspects: the direction of the route to and from the station, the distance from the subway station's outer wall, and others. By analyzing the vibration characteristics and propagation rules during train acceleration and deceleration, this study augments the research on environmental vibrations caused by subway trains in non-uniform operational states, thereby providing a more comprehensive scientific foundation for the investigation of vibration characteristics resulting from subway operations.

2. Materials and Methods

2.1. A Brief Review of the Experimental Site

The experimental site is located in a planning parking lot adjacent to a subway station in Foshan, Guangdong, China. During experiments, construction of the parking lot was underway, and the area had been smoothed over. The subway station, which commenced operations at the end of 2022, is an underground three-level island station with the first floor at a depth of 9.9 m, the second floor at 16.95 m, and the third floor at 24.1 m. The platform covers an area of 18,050 m^2, spans 120 m in length, and 11 m in width, and features three entrances and exits. The testing line is served by a type B train. Due to the dynamics of acceleration and deceleration as trains enter and exit the subway station, the trains' average speed ranges between 35 km/h and 45 km/h. No vibration mitigation measures are in place along the line segment leading from the station toward the urban area, whereas a medium level of vibration mitigation measures is applied in the direction of the suburbs. Specifically, no vibration reduction measures are implemented on the track adjacent to the experimental site. The map, and cross-sectional layouts of the test site are illustrated in Figures 1a and 1b, respectively.

(a) Map of the test site (b) Cross-sectional layout diagram

Figure 1. Map, and cross-sectional layouts of the experimental site. (**a**) Map of the test site; (**b**) Cross-sectional layout diagram.

2.2. Experimental Instruments

In this research, the JM3873 wireless vibration pickup, wireless gateway, and JM3873 wireless acquisition and analysis system were utilized for the acquisition and recording of signals, as depicted in Figure 2. Wireless sensors offer continuous deformation monitoring, while their non-intrusive deployment ensures minimal interference [33]. The JM3873 wireless vibration pickup is equipped with integrated horizontal and vertical 941 B vibration pickers, while requiring pre-setting before operation.

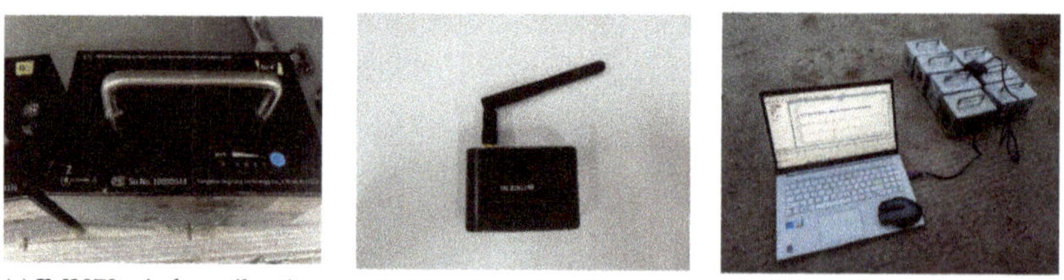

(a) JM3873 wireless vibration pickup (b) Wireless gateway (c) JM3873 wireless acquisition and analysis system

Figure 2. Experimental instruments. (**a**) JM3873 wireless vibration pickup; (**b**) Wireless gateway; (**c**) JM3873 wireless acquisition and analysis system.

By connecting all the vibration pickups to the computer via the wireless gateway, an oscilloscope was used to ascertain the suitable sampling range and frequency, as illustrated in Figure 3. To guarantee the integrity and reliability of the data, the vibration pickup's horizontal and vertical range was set to -0.01~$0.01\ \mathrm{m/s^2}$ for this test. The sampling frequency was set at 512 Hz, satisfying the Nyquist–Shannon criterion for frequencies up to 256 Hz. This ensured accurate capture of frequency components up to 256 Hz, as validated in similar studies. Continuous data acquisition was conducted during off-peak hours from 18:30 to 23:30 at night, encompassing a total duration of 5 h.

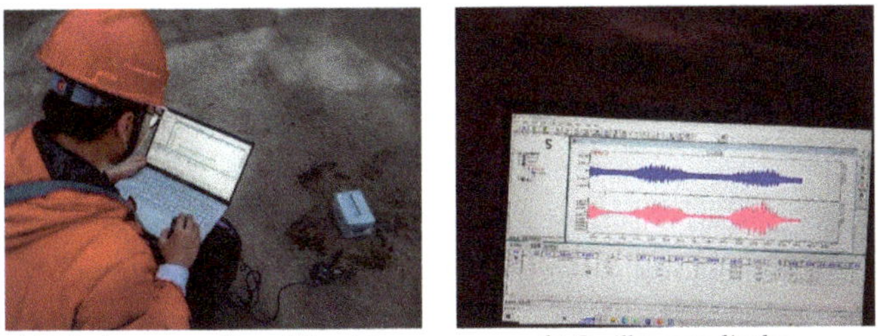

(**a**) Wireless connection of the pickup (**b**) Oscilloscope display

Figure 3. Instruments setup. (**a**) Wireless connection of the pickup; (**b**) Oscilloscope display.

2.3. Layout of Measuring Points

In the field test, there were 9 measuring points in total, established at three cross-sectional transects along the axis of the railway. Each transect was positioned at intervals of 20 m, all situated on the existing ground surface of the proposed site, as depicted in Figure 4. The closest measuring points are 10 m away from the subway station's enclosing structure, with the layout 1 m within the perimeter of the proposed site's enclosing structure. Section 1 is located on the left side of the parking lot, with two measuring points, K1 and K2, perpendicular to the direction of the railway. K1 is 15 m away from the outer wall of the subway station, and K2 is 20 m away from K1. Section 2 is situated centrally within the parking lot, perpendicular to the railway direction, with 4 measuring points: B1, B2, B3, and B4. The measuring point B1, which is nearest to the parking lot's outer wall, is set 10 m away from the subway station's outer wall, with B2 located 5 m from B1, and B3 and B4 placed 10 m apart from B2 and B3, respectively. Section 3 is positioned on the right side of the parking lot, perpendicular to the railway direction, with 3 measuring points: D1, D2, and D3. The measuring point D1 is situated 15 m away from the subway wall, with D2 and D3 placed 10 m away from D1 and D2, respectively. The main purpose of setting three sections is to evaluate the influence of train acceleration and deceleration-induced vibrations across different transects.

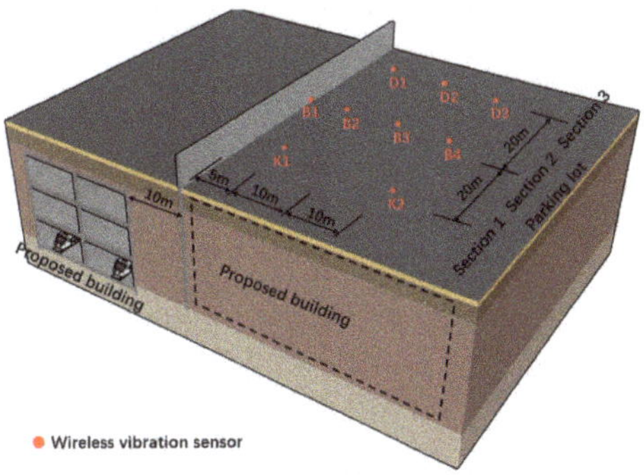

Figure 4. Layout of measuring points.

2.4. Vibration Evaluation Index

The vibration generated by train operation comprises a combination of vibrations at various frequencies. For such a complex vibration signal, the energy level is commonly used as a metric for assessment. Therefore, through an analysis of the time domain and frequency domain, the maximum vibration level of frequency division in JGJ/T170-2009 is utilized to evaluate the vibration and the limits of secondary radiation noise in buildings resulting from urban rail transit.

2.4.1. Vibration Acceleration Level

The vibration acceleration level (*VAL*) is the fundamental metric for characterizing the overall vibration environment's intensity, quantified in dB, as shown in Equation (1):

$$VAL = 20\log 10(\frac{a_e}{a_0}) \quad (1)$$

where a_e denotes the effective value of vibration acceleration, and a_0 represents the base acceleration, which is set to 10^{-6} m/s^2. The effective value of vibration acceleration (a_e) is defined, as shown in Equation (2):

$$a_e = \sqrt{\frac{1}{T}\int_0^T a_i(t)d(t)} \quad (2)$$

where $a_i(t)$ is the acceleration value at a specific time (*t*), $d(t)$ is the integration time, and *T* is the duration of the measuring time.

2.4.2. Maximum Vibration Level of Frequency Division

In the "Standard for limit and measuring method of building vibration and secondary noise caused by urban rail transit" (JGJ/T 170-2009) of China's current industry standard, the maximum vibration level within specific frequency divisions, denoted as VL_{max}, is used to access the environmental vibration of rail transit. This evaluation is determined through a 1/3 octave analysis of the vertical lead vibration acceleration, adjusted by a weight factor, as illustrated in Table 1. There are mainly three calculation methods: the linear average method, the peak-preserving method, and the maximum-preserving method [34–36]. Unlike root-mean-square (RMS)-based approaches, which average energy over time and may obscure short-duration peaks, the peak-preserving method retains maximum acceleration values within each 1/3 octave band. This is critical for assessing intermittent high-amplitude events, such as wheel–rail impacts during braking or startup, which are pivotal for structural fatigue analysis. The peak-preserving method was selected for vibration evaluation due to its efficacy in capturing transient vibration characteristics during train acceleration and deceleration. Based on comparative calculations of the Z-weighting factor for vibration acceleration across different algorithms, this study employs the peak-preserving method. For the experimental site, which is classified as a commercial mixed area, the vibration limit is set at 70 dB during daytime hours.

Table 1. Z-weighting factors for vibration acceleration at the center frequency of 1/3 octave bands, as specified in JGJ/T 170—2009.

1/3 Octave Center Frequency/Hz	Weighting Factor/dB	1/3 Octave Center Frequency/Hz	Weighting Factor/dB
4	0	31.5	−8
5	0	40	−10
6.3	0	50	−12
8	0	63	−14
10	0	80	−17

Table 1. *Cont.*

1/3 Octave Center Frequency/Hz	Weighting Factor/dB	1/3 Octave Center Frequency/Hz	Weighting Factor/dB
12.5	−1	100	−21
16	−2	125	−25
20	−4	160	−30
25	−6	200	−36

3. Results and Discussions

3.1. Analysis of Background Vibration Impact

To ensure that the interference from background vibration can be discounted, a comparative analysis of the background vibration and vehicle-induced vibration was conducted at each measuring point, both in the time domain and frequency domain. As all the measuring points were situated within the parking lot during the experiment, nighttime was chosen to ensure that no construction activities were taking place in the parking lot. During the analysis of background vibration, three measuring points located 15 m, 25 m, and 35 m from the subway station's outer wall were selected from each of the three sections. The results for the other measuring points were consistent with the analyzed data, as shown in Figures 5 and 6. The time-domain analysis revealed that the influence of background vibration was relatively insignificant in both the vertical and horizontal directions. Specifically, the amplitude of the background vibration acceleration remained below 3×10^{-4} m/s^2 in both directions, suggesting that the background vibration had a negligible effect on the vibration caused by the train. Furthermore, in the frequency-domain analysis, the primary frequency distribution of background vibration was within 0~10 Hz, while the main frequency distribution of vehicle-induced vibration was within 40~60 Hz, indicating that the background vibration had minimal impact on the dominant frequency of the vehicle-induced vibration. In conclusion, the background vibration at the experimental site had a minor influence on the experiment results.

3.2. Vibration Propagation Rules of Trains Running in Different Line Directions

An analysis was conducted on the vibration propagation dynamics of trains traveling in different line orientations, viz. urban-bound and suburban-bound directions, through the examination of both time-domain and frequency-domain perspectives. Specifically, the vehicle-induced vibration signals captured at the measurement points 15 m away from the subway station's outer wall across each test section were subjected to rigorous examination. The time-domain analysis examined the vibration signal amplitude changes with time, while the frequency-domain analysis employed the fast Fourier transform to transform the signal into a frequency-based representation, with the corresponding amplitude expressed as a functional dependency.

During the testing phase, a compendium of 99 vehicle-induced vibration signals was recorded, encompassing four types: inbound and outbound signals in the urban direction, as well as inbound and outbound signals in the suburban direction. Notably, the track structure oriented toward the urban district has not implemented vibration mitigation measures and is situated in close proximity to the proposed construction (proximate track). Conversely, the track structure directed toward the suburban district features an integrated track bed and utilizes medium vibration damping fasteners, positioning at a considerable distance from the proposed construction (distant track).

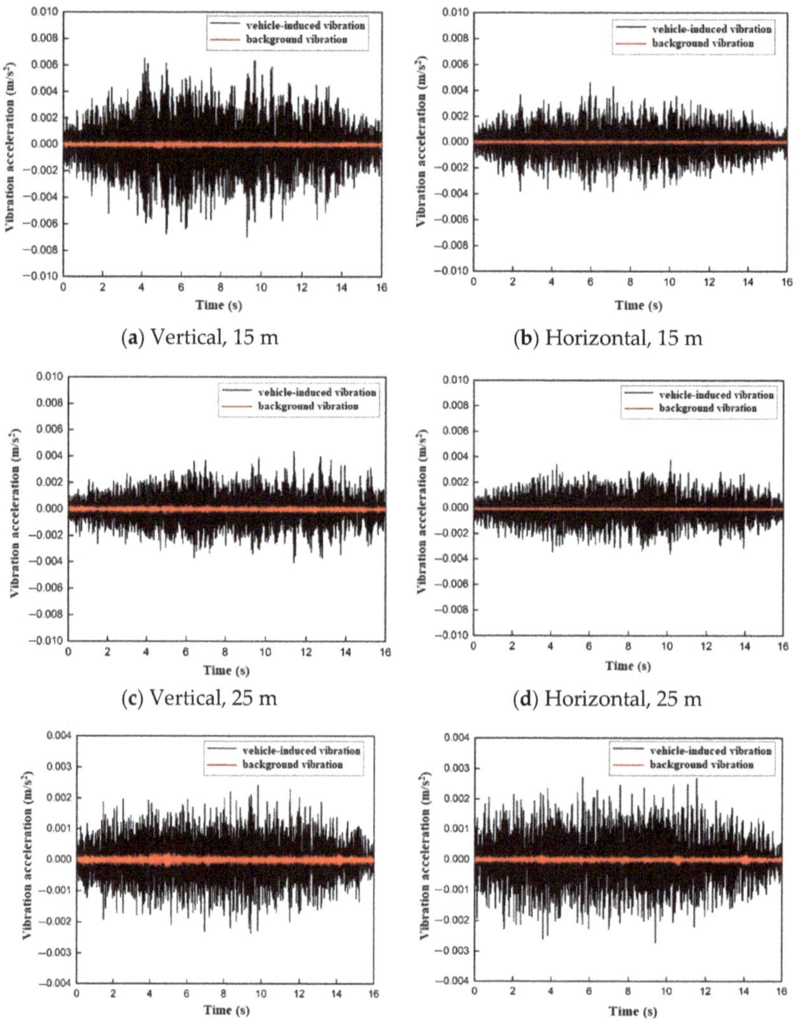

(a) Vertical, 15 m (b) Horizontal, 15 m (c) Vertical, 25 m (d) Horizontal, 25 m (e) Vertical, 35 m (f) Horizontal, 35 m

Figure 5. Time-domain comparison of vehicle-induced vibration and background vibration at various distances from the subway station's exterior wall. (**a**) Vertical, 15 m; (**b**) Horizontal, 15 m; (**c**) Vertical, 25 m; (**d**) Horizontal, 25 m; (**e**) Vertical, 35 m; (**f**) Horizontal, 35 m.

3.2.1. Time-Domain Analysis of Trains Running in Different Line Directions

A comparative analysis in the time domain of the characteristic ground vibration induced by trains proceeding in both urban and suburban directions is illustrated in Figure 7. For the urban-bound direction, the peak value of vertical vibration acceleration within Sections 1 to 3 varies between 0.004 m/s^2 and 0.008 m/s^2, whereas the peak horizontal vibration acceleration spans from 0.004 m/s^2 to 0.015 m/s^2. In contrast, for the suburban-bound direction, the peak vertical vibration acceleration for the same sections ranges from 0.001 m/s^2 to 0.005 m/s^2, and the peak horizontal vibration acceleration is between 0.003 m/s^2 and 0.005 m/s^2. Overall, irrespective of the orientation—vertical or horizontal—the vibration acceleration associated with the trains heading toward the urban direction exceeds that of trains heading toward the suburban direction. The principal factors contributing to this discrepancy include: (1) the suburban direction being situated on the far side of the rail, with a relatively farther distance that results in more significant vibration attenuation along the propagation path; (2) the installation of medium vibration-damping fasteners on the suburban track, adjacent to the residential buildings on the opposite side of the subway station, which serves to attenuate the vibration at the source to some extent. The combined effect of these two factors results in the vibration acceleration being higher in the urban direction compared to the suburban direction.

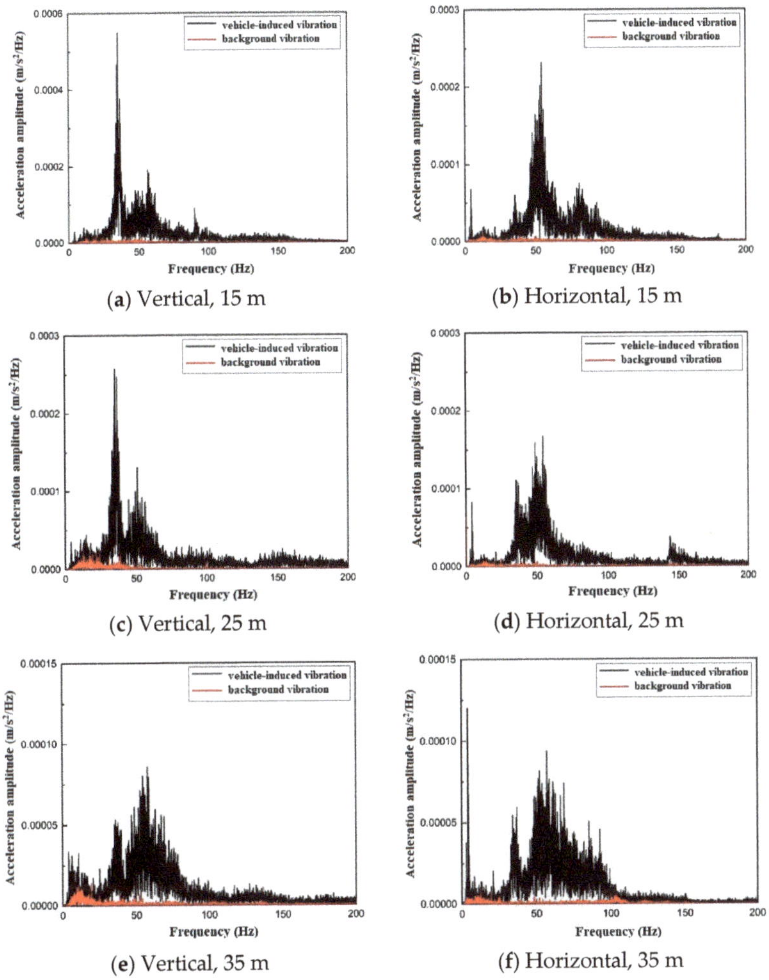

Figure 6. Frequency-domain comparison of vehicle-induced vibration and background vibration at various distances from the exterior wall of the subway station. (**a**) Vertical, 15 m; (**b**) Horizontal, 15 m; (**c**) Vertical, 25 m; (**d**) Horizontal, 25 m; (**e**) Vertical, 35 m; (**f**) Horizontal, 35 m.

3.2.2. Frequency-Domain Analysis of Trains Running in Different Line Directions

Figure 8 illustrates the spectrogram of typical vehicle-induced vibration signals, caused by trains traveling in both urban and suburban directions, subsequent to the fast Fourier transform. This representation corresponds to the time-domain plot. Typically, the acceleration amplitude within the urban direction is higher than that in the suburban direction, with the predominant influence frequency band localized within the range of 40 to 60 Hz. To elaborate further, in the urban direction, two distinct peaks are observed in the vertical direction across three sections, situated at 40 Hz and 60 Hz, respectively. The peak frequency in the horizontal direction is predominantly at 60 Hz. Moreover, the frequency response bands for Sections 2 and 3 are broader in comparison to Section 1. While in the suburban direction, the vertical vibration frequencies for Sections 1 and 2 are distributed between 40 and 60 Hz, featuring multiple peaks, whereas the peak frequency for Section 3 is confined to 40 Hz. In comparison, the horizontal frequencies in the suburban direction for Sections 2 and 3 range from 40 to 60 Hz, with multiple peaks, and Section 1's peak frequency is centered at 50 Hz.

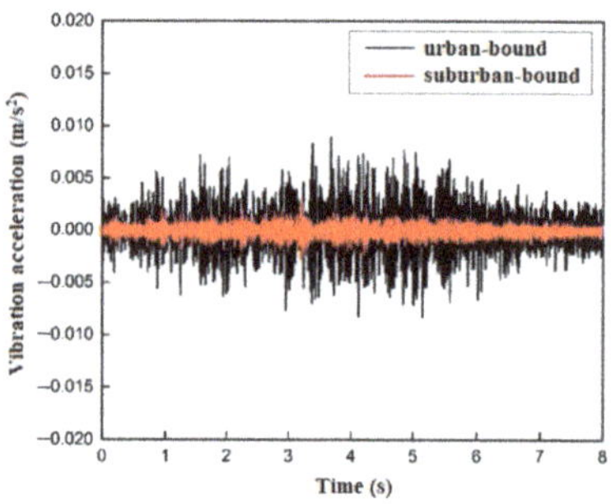

(**a**) Vertical, Section 1

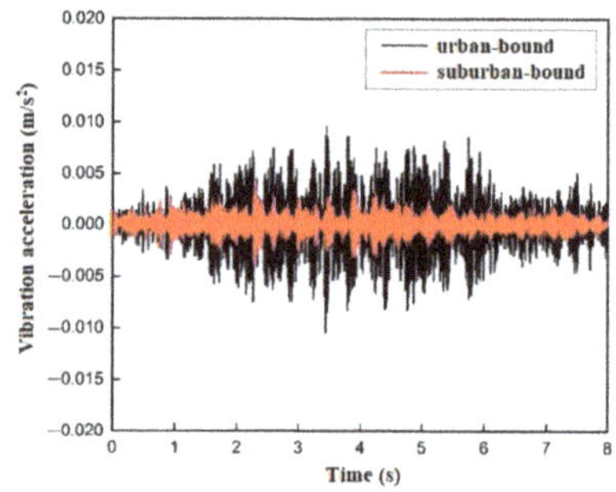

(**b**) Horizontal, Section 1

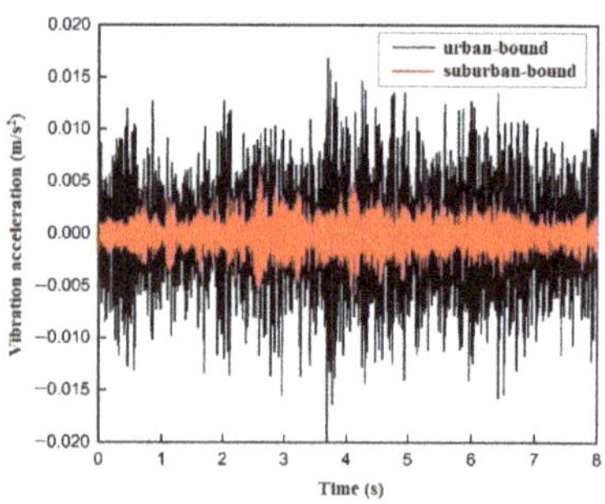

(**c**) Vertical, Section 2

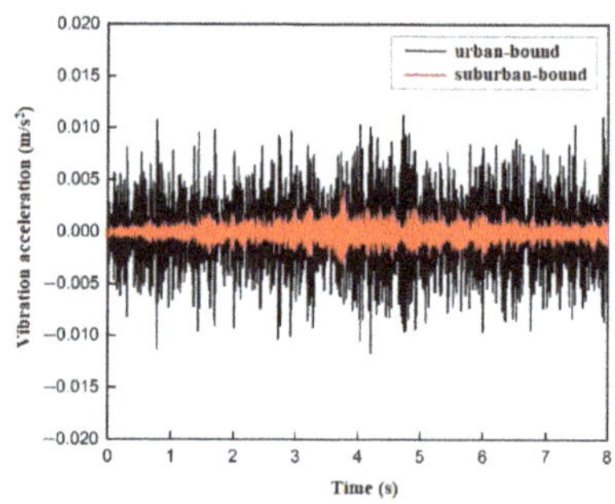

(**d**) Horizontal, Section 2

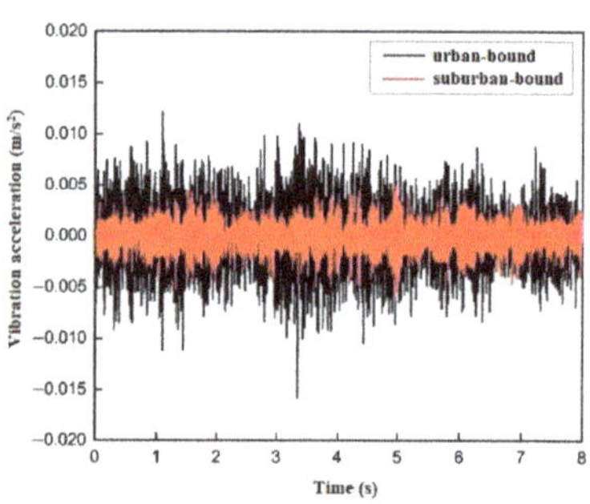

(**e**) Vertical, Section 3

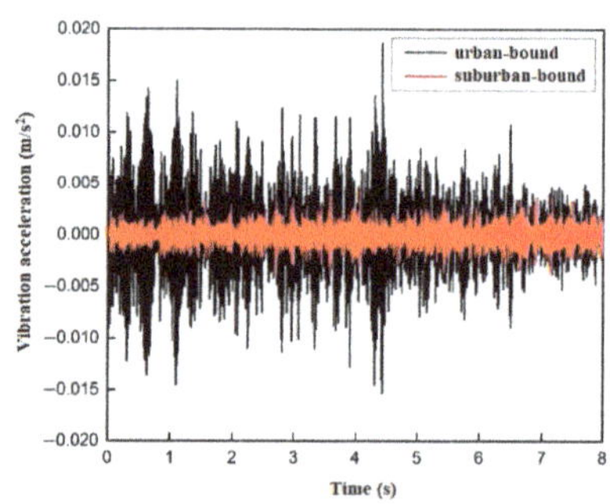

(**f**) Horizontal, Section 3

Figure 7. Time-domain plot of vehicle-induced vibration in urban-bound and suburban-bound directions. (**a**) Vertical, Section 1; (**b**) Horizontal, Section 1; (**c**) Vertical, Section 2; (**d**) Horizontal, Section 2; (**e**) Vertical, Section 3; (**f**) Horizontal, Section 3.

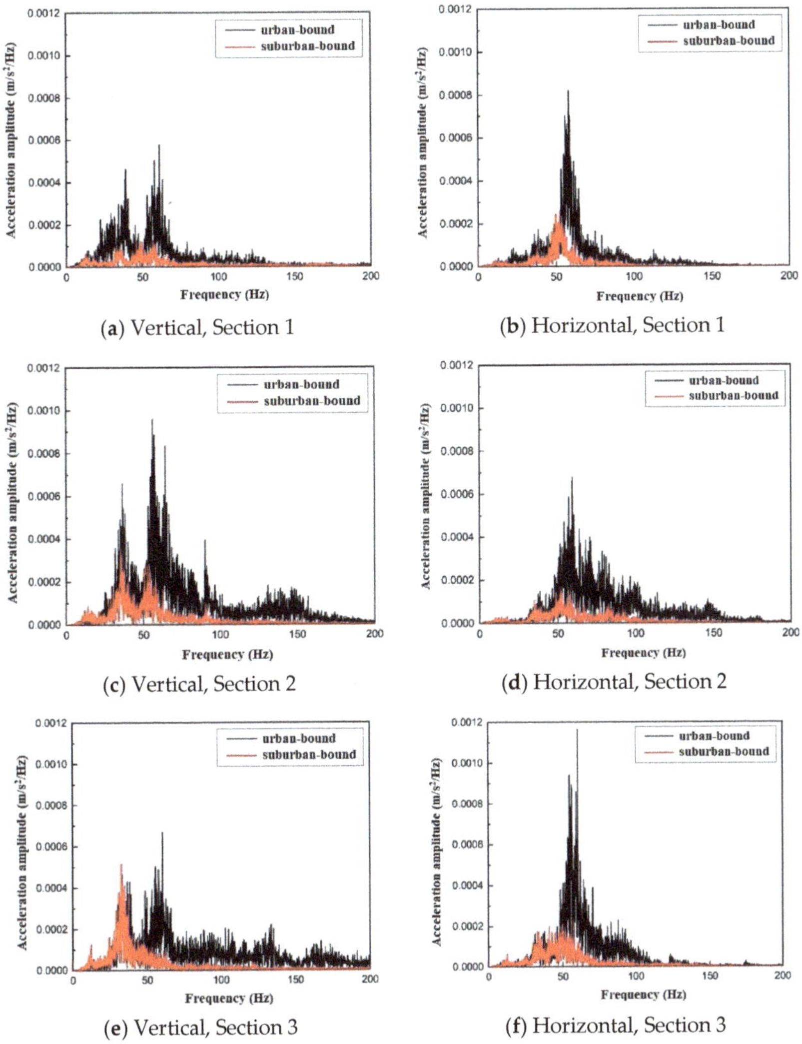

Figure 8. Frequency-domain plot of vehicle-induced vibration in urban-bound and suburban-bound directions. (**a**) Vertical, Section 1; (**b**) Horizontal, Section 1; (**c**) Vertical, Section 2; (**d**) Horizontal, Section 2; (**e**) Vertical, Section 3; (**f**) Horizontal, Section 3.

The calculation of the 1/3 octave band vibration acceleration level for each train, in both urban and suburban directions of every section, has yielded insights into the influence range and average vibration values, as depicted in Figure 9. The analysis reveals that, in the vertical direction, the average peak vibration acceleration levels for the three sections in the urban direction are 65 dB, 64 dB, and 69 dB, with corresponding peak frequencies at 40 Hz, 40 Hz, and 63 Hz. In the suburban direction, the vibration acceleration levels range around 61 dB, with peak frequencies that are similarly aligned. For the horizontal direction, the average peak vibration acceleration levels for the three sections are 69 dB, 62 dB, and 72 dB, respectively, with peak frequencies at 63 Hz, 63 Hz, and 50 Hz. In the suburban direction, the average peak vibration acceleration levels for the three sections are 67 dB, 61 dB, and 65 dB, with peak frequencies at 63 Hz, 63 Hz, and 50 Hz. The comparison between urban and suburban directions indicates a higher peak vibration acceleration level in urban areas, aligning with the previously mentioned influencing factors. Specifically, the suburban direction exhibits a higher average vibration acceleration level at lower frequencies, whether in the vertical or horizontal direction. Conversely, at higher frequencies, the urban direction demonstrates elevated vibration acceleration levels. To mitigate vibration effects, the installation of vibration-reducing fasteners on the track in the suburban direction aims to lower the track's natural frequency, thereby achieving a vibration reduction effect. This intervention, however, may result in amplified vibrations at lower frequencies, although the peak values are expected to decrease. Urban-bound trains exhibited vertical VALs ranging from 65 to 69 dB, whereas suburban-bound trains showed a 15–20% reduction (55–61 dB), attributable to medium-damping fasteners installed on suburban tracks.

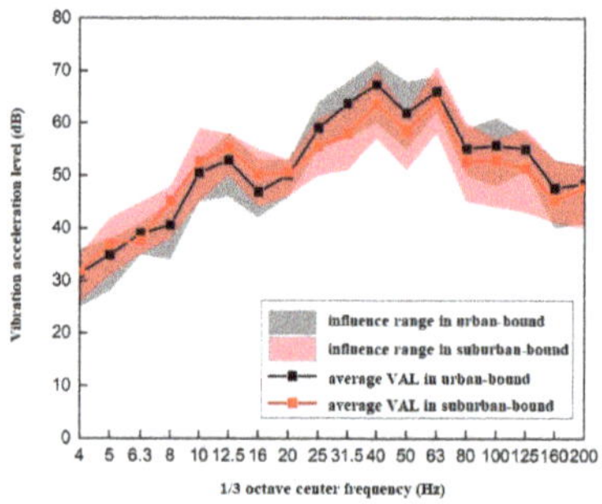

(**a**) Vertical, Section 1

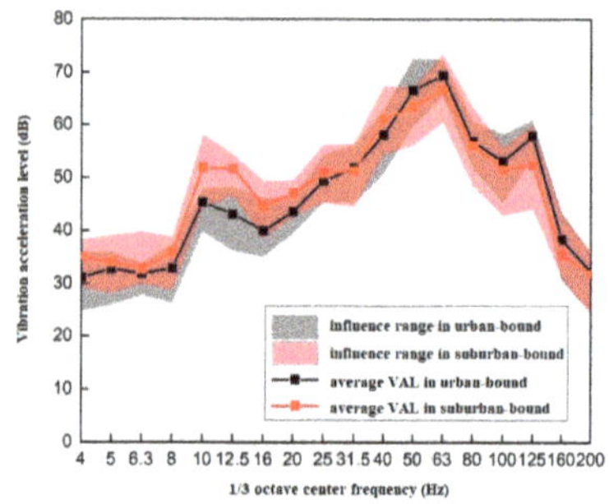

(**b**) Horizontal, Section 1

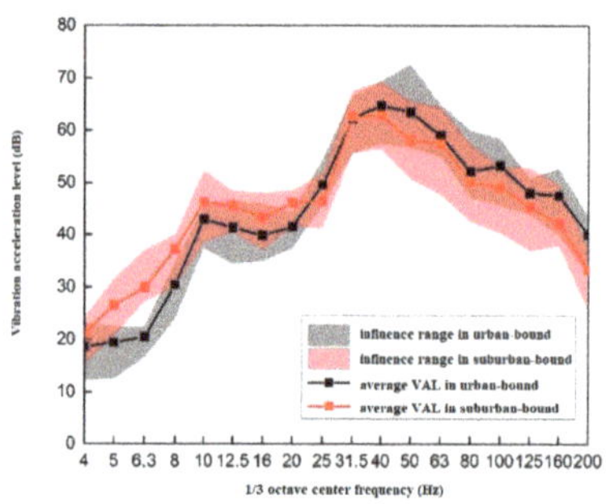

(**c**) Vertical, Section 2

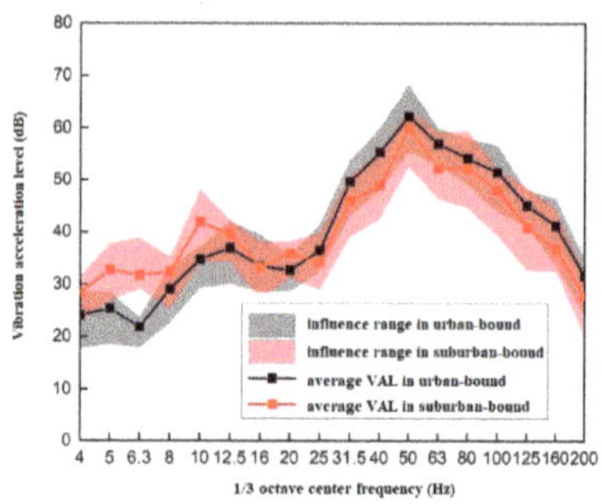

(**d**) Horizontal, Section 2

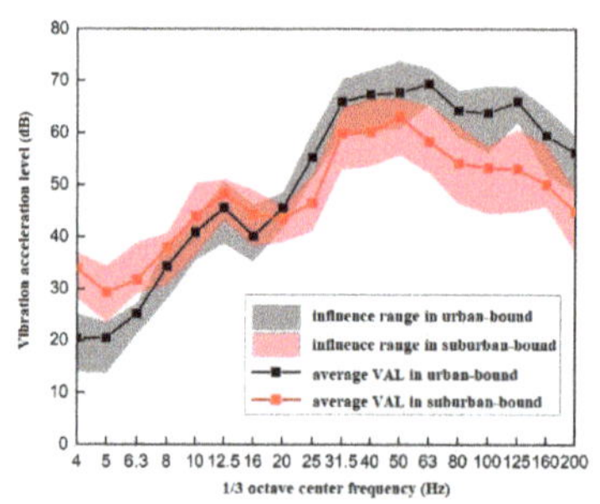

(**e**) Vertical, Section 3

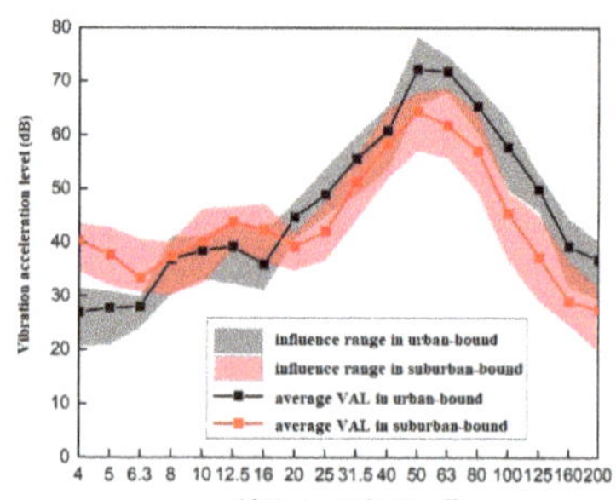

(**f**) Horizontal, Section 3

Figure 9. Train-induced soil vibration acceleration in 1/3 octave bands in urban-bound and suburban-bound directions. (**a**) Vertical, Section 1; (**b**) Horizontal, Section 1; (**c**) Vertical, Section 2; (**d**) Horizontal, Section 2; (**e**) Vertical, Section 3; (**f**) Horizontal, Section 3.

3.3. Influence of Acceleration and Deceleration in and out of Subway Station on Vibration Propagation Law

3.3.1. Time-Domain Analysis of Trains Entering and Leaving the Subway Station

Upon conducting time-domain analysis, it is observed that the vehicle-induced vibration signals of trains entering and leaving the station typically manifest in pairs. For a detailed examination, two representative sets of complete inbound and outbound vibration signals recorded at measuring point B2 are presented in Figure 10. The pairing of these signals arises from the fact that, as the train completely traverses the subway station, the vibration pickup initially captures the inbound signal, followed by the outbound signal after an interval of approximately 30 to 40 s subsequent to the train's cessation. Correlation with the train schedule facilitates the identification of the former as the inbound train's time-domain waveform and the latter as the outbound train's time-domain waveform.

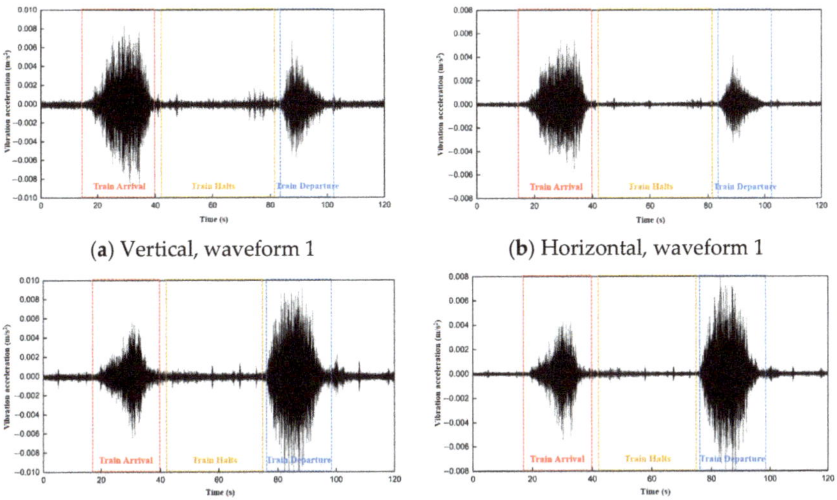

(a) Vertical, waveform 1 (b) Horizontal, waveform 1

(c) Vertical, waveform 2 (d) Horizontal, waveform 2

Figure 10. Complete time-domain plot of a train passing through the subway station. (**a**) Vertical, waveform 1; (**b**) Horizontal, waveform 1; (**c**) Vertical, waveform 2; (**d**) Horizontal, waveform 2.

Moreover, waveform 1 discloses that the inbound train exhibits peak accelerations of approximately 0.007 m/s^2 vertically and 0.004 m/s^2 horizontally, whereas the outbound train demonstrates peak accelerations of around 0.005 m/s^2 vertically and 0.002 m/s^2 horizontally. This discrepancy is attributable to the fact that the experiment was carried out during the evening peak hours between 18:30 and 23:30, during which the volume of passengers fluctuates. After the train halts at the station, the number of alighting passengers exceeds that of boarding passengers, leading to a reduction in the train's overall mass and a resultant attenuation of vibration at the train's departure. In waveform 2, the inbound train's peak acceleration is noted as approximately 0.004 m/s^2 vertically and 0.003 m/s^2 horizontally. Conversely, the outbound train's peak acceleration escalates to roughly 0.008 m/s^2 vertically and 0.006 m/s^2 horizontally. During this time, the number of passengers disembarking is lower than those embarking, thus increasing the train's overall mass and intensifying the vibration at the train's departure.

Upon dissecting the independent complete time-domain waveform of the inbound train, it is segmented into three distinct phases, as illustrated in Figure 11. The initial phase pertains to the train's approach to the measuring point, where the wheelset is nearing but has yet to surpass the measuring point, which is indicated by deceleration. It is evident from the figure that as the train approaches closer, the vibration collected by the sensor progressively intensifies. The second phase involves the train decelerating through the measuring point, with the waveform acceleration peaking as the wheelset passes. The third phase constitutes the braking segment, where the transition from deceleration to braking is discernible in the time domain, marked by an abrupt vibration due to the braked-induced impact load, followed by a rapid decay of the vibration.

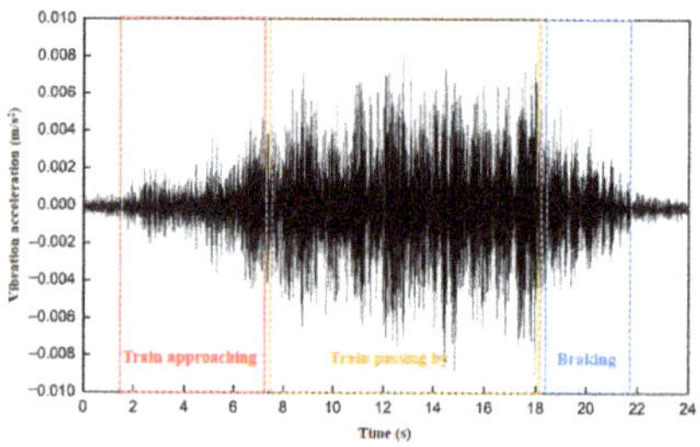

(**a**) Vertical

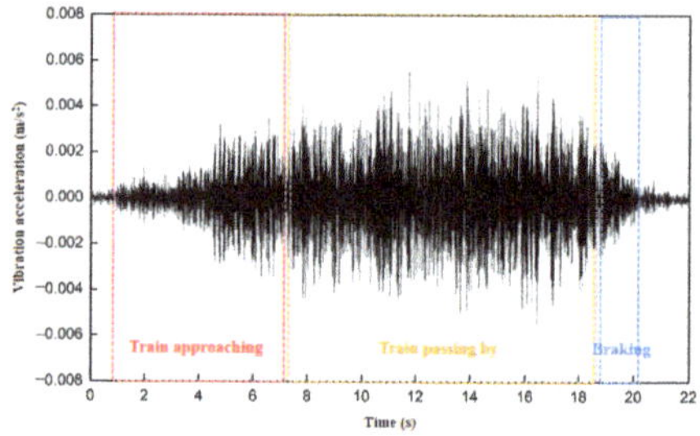

(**b**) Horizontal

Figure 11. Complete time-domain plot of train entering the subway station. (**a**) Vertical; (**b**) Horizontal.

The analysis of the independent complete time-domain waveform of the outbound train reveals a bipartite structure, as shown in Figure 12. The first part corresponds to the acceleration phase during the train's startup, with the waveform acceleration reaching a peak in correlation with the wheelset's passage. The second part represents the train's departure, during which the wheelset moves away from the measuring point and the train accelerates. As depicted in the figure, the vibration recorded by the sensor progressively diminishes as the train departs.

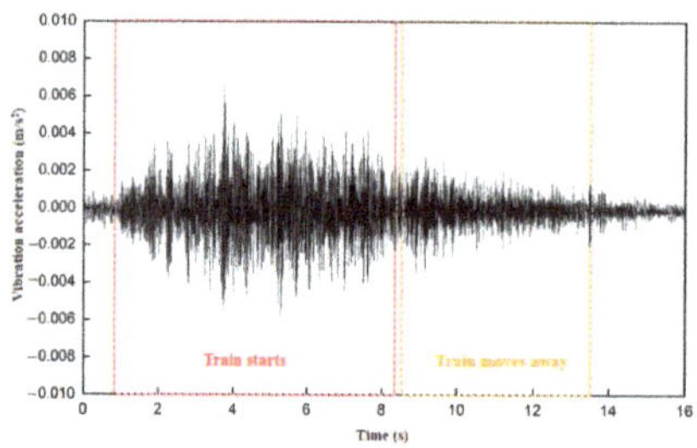

(**a**) Vertical

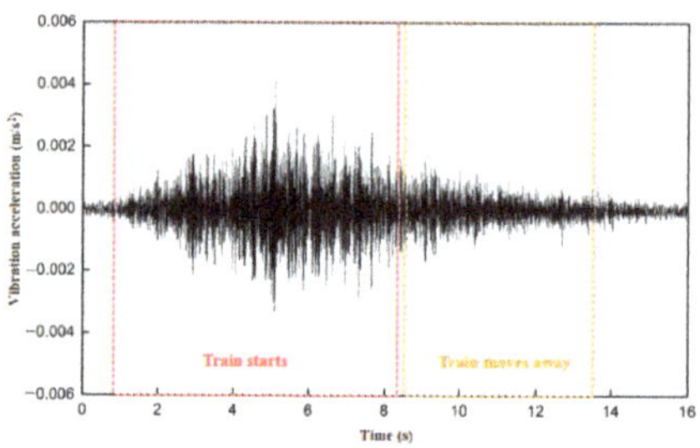

(**b**) Horizontal

Figure 12. Complete time-domain plot of a train exiting the subway station. (**a**) Vertical; (**b**) Horizontal.

3.3.2. Time-Domain Analysis of Train Acceleration and Deceleration

Upon an initial examination of the inbound and outbound train vibration waveforms within the time domain, a subsequent analysis is conducted to assess the impact of velocity on vehicle-induced vibrations. Specifically, the characteristic vehicle-induced vibrations of trains inbound and outbound toward the urban area are delineated for examination, as illustrated in Figures 13 and 14. Notably, the chosen measuring point is situated 15 m away from the subway station's outer perimeter.

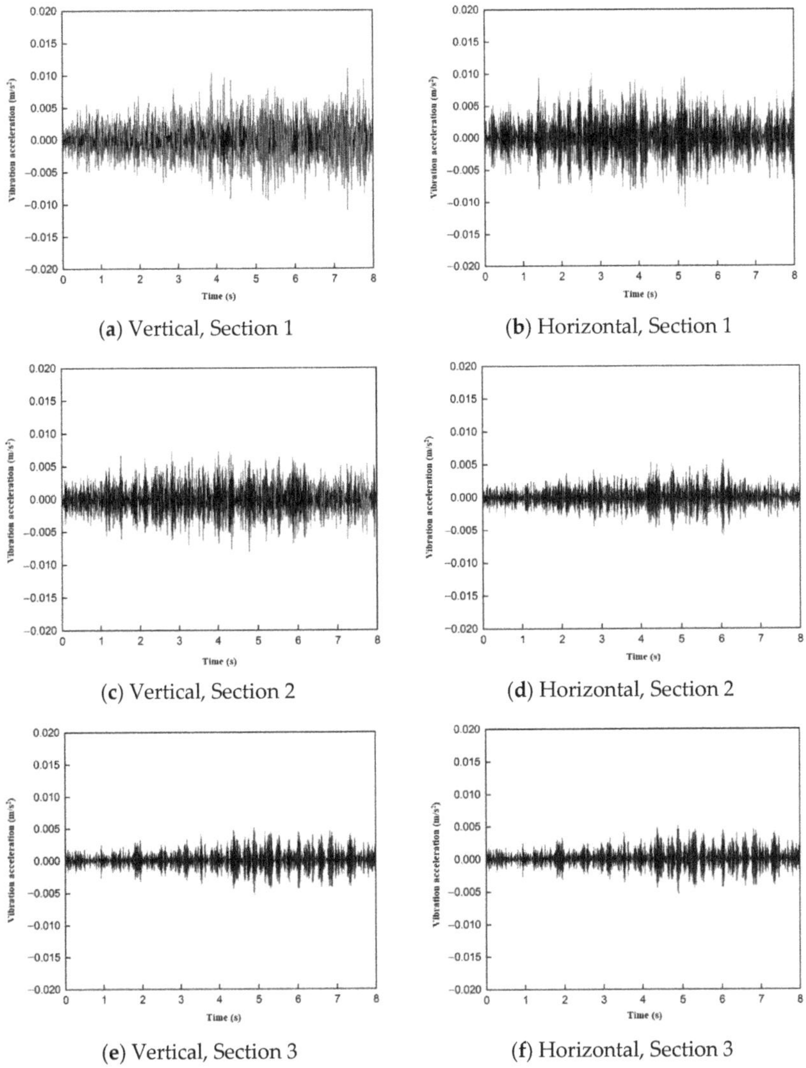

Figure 13. Time-domain plot of trains deceleration. (**a**) Vertical, Section 1; (**b**) Horizontal, Section 1; (**c**) Vertical, Section 2; (**d**) Horizontal, Section 2; (**e**) Vertical, Section 3; (**f**) Horizontal, Section 3.

An analysis of the time-domain diagram of the train's deceleration-induced vibration reveals that both the vertical and horizontal components of the train's vibration acceleration diminish progressively from Sections 1 to 3. This decrement is attributed to the train's deceleration phase, during which its velocity peaks as it traverses Section 1 before progressively diminishing. Consequently, the reduction in velocity correlates with a diminished vibration acceleration.

Conversely, the time-domain diagram of the train's acceleration-induced vibration illustrates an incremental trend in vibration acceleration in both vertical and horizontal directions from Sections 1 to 3. During this interval, the train transitions from launching to an accelerated state, with the minimum velocity recorded at Section 1, and the maximum velocity at Section 3. The train's vibration acceleration is observed to escalate in correlation with its increasing velocity.

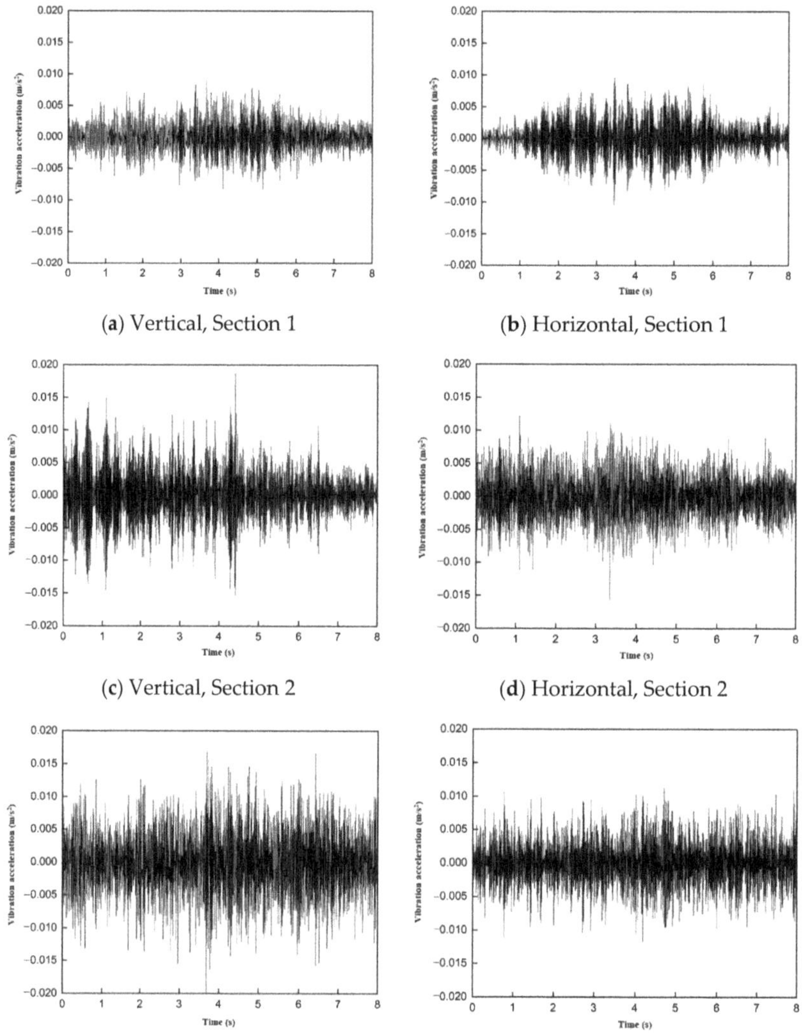

(**a**) Vertical, Section 1 (**b**) Horizontal, Section 1

(**c**) Vertical, Section 2 (**d**) Horizontal, Section 2

(**e**) Vertical, Section 3 (**f**) Horizontal, Section 3

Figure 14. Time-domain plot of train acceleration. (**a**) Vertical, Section 1; (**b**) Horizontal, Section 1; (**c**) Vertical, Section 2; (**d**) Horizontal, Section 2; (**e**) Vertical, Section 3; (**f**) Horizontal, Section 3.

A comparative examination of the train's deceleration and acceleration phases discloses that the vibration acceleration is affected by the train's mass, which fluctuates with the volume of passengers as the train enters and exits from the station. Furthermore, vibration acceleration is susceptible to the train's velocity, with discrepancies in the magnitude of acceleration during deceleration and acceleration phases leading to inconsistent variations in train velocity. These two factors exert an influence on the vibration acceleration of the ground, precluding an accurate assessment of the train's deceleration or acceleration impact on the ground vibration acceleration.

3.3.3. Frequency-Domain Analysis of Trains Acceleration and Deceleration

Upon analyzing the inbound and outbound train vibration spectra at urban stations, as depicted in Figures 15 and 16, it is observed that as the train enters the station, Sections

1 and 2 exhibit vertical direction peaks at 40 Hz and 60 Hz, respectively, while Section 3 demonstrates a peak at 70 Hz. In the horizontal direction, all three sections peak at 60 Hz, with Section 1 exhibiting a broader response frequency band. When the train departs the station, all three sections in the vertical direction display two peak frequencies at 40 Hz and 60 Hz. Section 1 has an additional peak at 40 Hz. The peak frequency for all three sections in the horizontal direction remains at 60 Hz, with Sections 2 and 3 generally displaying a broader frequency band.

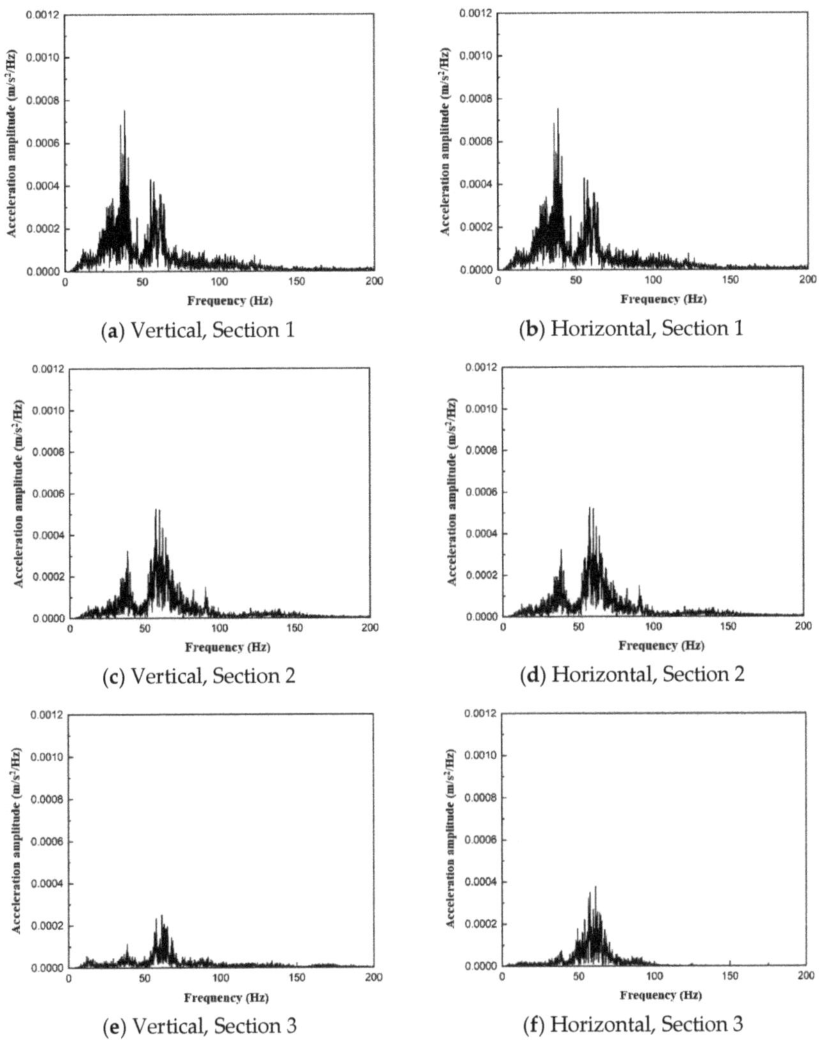

(a) Vertical, Section 1 (b) Horizontal, Section 1

(c) Vertical, Section 2 (d) Horizontal, Section 2

(e) Vertical, Section 3 (f) Horizontal, Section 3

Figure 15. Frequency spectrum plot of vibration induced by incoming trains. (**a**) Vertical, Section 1; (**b**) Horizontal, Section 1; (**c**) Vertical, Section 2; (**d**) Horizontal, Section 2; (**e**) Vertical, Section 3; (**f**) Horizontal, Section 3.

The 1/3 octave vibration acceleration level has been computed for the areas adjacent to the station as the train approaches and for those distant from the station as the train departs, with the average values and influence ranges illustrated in Figure 17. The peak value of the average vibration acceleration level for both vertical and horizontal directions is found at 50 Hz, with vertical direction peaks as the train approaches and departs at 59 dB and 56 dB, respectively, and horizontal peaks at 58 dB and 50 dB. Notably, at higher frequencies, the average vibration acceleration level of the train approaching the station is higher in both vertical and horizontal directions compared to that of the train departing, aligning with the Doppler effect. This effect posits that the receiving frequency of a wave increases as the wave source approaches the measuring point and decreases as it recedes. Consequently, the high-frequency vibration acceleration level near the measuring point for the train entering the station is higher than that for the train leaving the station. At lower frequencies, the difference between approaching and departing trains is negligible.

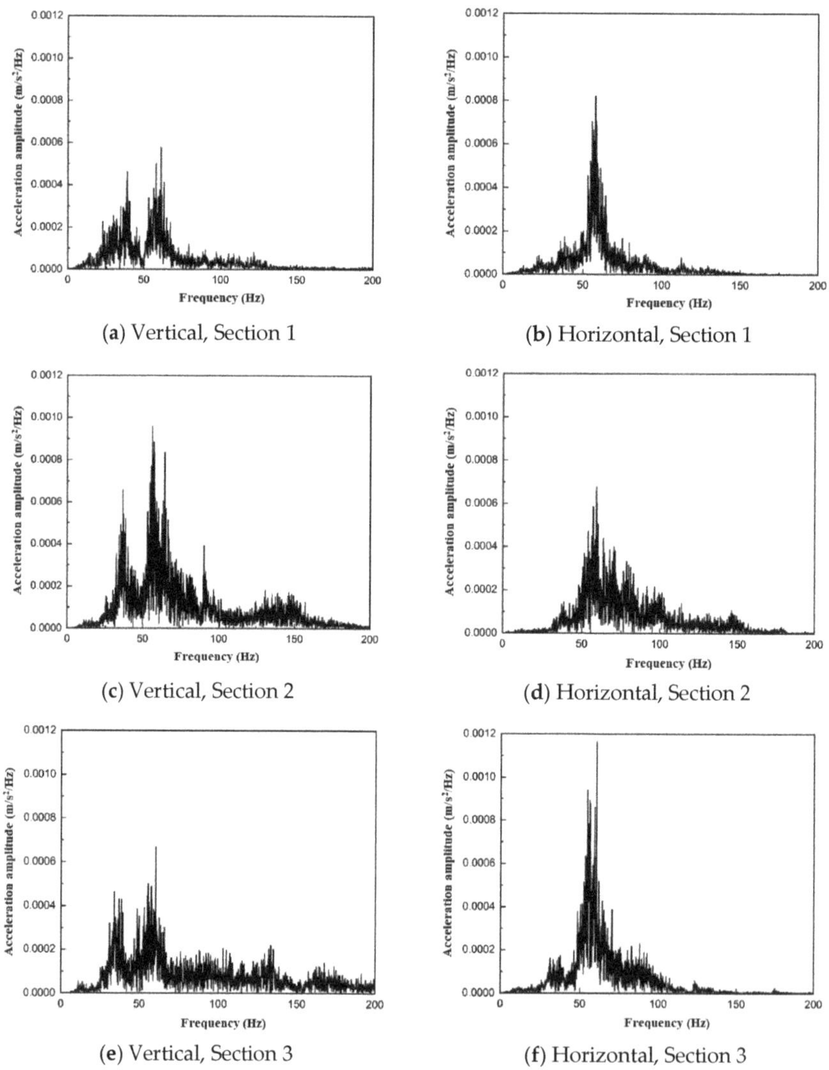

Figure 16. Frequency spectrum plot of vibration induced by outgoing trains. (**a**) Vertical, Section 1; (**b**) Horizontal, Section 1; (**c**) Vertical, Section 2; (**d**) Horizontal, Section 2; (**e**) Vertical, Section 3; (**f**) Horizontal, Section 3.

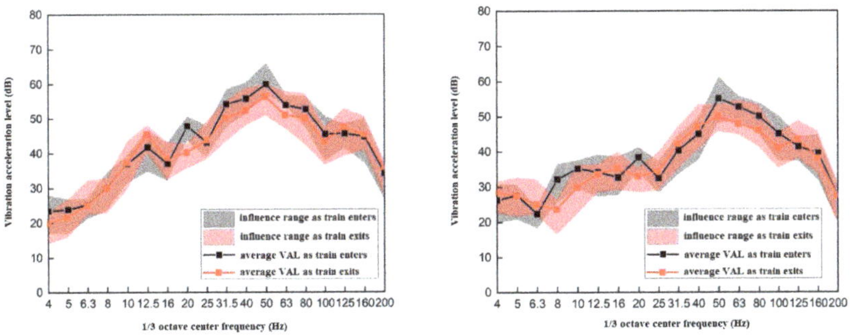

(a) Vertical (b) Horizontal

Figure 17. Train approach/departure vibration acceleration in 1/3 octave bands. (a) Vertical; (b) Horizontal.

Figure 18 presents the 1/3 octave vibration acceleration level for the vibration caused by the train's acceleration and deceleration in each section, with the influence range and average values obtained. In the vertical direction, the peak average vibration acceleration levels for the three sections during train deceleration are 66 dB, 68 dB, and 60 dB, respectively, with peak frequencies at 40 Hz, 63 Hz, and 63 Hz. For the train acceleration phases, the average peak vibration acceleration levels are 61 dB, 68 dB, and 60 dB, with peak frequencies at 40 Hz, 63 Hz, and 63 Hz, respectively. In the horizontal direction, the average peak vibration acceleration levels during train deceleration are 71 dB, 69 dB, and 60 dB, with peak frequencies at 63 Hz, 50 Hz, and 63 Hz, respectively. For the acceleration phases, the average peak values are 65 dB, 70 dB, and 71 dB, with peak frequencies consistently at 63 Hz.

When comparing the 1/3 octave vibration acceleration level diagrams for the urban and suburban directions, it is evident that the difference in vibration acceleration levels during train acceleration and deceleration across the three sections is not significant in the low-frequency range. However, in the high-frequency range, from Sections 1 to 3, the vibration acceleration level during train deceleration progressively decreases, while that during acceleration progressively increases. This is consistent with the earlier observation that train vibration is influenced by speed. Specifically, during deceleration, Section 1 is at the position of maximum operating speed, which then gradually decreases. Conversely, during acceleration, Section 1 is at the position of lowest operating speed when the train passes through it and then gradually increases. This indicates that train speed affects vibration, primarily within the high-frequency range, with minimal impact within the low-frequency range.

3.4. Characteristics of Vibration Propagation at Different Distances from the Outer Walls of Subway Stations

In order to explore the impact of the proximity to the vibration source on the dissemination of vibrations, this segment isolates a representative vehicle-induced vibration signal derived from the outbound train's vibrations in the urban direction. Subsequently, the propagation dynamics are examined through a comprehensive analysis of the time-domain and frequency-domain characteristics, along with the proximity to the vibration source across three distinct sections.

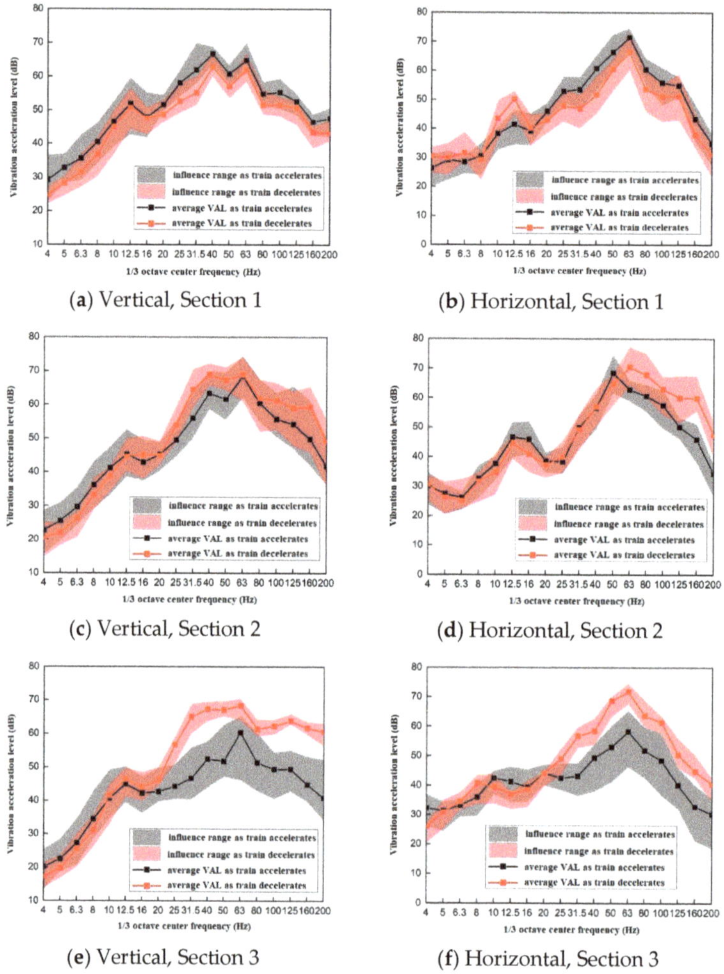

Figure 18. Train acceleration/deceleration vibration levels in 1/3 octave bands. (**a**) Vertical, Section 1; (**b**) Horizontal, Section 1; (**c**) Vertical, Section 2; (**d**) Horizontal, Section 2; (**e**) Vertical, Section 3; (**f**) Horizontal, Section 3.

3.4.1. Time-Domain Analysis of Trains at Different Distances from the Outer Walls of Subway Stations

The integration of a time-frequency diagram into the time-domain analysis facilitates the examination of the time and frequency attributes of vibration signals at varying distances from the external walls of subway stations. The measuring points K1 and K2 within Section 1 are positioned 20 m apart. As depicted in Figures 19 and 20, in the vertical direction, the peak acceleration diminishes from 0.005 m/s^2 to 0.0025 m/s^2, with the predominant frequency bands confined to 30~70 Hz, whereas in the horizontal direction, the peak acceleration decreases from 0.005 m/s^2 to 0.001 m/s^2, with the predominant frequency bands concentrated from 30 to 60 Hz. Generally, K1 exhibits a richer presence of both high- and low-frequency components compared to K2.

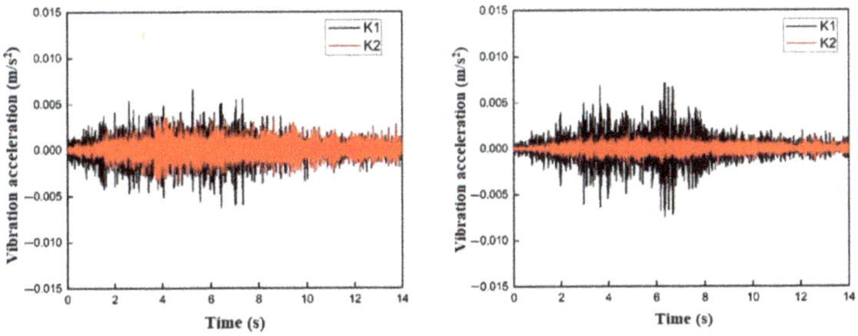

(a) Vertical (b) Horizontal

Figure 19. Time-domain plot for Section 1 at varying distances from the subway station wall. (**a**) Vertical; (**b**) Horizontal.

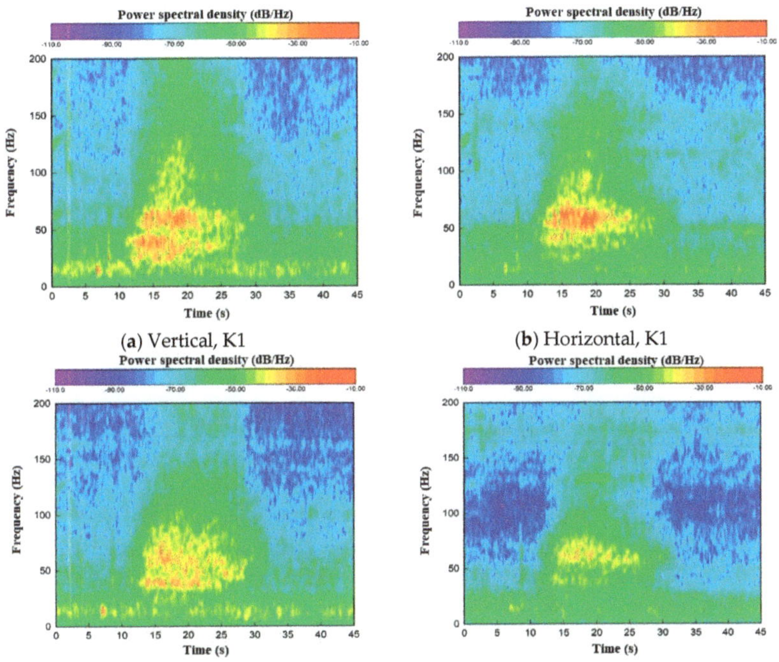

(c) Vertical, K2 (d) Horizontal, K2

Figure 20. The frequency-domain plot of Section 1 at different distances from the exterior wall of the subway station. (**a**) Vertical, K1; (**b**) Horizontal, K1; (**c**) Vertical, K2; (**d**) Horizontal, K2.

Within Section 2, the measuring points B1 and B2 are separated by 5 m, while B2, B3, and B4 are 10 m apart from each other. As illustrated in Figures 21 and 22, in the vertical direction, the peak acceleration of B2, B3, and B4 diminishes from 0.008 m/s^2 to 0.005 m/s^2 and 0.004 m/s^2, respectively. However, due to B1's proximity to the parking lot's external wall, its vibrations are restricted, resulting in a reduced peak acceleration. Consequently, despite being closer to the subway station's external wall, B1's peak acceleration is 0.005 m/s^2, lower than B2. Horizontally, the peak acceleration of B2, B3, and B4 decays from 0.005 m/s^2 to 0.004 m/s^2. The principal frequency bands in both vertical and horizontal directions are concentrated within the range of 50 Hz to 60 Hz, with B1, B2, and B3 demonstrating a more copious high-frequency response than B4.

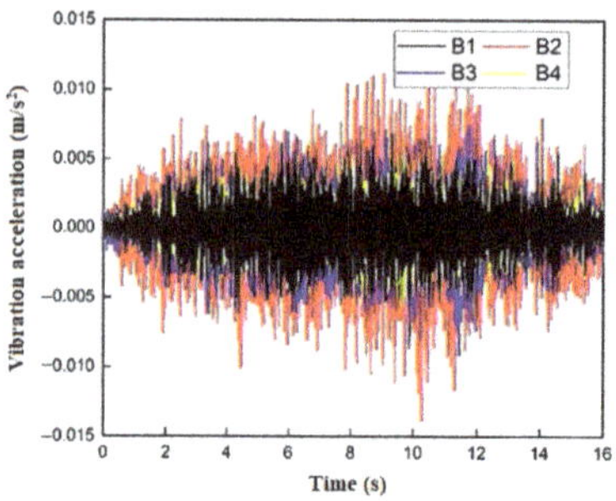

(a) Vertical

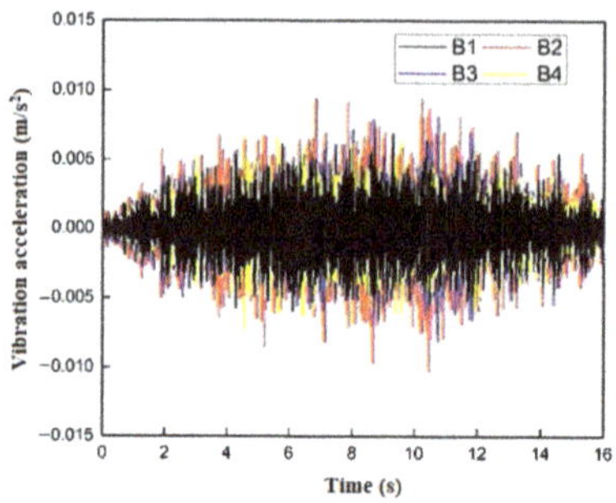

(b) Horizontal

Figure 21. Time-domain plot for Section 2 at various distances from the subway station wall. (**a**) Vertical; (**b**) Horizontal.

Within Section 3, the measuring points D1, D2, and D3 are each spaced 10 m apart. As depicted in Figures 23 and 24, in the vertical direction, the peak acceleration of D1, D2, and D3 decreases from 0.006 m/s^2 to 0.005 m/s^2. The main frequency bands are all within the 40 Hz to 60 Hz spectrum, with the high-frequency component progressively diminishing. In the horizontal direction, the peak acceleration of D1, D2, and D3 falls from 0.009 m/s^2 to 0.008 m/s^2 and 0.003 m/s^2, with the main frequency band ranging from 50 Hz to 60 Hz, and the high-frequency component also decreases progressively. Moreover, for measuring points D1 and D2 in the same section, the horizontal acceleration peak is higher than that in the vertical direction. This suggests that as the train traverses the section, the horizontal track irregularity exceeds that of the vertical, with the vibration attenuation in the horizontal direction occurring more rapidly beyond a 25 m distance, and the attenuation at D3 being particularly pronounced.

According to the analysis of these three sections, the vibration acceleration typically diminishes with increasing distance. At distances of 15 m and 35 m from the subway station's external wall, the vibration attenuation in the vertical direction of Sections 1, 2, and 3 is 50%, 50%, and 16.7%, respectively, while in the horizontal direction, the attenuation is 80%, 20%, and 66.7%. It is observed that the horizontal vibration attenuation rate of Sections 1 and 3 exceeds that of the vertical direction, and the vibration attenuation rate in Section 1 is higher than that in Sections 2 and 3, which correlates with the characteristics of the soil layer beneath each section. At distances of 15 m and 25 m from the subway station's external wall, Sections 2 and 3 exhibit vertical attenuation of 37.5% and 16.7%, and horizontal attenuation of 25% and 11.1%, respectively. At distances of 25 m and 35 m from the subway station's external, Section 2 shows a 20% vertical attenuation, whereas Section 3 does not. In the horizontal direction, Section 2 shows no attenuation, while Section 3's attenuation reaches 62.5%. The slow attenuation in the vertical direction at distances of 25 m to 35 m from the subway station's external wall within both sections is attributed to the soil layer properties. Considering that the soil's elastic modulus at this distance is affected by the surface layer's hardening, it hinders vibration attenuation, indicating a non-linear relationship between vibration attenuation and increasing distances.

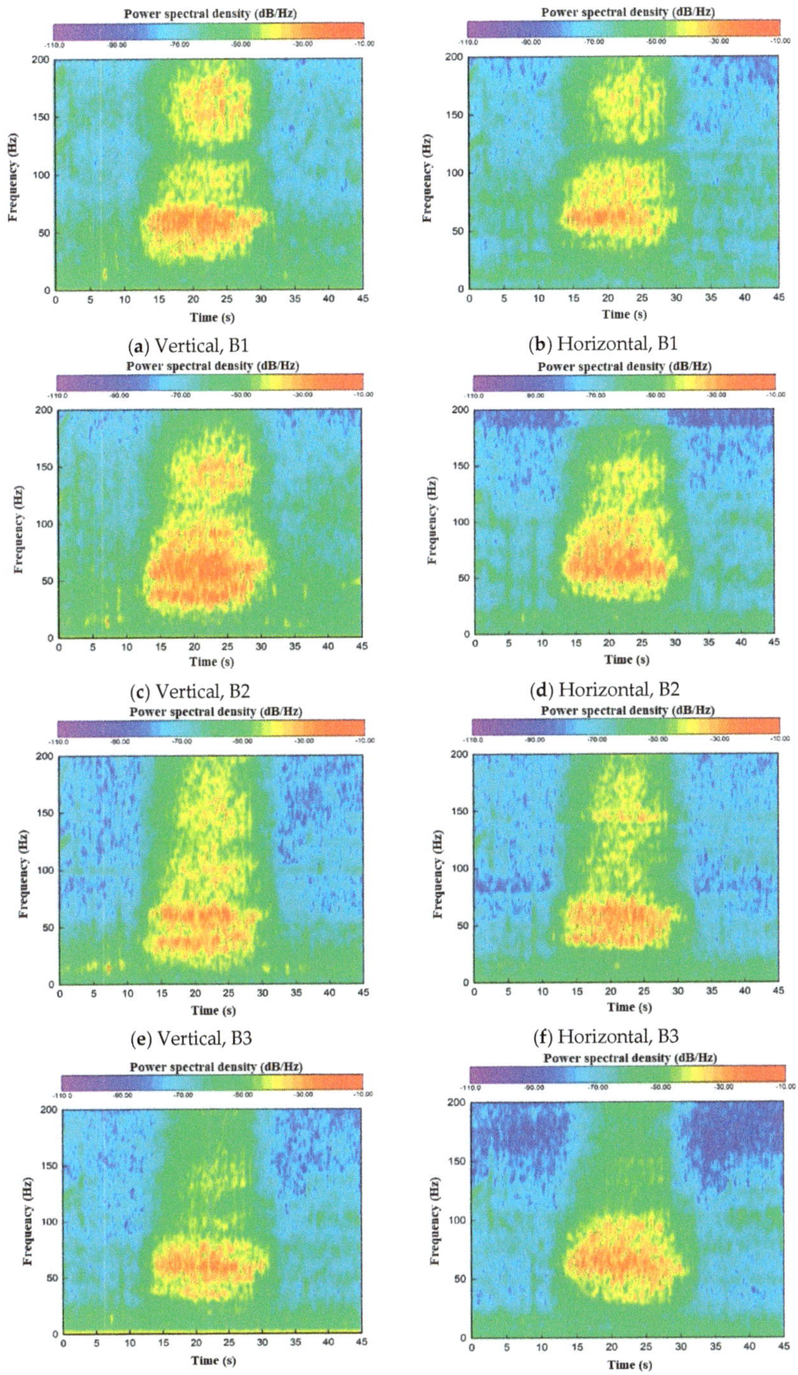

(a) Vertical, B1
(b) Horizontal, B1
(c) Vertical, B2
(d) Horizontal, B2
(e) Vertical, B3
(f) Horizontal, B3
(g) Vertical, B4
(h) Horizontal, B4

Figure 22. The frequency-domain plot of Section 2 at different distances from the subway station's exterior wall. (**a**) Vertical, B1; (**b**) Horizontal, B1; (**c**) Vertical, B2; (**d**) Horizontal, B2; (**e**) Vertical, B3; (**f**) Horizontal, B3; (**g**) Vertical, B4; (**h**) Horizontal, B4.

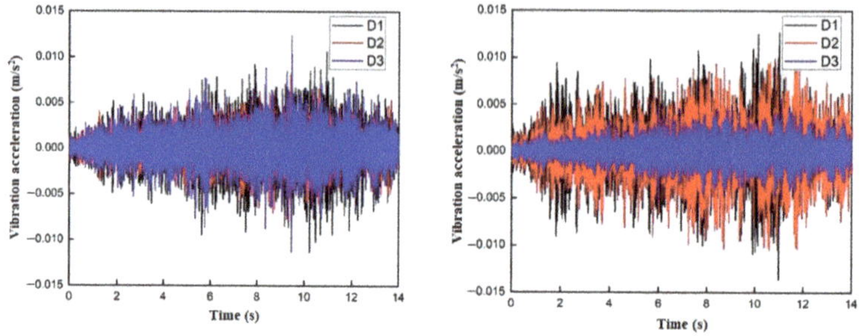

(a) Vertical (b) Horizontal

Figure 23. Time-domain plot of Section 3 at different distances from the exterior wall of the subway station. (a) Vertical; (b) Horizontal.

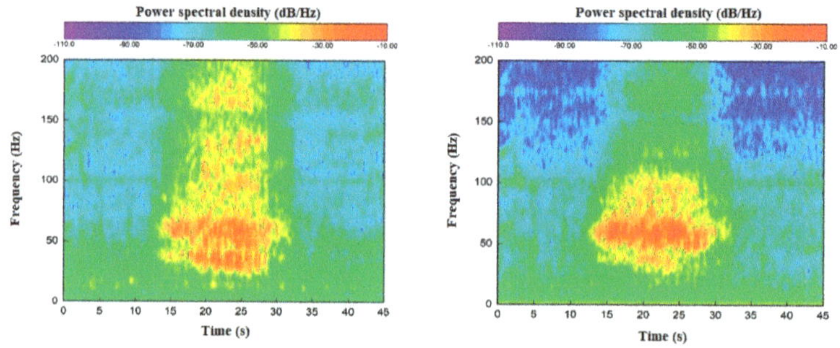

(a) Vertical, D1 (b) Horizontal, D1

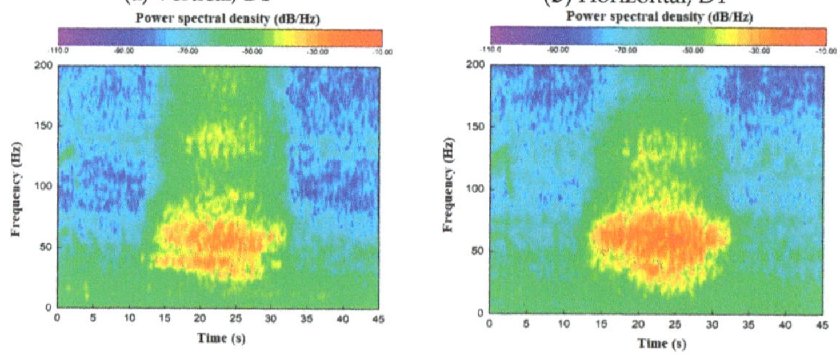

(c) Vertical, D2 (d) Horizontal, D2

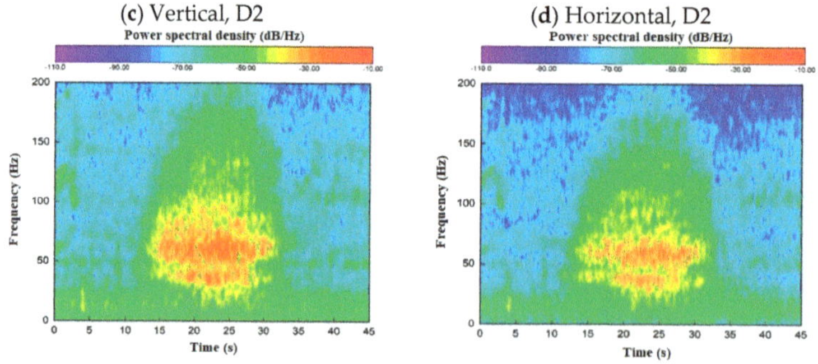

(e) Vertical, D3 (f) Horizontal, D3

Figure 24. The frequency-domain plot of Section 3 at different distances from the exterior wall of the subway station. (a) Vertical, D1; (b) Horizontal, D1; (c) Vertical, D2; (d) Horizontal, D2; (e) Vertical, D3; (f) Horizontal, D3.

3.4.2. Frequency-Domain Analysis of Trains at Different Distances from the Outer Walls of Subway Stations

Figure 25 delineates the typical vehicle-induced vibration spectra at each measuring point across three sections along the urban departure direction. In the vertical direction of Section 1, the spectra for K1 and K2 exhibit dual peaks at 40 Hz and 60 Hz. Conversely, in the horizontal direction, a singular peak occurs at 60 Hz. As the distance from the source increases, the amplitude of acceleration diminishes, and this effect becomes more pronounced in the horizontal direction. Considering the characteristics of the soil layer in Section 1, the vibration in the horizontal direction is easier to propagate and be affected than that in the vertical direction, and the attenuation is obvious. The spectrum of Section 2 is close to each other in vertical and horizontal directions, and the frequency bands of B1, B2, B3, and B4 are wide. B3 has two peaks at 40 and 60 Hz, B1 and B4 have a peak at 60 Hz, and the main frequency band of B2 in the vertical direction ranges from 30 Hz to 90 Hz with multiple peaks. In the horizontal direction, the peak appears at 60 Hz. With the increase in distance, the number of small peaks in the high-frequency band gradually decreases, and the peak frequency gradually concentrates at 60 Hz. Because B1 is close to the wall, the wall is affected by train vibration, and its own vibration leads to the abundance of the B1 high-frequency band. Section 3 Spectrum in the vertical direction, D1, D2, and D3 each have two peaks at 40 and 60 Hz, as displayed in Figure 25. In the horizontal direction, the peak value appears at 60 Hz, and the peak value appears at 40 Hz, but the acceleration amplitude is small. With the increase in distance, there is little difference in the acceleration amplitude in the vertical direction, and D3 attenuates obviously when the distance of Section 3 is greater than 70 Hz. It is considered that the distance of Section 3 has little effect on the attenuation of vertical vibration, while the soil attenuates at a high frequency in a small amplitude. In the horizontal direction, the vibration attenuation is obvious.

Upon analyzing the vehicular-induced vibrations recorded at various measuring points across different sections, a dataset comprising 20 vehicle traversal instances was utilized to ascertain the 1/3 octave vibration acceleration levels. As depicted in Figure 26, the influence range and average vibration values were determined. Vertically for Section 1, the maximum average vibration acceleration level observed at 15 m from the subway station's external wall was 65 dB, with a peak frequency of 40 Hz. At a distance of 35 m, the peak frequency decreases to 61 dB, with the peak frequency shifting to 63 Hz. In the horizontal direction, the maximum average vibration acceleration level was 62 dB at a distance of 15 m, with a peak frequency of 40 Hz, while at 35 m, it reduced to 55 dB, maintaining the same peak frequency. Vertically for Section 2, the maximum average vibration acceleration level was 68 dB at 15 m from the subway station's outer wall, declining to 65 dB at 25 m, 62 dB at 35 m, and 63 dB at 10 m, with the peak frequency consistent at 63 Hz across all distances. Horizontally, the peak average vibration acceleration level was 70 dB at 15 m, decreasing to 68 dB at 25 m, and 67 dB at both 10 m and 35 m, with the peak frequency for all four distances also remaining at 63 Hz. In Section 3, vertically, the peak average vibration acceleration level was 69 dB at 15 m from the subway station's external wall, decreasing to 67 dB at 25 m, and further to 66 dB at 35 m, with the peak frequency consistently at 63 Hz. In the horizontal direction, the peak vibration acceleration level was 76 dB at 15 m, with an attenuation of 72 dB at 25 m, and 65 dB at 35 m, with a peak frequency maintained at 63 Hz. At measuring points D1 and D2 within this section, the average horizontal vibration acceleration levels were found to be higher than those in the vertical direction, also with a higher peak frequency, which was influenced by the ground's reflection and scattering, as well as the site wall structure, facilitating greater propagation of vibrations in the horizontal direction.

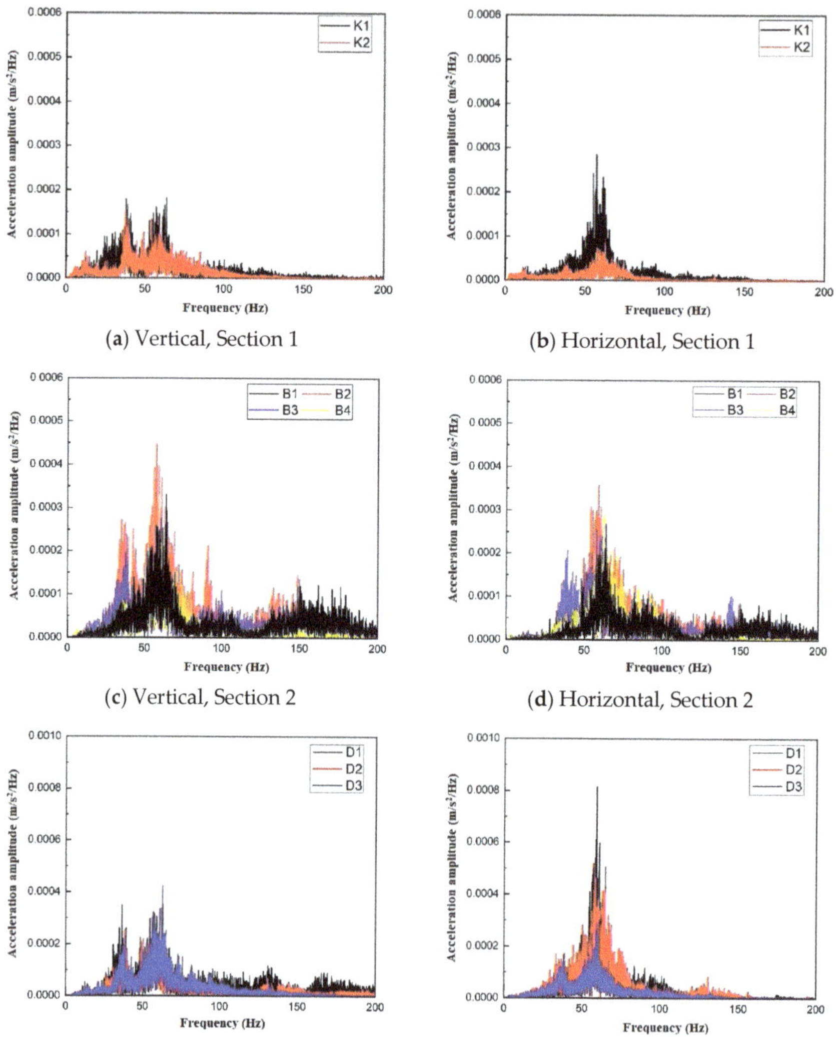

(**a**) Vertical, Section 1 (**b**) Horizontal, Section 1

(**c**) Vertical, Section 2 (**d**) Horizontal, Section 2

(**e**) Vertical, Section 3 (**f**) Horizontal, Section 3

Figure 25. Frequency spectrum plot of vehicle-induced vibration at different distances from Section 1, 2, and 3, respectively. (**a**) Vertical, Section 1; (**b**) Horizontal, Section 1; (**c**) Vertical, Section 2; (**d**) Horizontal, Section 2; (**e**) Vertical, Section 3; (**f**) Horizontal, Section 3.

The former results of the three sections indicate that the vertical vibration levels at the nine measuring points, located at 15 m, 25 m, and 35 m from the subway station's outer wall, do not exceed the specified daytime limit of 70 dB. Notably, measuring points in Sections 2 and 3, located 15 m from the subway station's external wall, reach 69 dB, which is close to the specified threshold. At the distances of 15 m and 35 m from the subway station's outer wall, the peak average vibration acceleration level in the vertical direction of all three sections experiences respective attenuations of 4 dB, 6 dB, and 3 dB, while the peak average vibration acceleration level in the horizontal direction attenuates by 7 dB, 3 dB, and 11 dB, respectively. When examining the attenuation from 15 m to 25 m and from 25 m to 35 m in Sections 2 and 3, it is evident that the vibration diminishes progressively within the range from 15 m to 35 m. In the vertical direction, the vibration attenuation observed in

Sections 1 and 2 is more pronounced than that of cross-Section 3. The variation in vibration attenuation is correlated with the train's speed and the duration of its passage. As the train accelerates out of the station, Section 3 experiences the longest exposure time, and the train's speed upon exiting the section is also the largest, yielding a more comprehensive collection of vibration signals that reflect the train's attenuation. Conversely, sections with shorter train passage time and slower speed upon exit result in greater attenuation of the collected vibration signals.

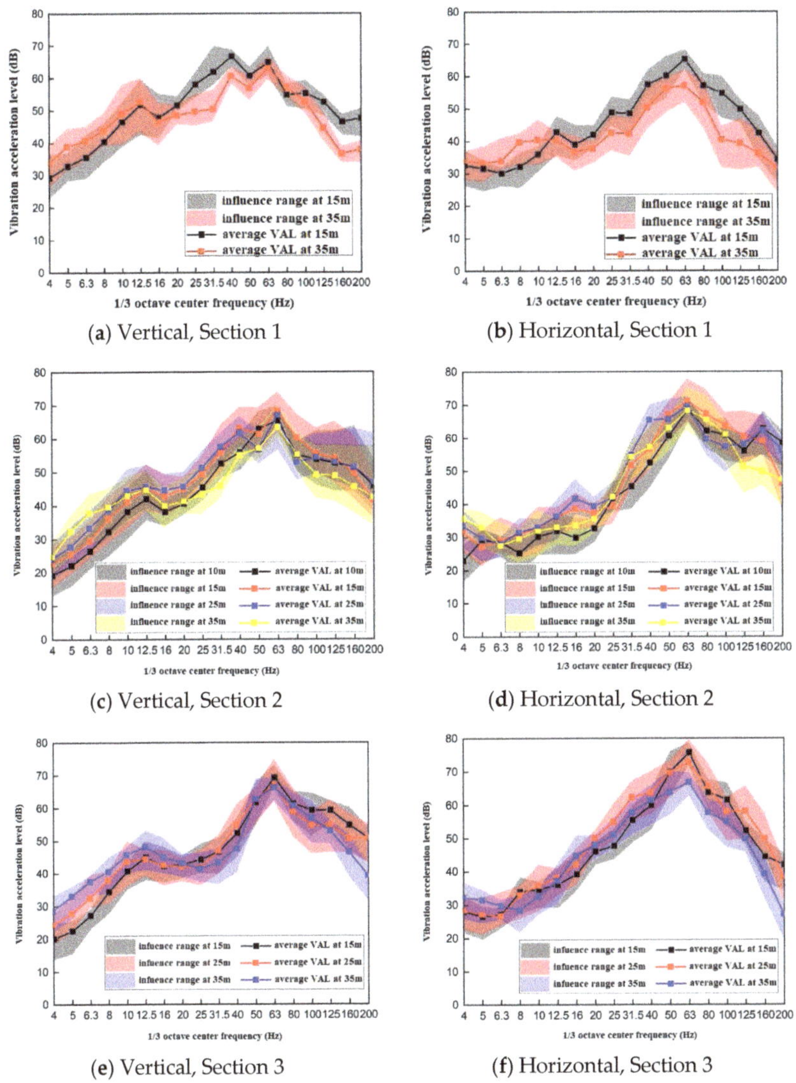

Figure 26. Acceleration levels of vibration in 1/3 octave bands at different distances from Sections 1, 2, and 3, respectively. (**a**) Vertical, Section 1; (**b**) Horizontal, Section 1; (**c**) Vertical, Section 2; (**d**) Horizontal, Section 2; (**e**) Vertical, Section 3; (**f**) Horizontal, Section 3.

4. Conclusions

In this study, field measurements of soil vibration adjacent to an operational railway station in Foshan were conducted to investigate the propagation behavior of vibrations caused by subway train operations. The analysis focused on three key aspects: the direction of train travel, the acceleration and deceleration of trains as they enter and exit the station, and the distance from the subway station. The response characteristics of the vibration were examined in both the time and frequency domains for both vertical and horizontal directions. The principal findings are as follows:

In the time domain, the amplitude of background vibration acceleration remained below 3×10^{-4} m/s^2, significantly lower than that of vehicle-induced vibration. In the frequency domain, the primary frequency of background vibration ranged from 0 to 10 Hz, which is inconsistent with the 40 to 60 Hz range of vehicle-induced vibration. Therefore, the background vibration at the experiment site minimally impacted the experiment results.

Due to the effect of vibration-damping fastenings, the low-frequency vibration of trains in the suburban direction was found to be greater than that in the urban direction. Moreover, as the suburban direction is located on the side farther from the rail, the peak vibration acceleration in the vertical direction within each section was within the range of 0.001 m/s^2 to 0.005 m/s^2, lower than that of the 0.004 m/s^2 to 0.008 m/s^2 range observed in the urban direction.

When the train decelerated to enter the station and accelerated to exit, the resulting vibration waves exhibited the Doppler effect. The ground soil vibration acceleration caused by the train was influenced by the train's weight and speed, positively correlating with both. However, the currently available measured data were insufficient to accurately ascertain the impact of train deceleration or acceleration on ground vibration acceleration.

When compared with the daytime limit of 70 dB vertically in China's current industry standard, the nine measuring points at distances of 15 m, 25 m, and 35 m from the subway station's outer wall within the three sections did not surpass the limit. Specifically, Section 2 and 3 reached 69 dB at a distance of 15 m from the outer wall, which was close to the limit.

As the distance increased, the vibration at the measuring points progressively attenuated, although the attenuation pattern was non-linear. The attenuation rate was influenced by the hardening of the adjacent soil layer, with a higher elastic modulus after hardening resulting in a slower vibration attenuation rate. Vibration attenuation was also affected by factors such as train speed and passage time. Specifically, as the train spent a short time passing through a section and exiting at a low speed, fewer vibration signals were collected, leading to a perceived larger vibration attenuation.

5. Future Work

While this study provides critical insights into the vibration propagation behaviors of non-uniform speed subway trains, several limitations must be acknowledged. First, the experimental data were collected at a single site with a limited sample size over a short-term period. While the methodology aligns with localized vibration studies [15,24], site-specific geological and structural characteristics may limit the generalizability of conclusions. Future work should expand measurements to diverse geotechnical environments and structural configurations, as well as long-term monitoring to evaluate the universality of vibration propagation patterns.

Second, the analysis did not account for seasonal climatic impacts on geological properties, which may modulate vibration propagation pathways. Seasonal variations in soil properties—such as moisture content and temperature—could significantly alter wave propagation dynamics. Future research should incorporate year-round measurements to quantify these seasonal effects and validate the robustness of the proposed models.

Additionally, the current dataset lacks granularity to decouple speed-dependent effects during transient operational phases (acceleration/deceleration), as speed variations were conflated with confounding factors like passenger load fluctuations. Future work should integrate high-resolution train telemetry with vibration monitoring to isolate speed-dependent effects. In addition, a phased analysis framework would differentiate vibration characteristics between transient operational phases and steady-state operations, supported by numerical simulations validated against field data.

Author Contributions: Methodology, Y.S.; Investigation, Y.S. and N.C.; Writing—original draft, Y.S.; Writing—review and editing, N.C. and Y.C.; Supervision, Y.C.; Funding acquisition, Y.C. All authors have read and agreed to the published version of the manuscript.

Funding: This research was funded by the education bureau of Guangzhou municipality, grant number 2024312560, and Guangzhou Railway Polytechnic, grant number GTXYR2210.

Data Availability Statement: Dataset available on request from the authors.

Conflicts of Interest: Author Na Cai is employed by the company Zhejiang Taicheng Environmental Technology Co., Ltd. The authors declare that the research was conducted in the absence of any commercial or financial relationships that could be construed as a potential conflict of interest.

References

1. Zhang, X. Forty years of highly synchronous development in China's scientific research and economy. *Natl. Sci. Rev.* **2018**, *5*, 796–798.
2. Sun, M.; Liu, Y. Economic growth and urban rail transit development: Evidence from China's megacities. *Transp. Policy* **2021**, *98*, 62–72.
3. Chen, J.; Xu, G. Urban rail transit systems in China: Current trends and future challenges. *Cities* **2023**, *134*, 104021.
4. Wang, J.; Zhang, L. The impact of economic growth on the development of urban rail transit in China. *J. Urban Plan. Dev.* **2021**, *147*, 04021025.
5. Zhang, Y.; Chen, Z. Urbanization and its impact on public transportation: Evidence from major Chinese cities. *Transp. Res. Part A Policy Pract.* **2022**, *156*, 262–274.
6. Liu, W.; Lund, H.; Mathiesen, B.V. Modelling the transport system in China and evaluating the current strategies towards the sustainable transport development. *Energy Policy* **2013**, *58*, 347–357.
7. Xu, Q.-Y.; Ou, X.; Au, F.; Lou, P.; Xiao, Z.-C. Effects of track irregularities on environmental vibration caused by underground railway. *Eur. J. Mech.-A/Solids* **2016**, *59*, 280–293.
8. Tao, Z.; Wang, Y.; Sanayei, M.; Moore, J.A.; Zou, C. Experimental study of train-induced vibration in over-track buildings in a metro depot. *Eng. Struct.* **2019**, *198*, 109473. [CrossRef]
9. Zou, C.; Li, X.; He, C.; Zhou, S. An efficient method for estimating building dynamic response due to train operations in tunnel considering transmission path from source to receiver. *Comput. Struct.* **2024**, *305*, 107555. [CrossRef]
10. Liu, Q.; Li, X.; Zhang, X.; Zhou, Y.; Chen, Y.F. Applying constrained layer damping to reduce vibration and noise from a steel-concrete composite bridge. *J. Sandw. Struct. Mater.* **2020**, *22*, 1743–1769. [CrossRef]
11. Li, X.; Chen, Y.; Zou, C.; Wang, H.; Zheng, B.; Chen, J. Building structure-borne noise measurements and estimation due to train operations in tunnel. *Sci. Total Environ.* **2024**, *926*, 172080. [CrossRef]
12. Huang, S.; Chen, Y.; Zou, C.; Jian, S. Train-induced environmental vibrations by considering different building foundations along curved track. *Transp. Geotech.* **2022**, *35*, 100785. [CrossRef]
13. Qu, S.; Yang, J.; Zhu, S.; Zhai, W.; Kouroussis, G.; Zhang, Q. Experimental study on ground vibration induced by double-line subway trains and road traffic. *Transp. Geotech.* **2021**, *29*, 100564.
14. Gutowski, T.G.; Dym, C.L. Propagation of ground vibration: A review. *J. Sound Vib.* **1976**, *49*, 179–193.
15. Zou, C.; Wang, Y.; Moore, J.A.; Sanayei, M. Train-induced field vibration measurements of ground and over-track buildings. *Sci. Total Environ.* **2017**, *575*, 1339–1351. [PubMed]
16. Di, H.; Su, G.; Yu, J.; Xu, Y.; Xu, P.; Zhou, S. Field measurement and evaluation of vibrations inside buildings above metro tunnels. *Soil Dyn. Earthq. Eng.* **2023**, *166*, 107767.
17. Forrest, J.A.; Hunt, H.E.M. A three-dimensional tunnel model for calculation of train-induced ground vibration. *J. Sound Vib.* **2006**, *294*, 678–705.
18. He, C.; Jia, Y.; Zhou, S. Semi-analytical method for calculating ground vibrations from a tunnel in a homogeneous half-space with an irregular surface. *J. Sound Vib.* **2024**, *591*, 118615.
19. He, C.; Zhou, S.; Guo, P.; Di, H.; Zhang, X. Analytical model for vibration prediction of two parallel tunnels in a full-space. *J. Sound Vib.* **2018**, *423*, 306–321.

20. Qiu, Y.; Zou, C.; Hu, J.; Chen, J. Prediction and mitigation of building vibrations caused by train operations on concrete floors. *Appl. Acoust.* **2024**, *219*, 109941.
21. Zhang, J.; Xi, C.; Sun, Y.; Li, G.; Liu, J.; Wang, M. Analysis of antideformation and antivibration effect of different track beds under subway train vibration. *Shock Vib.* **2022**, *2022*, 4814039.
22. Tao, Z.Y.; Zou, C.; Yang, G.R.; Wang, Y.M. A semi-analytical method for predicting train-induced vibrations considering train-track-soil and soil-pile-building dynamic interactions. *Soil Dyn. Earthq. Eng.* **2023**, *167*, 107822.
23. Hu, J.; Zou, C.; Liu, Q.; Li, X.; Tao, Z. Floor vibration predictions based on train-track-building coupling model. *J. Build. Eng* **2024**, *89*, 109340.
24. Shi, W.; Bai, L.; Han, J. Subway-induced vibration measurement and evaluation of the structure on a construction site at curved section of metro line. *Shock Vib.* **2018**, *2018*, 5763101.
25. Huang, Q.; Li, P.; Zhang, D.; Huang, H.; Zhang, F. Field measurement and numerical simulation of train-induced vibration from a metro tunnel in soft deposits. *Adv. Civ. Eng.* **2021**, *2021*, 6688746.
26. He, C.; Zhou, S.; Guo, P. An efficient three-dimensional method for the prediction of building vibrations from underground railway networks. *Soil Dyn. Earthq. Eng.* **2020**, *139*, 106269.
27. Lei, X.; Jiang, C. Analysis of vibration reduction effect of steel spring floating slab track with finite elements. *J. Vib. Control* **2016** *22*, 1462–1471.
28. Zou, C.; A Moore, J.; Sanayei, M.; Wang, Y.; Tao, Z. Efficient impedance model for the estimation of train-induced vibrations in over-track buildings. *J. Vib. Control* **2021**, *27*, 924–942.
29. Jin, H.; Wang, H.; Li, Z.; Zhou, X. Vibration-reduction optimization of the point-supporting floating-slab track based on local resonance mechanism. *J. Vib. Control* **2023**, *29*, 1176–1190.
30. Kong, G.; Sun, G.; Liu, H.; Li, J. Dynamic Response of Ballastless Track XCC Pile-Raft Foundation under Train Axle Loads. *J. Test Eval.* **2021**, *49*, 1691–1704.
31. Tao, Z.; Zhang, D.; Tu, D.; He, L.; Zou, C. Prediction of train-induced ground-borne vibration transmission considering parametric uncertainties. *Probabilistic Eng. Mech.* **2025**, *76*, 103731. [CrossRef]
32. Li, X.; Chen, Y.; Zou, C.; Chen, Y. Train-induced vibration mitigation based on foundation improvement. *J. Build. Eng.* **2023**, *76* 107106. [CrossRef]
33. Xu, D.S.; Zhao, Y.M.; Liu, H.B.; Zhu, H.H. Deformation Monitoring of Metro Tunnel with a New Ultrasonic-Based System. *Sensors* **2017**, *17*, 1758. [CrossRef] [PubMed]
34. Zhou, S.; He, C.; Guo, P.; Yu, F.; Di, H. Dynamic response of a segmented tunnel in saturated soil using a 2.5-D FE-BE methodology *Soil Dyn. Earthq. Eng.* **2019**, *120*, 386–397.
35. Deng, L.; Liu, W.; Liu, W.; Meng, M. Computation and analysis of evaluation indicators of environmental vibration induced by urban rail transit. *Urban Rapid Rail Transit* **2017**, *30*, 40–45.
36. Li, X.; Zheng, B.; Chen, Y.; Zou, C. A hybrid methodology for estimating train-induced rigid foundation building vibrations *Constr. Build. Mater.* **2025**, *460*, 139852. [CrossRef]

buildings

Article

Experimental Study on Noise Reduction Performance of Vertical Sound Barrier in Elevated Rail Transit

Lizhong Song [1,2], Yisheng Zhang [1,2], Quanmin Liu [1,2,*], Yunke Luo [3] and Ran Bi [4]

1 State Key Laboratory of Safety and Resilience of Civil Engineering in Mountain Area, East China Jiaotong University, Nanchang 330013, China; songlizhong@ecjtu.edu.cn (L.S.); 2022138086100050@ecjtu.edu.cn (Y.Z.)
2 MOE Engineering Research Center of Railway Environmental Vibration and Noise, East China Jiaotong University, Nanchang 330013, China
3 Department of Civil and Environmental Engineering, The Hong Kong Polytechnic University, Hong Kong 999077, China; yunke.luo@connect.polyu.hk
4 State Key Laboratory of Bridge Intelligent and Green Construction, Southwest Jiaotong University, Chengdu 611756, China; ran_bi@my.swjtu.edu.cn
* Correspondence: liuquanmin@ecjtu.edu.cn

Abstract: With the large-scale construction of rail transit in mainland China, the noise problem caused by passing trains has become increasingly prominent. The vertical sound barrier is currently the most effective noise control measure for rail transit. However, the noise reduction performance of the vertical sound barrier at different train speeds remains unclear. This study focuses on the box-girder cross-sections of an elevated urban rail transit line with and without vertical sound barriers, conducting field tests during train passages. Based on the test results, the influence of train speed on noise levels at both cross-sections was investigated, the sound source characteristics were analyzed, and the noise reduction performance of the vertical sound barriers at different speeds was explored. The findings indicate the following: Regardless of the presence of sound barriers, within the speed range of 20 to 80 km/h, the linear sound pressure levels at the track-side and beam-side measurement points exhibit a strong linear correlation with speed, while the correlation is weaker at the beam-bottom measurement points. As speed increases, the wheel–rail noise increases by approximately 1.5 dB compared to the structural noise at the same speed. Vertical sound barriers significantly reduce mid-to-high-frequency noise, but in the low frequency band between 20 and 63 Hz, the noise increases, likely due to secondary structural noise radiated by the self-vibration of the barriers when trains pass. At speeds of 20, 40, 60, and 80 km/h, the insertion loss at measurement points located 7.5 m from the track centerline ranges from 6.5 to 9.0, 8.5 to 10.5, 7.5 to 9.5, and 7.5 to 10.2 dB, respectively. At 25 m from the track centerline, the insertion loss ranges from 1.5 to 2.5, 6.0 to 6.5, 5.5 to 6.0, and 5.0 to 6.0 dB, respectively. The noise reduction capability of the vertical sound barrier initially increases and then decreases with higher speeds, and the rate of reduction slows as speed increases. This research will provide a reference and basis for determining speed limits in the rail transit sections equipped with sound barriers.

Keywords: elevated rail transit; different speeds; vertical sound barrier; field test; noise reduction

1. Introduction

In recent years, elevated urban rail transit lines have developed rapidly due to their advantages, such as low construction costs and minimal land use. However, the proportion of elevated lines in urban rail transit networks remains relatively low [1]. One of the

https://doi.org/10.3390/buildings15101621

significant factors limiting their widespread application is the vibration and noise issues caused by trains passing through elevated sections, which severely affect the lives of residents along the lines [2].

When trains pass over railways, wheel–rail noise is generated due to the interaction between the wheels and rails caused by irregularities in the track [3–7]. Additionally, rail vibrations are transmitted to the bridge structure, leading to bridge-structure noise [8,9]. Furthermore, aerodynamic noise is produced by the friction between the train and the air [10,11], and pantograph–catenary noise arises from the friction between the pantograph and the overhead contact line [12,13]. However, in urban rail transit systems, where operational speeds are generally low (typically not exceeding 120 km/h), the primary sources of noise are wheel–rail noise and bridge-structure noise, with wheel–rail noise being the dominant contributor [14].

Installing sound barriers along the tracks is an effective measure to mitigate the impact of wheel–rail noise on nearby residents. Numerous studies have been conducted by domestic and international scholars on the noise reduction effectiveness of such sound barriers [15–18]. Wang et al. [19] established a numerical prediction model for wheel–rail noise using finite element-boundary element and statistical energy analysis methods, comparing the noise reduction effects of vertical sound barriers and low-height near-rail barriers. Xin et al. [20] analyzed the noise reduction performance of fully enclosed sound barriers on the Jing Xiong Intercity Railway through field tests and numerical simulations, finding a noise reduction of 19.8 to 20.1 dB(A) when trains passed at 350 km/h. Li et al. [21] conducted field tests to study the noise reduction effects of semi-enclosed sound barriers on high-speed railways, showing that the sound insulation of semi-enclosed barriers increases with frequency, providing a noise reduction of 15 to 17 dB(A) within 25 m of the track centerline. Sun et al. [22,23] investigated the noise reduction effects of barriers with different top shapes through surveys and empirical formulas, offering references for sound barrier selection. Jung et al. [24] compared the noise reduction performance of sound barriers with two different top shapes using field tests and scaled model experiments, concluding that cylindrical-top barriers perform better than vertical reflective barriers of the same height. Zhang et al. [25] analyzed the acoustic performance of vertical and semi-enclosed sound barriers through field tests and numerical simulations, demonstrating that semi-enclosed sound barriers provide better noise reduction, outperforming vertical sound barriers by 6 dB(A). Based on field tests, Liu et al. [26] studied the noise reduction effects of sound barriers with different heights and distances, finding that sound pressure levels increase with distance from the barrier at the same height and decrease with proximity to the ground at the same distance. Song et al. [27] used a 3D dynamic model and a 2.5D acoustic model to analyze the noise reduction effects of vertical sound barriers on U-shaped girder bridges, showing significant effectiveness in reducing wheel–rail noise.

However, testing research on the noise reduction performance of vertical sound barriers on elevated urban rail transit sections under different train speeds is rarely reported. Therefore, this study focuses on the box-girder cross-sections of an elevated urban rail transit line with and without vertical sound barriers, conducting field noise tests during train passages. Based on the test results, the noise reduction performance of vertical sound barriers under different train speeds is analyzed.

2. Field Test

2.1. Overview of the Test Section

On an elevated section of a urban rail transit line, sections without sound barriers and sections with vertical sound barriers were selected to conduct field vibration and noise tests caused by train passages. The elevated line uses a six-carriage Type B train with a maximum operating speed of 80 km/h. Figure 1 shows the cross-section of this test box beam and test photos. The elevated section employs a 30 m span double-track simply supported concrete box girder, the panel width of the box beam bridge is 9.3 m, the floor width is 4.03 m, the beam height is 1.8 m, the thickness of the top plate within the span is 0.25 m, the thickness of the web is about 0.408 m, the thickness of the bottom plate is 0.25 m, the line spacing is 4 m, and the distance between the beam bottom and the ground is about 2.3 m.

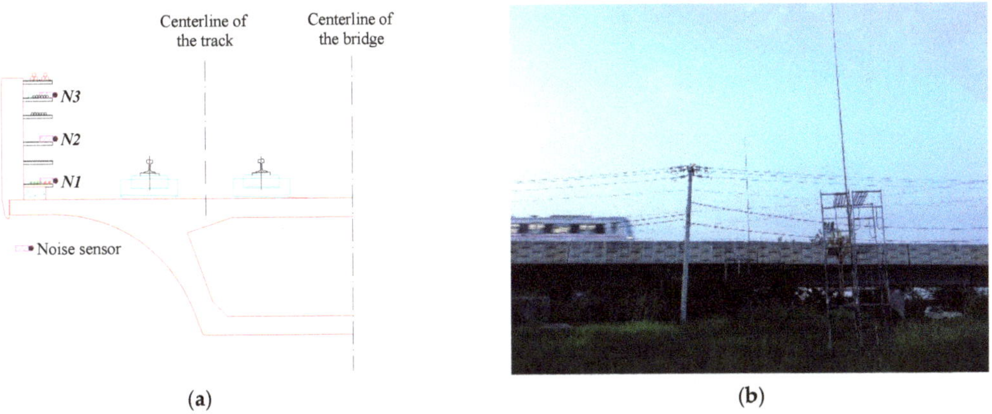

Figure 1. Cross-section of the viaduct box girder section and test photo: (**a**) cross-section of the box girder section; (**b**) photo of the test section.

A 1.6-meter-high retaining wall is installed on both sides of the box girder, with a 2.4-meter-high vertical sound barrier mounted on top of the retaining wall, as shown in Figure 2. The sound barrier primarily consists of H-shaped steel columns, sound-absorbing and insulating panels, and PC transparent sound-insulating panels. The H-shaped steel columns are spaced at intervals of 2.0 m, the sound-absorbing and insulating panels of the barrier are 1.723 m in length and 0.4 m in width, while the PC transparent sound-insulating panels are 1.723 m in length and 1.0 m in width. Both sides of the unit panels are fixed with 2 mm thick soft rubber pads and connected to the H-shaped steel columns via socket joints. The noise barrier is connected to the bridge retaining wall using high-strength bolts.

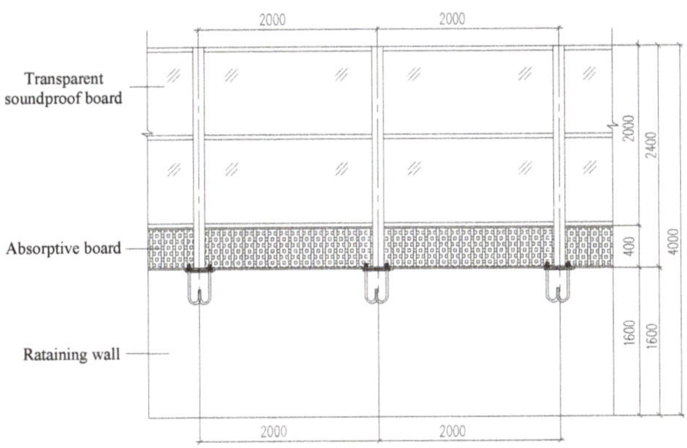

Figure 2. Vertical sound barrier (unit: mm).

2.2. Test Plan

2.2.1. Arrangement of Test Measurement Points

For ease of comparison, noise measurement points were set up at identical locations in both cross-sections to measure the noise caused by trains passing through the elevated urban rail transit section. In strict compliance with ISO 3095: 2013 standards [28], nine noise measurement points were systematically deployed in each of the two cross-sections, with microphones placed beside the outer rail, under the girder, and on the side of the girder to measure the track-side noise, under-girder noise, and side-girder noise, respectively. On the section without a sound barrier, measurement points N1, N2, and N3 are located beside the outer rail, 2.06 m from the track centerline, and at heights of 0.2183 m, 0.7648 m, and 1.3416 m above the bridge deck, (as shown in Figure 3). Measurement points N5 and N6 are located under the girder, 0.3 m from the center of the bottom slab and 1.2 m above the ground, respectively. Measurement points N8, N9, N13, and N14 are located 7.5 m and 25 m from the outer track centerline, at heights of 1.2 m and 3.5 m above the rail surface, respectively. On the section with the vertical sound barrier, the measurement points N1′–N3′, N5′–N6′, N8′–N9′, and N13′–N14′ correspond exactly to the locations of points N1–N3, N5–N6, N8–N9, and N13–N14 on the section without the sound barrier.

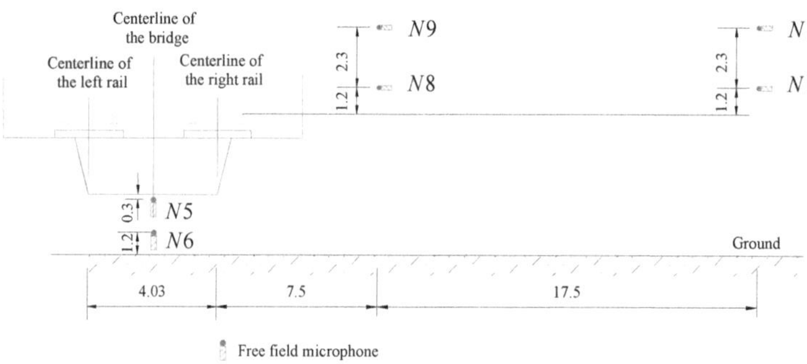

Figure 3. Schematic diagram of test measurement points arrangement.

2.2.2. Test Conditions

The testing was conducted on 2 August 2023 and 9 August 2023. Both tests were conducted at night under similar weather conditions, with a temperature of 25 °C, relative humidity of 81%, no rainfall, windless conditions, etc. Trains passed through the two test sections at speeds of 20 km/h, 40 km/h, 60 km/h, and 80 km/h. To ensure the validity of the test data, five measurements were taken for each speed condition. On the 2nd, 20 sets of valid data were collected for the section without a sound barrier at constant speeds, and on the 9th, 20 sets of valid data were collected for the section with the vertical sound barrier. The measurement points for test data were positioned on the bridge section proximate to the train's travel route.

2.3. Test Equipment

For this test, noise measurements were conducted using GRAS sound sensors. The data acquisition system employed was the DATaRec 4 DIC24 from Head Acoustics, Germany, and ArtemiS data acquisition and analysis software was used for data collection and analysis. This setup is illustrated in Figure 4.

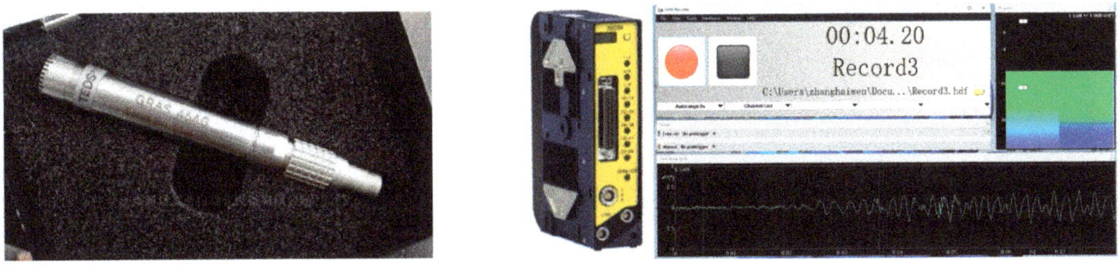

(a) (b)

Figure 4. Test equipment: (**a**) GRAS sound sensor; (**b**) DATaRec 4 DIC24 data acquisition instrument.

3. Analysis of Test Results

3.1. Impact of Speed on Noise in Both Sections

3.1.1. Impact of Speed on Noise at the Section Without a Sound Barrier

To investigate the influence of train speed on the noise levels at the section without a sound barrier, noise radiation was measured as trains passed through the elevated urban rail transit section at speeds of 20 km/h, 40 km/h, 60 km/h, and 80 km/h. The linear sound pressure level spectra for typical measurement points N1, N5, and N8 under different speed conditions are shown in Figure 5. From Figure 5, the following observations can be made: In general, as the train speed increases, the noise levels at all typical measurement points also increase. Noise levels tend to decrease as the distance from the measurement points to the track increases. In addition, the peak noise frequencies at each measurement point are relatively similar across different speeds. The noise measurement point N1 is located close to the rail and is primarily influenced by wheel–rail noise, with a peak frequency around 800 Hz. A peak appears near 63 Hz in the low-frequency range, which may be attributed to partial background noise interference during the testing process. The noise measurement point N5 is positioned close to the bridge bottom slab and is mainly influenced by bridge-structure noise, with a peak frequency around 63 Hz. The noise measurement point N8 is located on the side of the girder, 7.5 m from the track centerline and 1.2 m above the rail surface, and it is influenced by both wheel–rail noise and bridge-structure noise. The peak frequency in the low-frequency range is around 80 Hz, while in the mid-to-high-frequency range, the peak frequency is around 800 Hz.

To further quantitatively analyze the influence of train speed on the noise levels at the section without a sound barrier, the variation in the linear overall sound pressure level with speed at typical measurement points N1, N5, and N8 is presented. More than five noise tests were conducted under each train speed condition, from which test groups containing background noise (e.g., aircraft noise or dog barking) were removed, retaining five valid test sets, as shown in Figure 6. From Figure 6, the following observations can be made: The linear overall sound pressure levels obtained from the noise tests under the same speed condition at each typical measurement point show little variation, indicating the validity of the noise test data. When the train speed varies between 20 km/h and 80 km/h, the linear overall sound pressure levels at measurement points N1 and N8 exhibit a strong linear correlation with speed. In contrast, the linear overall sound pressure level at measurement point N5, which is more influenced by bridge-structure noise, shows a weaker linear correlation with speed. The noise levels at all typical measurement points increase with higher train speeds. For every 10 km/h increase in speed, the linear overall sound pressure levels at measurement points N1, N5, and N8 increase by approximately 2.3 dB, 1.3 dB, and 1.8 dB, respectively. From the comparison between measurement points

N1 and N5, it can be concluded that, as the train speed increases, wheel–rail noise increases faster than structure noise.

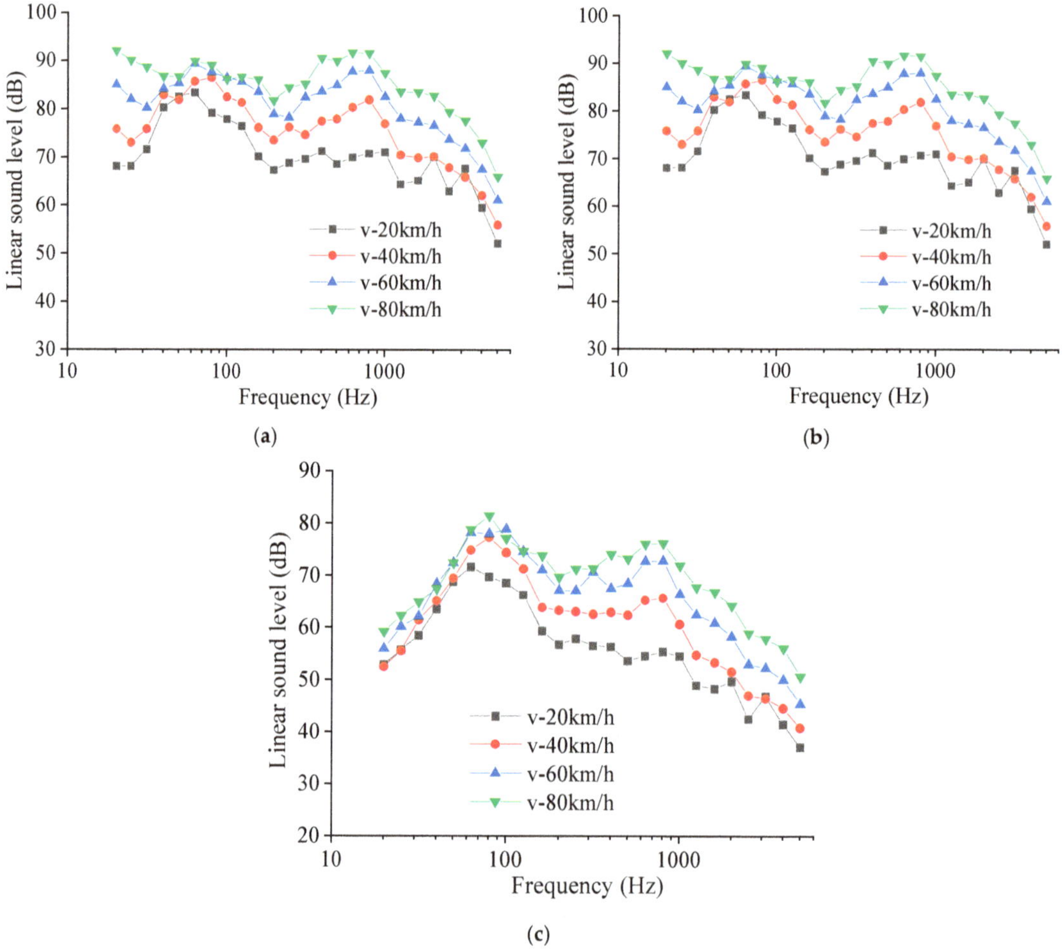

Figure 5. Linear sound-pressure-level spectrum diagram of typical noise measurement points at different speeds/dB: (**a**) measurement point N1; (**b**) measurement point N5; (**c**) measurement point N8.

3.1.2. Impact of Speed on Noise at the Section with a Vertical Sound Barrier

To investigate the influence of train speed on the noise levels at the section with a vertical sound barrier, the noise radiation was measured as trains passed through the elevated urban rail transit section at speeds of 20 km/h, 40 km/h, 60 km/h, and 80 km/h. The linear sound pressure level spectra for typical measurement points N1′, N5′, and N8′ under different speed conditions are shown in Figure 7. From Figure 7, the following observations can be made: In general, as the train speed increases, the noise levels at all typical measurement points also increase. Noise levels tend to decrease as the distance from the measurement points to the track increases. In addition, the peak noise frequencies at each measurement point are relatively similar across different speeds. The noise measurement point N1′ is positioned close to the rail and is mainly influenced by wheel–rail noise, with a peak frequency around 400 Hz in the mid-to-high-frequency range. The noise measurement point N5′ is positioned close to the bridge deck and is primarily influenced by the structural noise of the bridge, with a peak frequency in the low-frequency range at 80 Hz. The noise measurement point N8′ is situated on the side of the beam and is affected by both wheel–rail noise and bridge-structure noise. It exhibits peaks in both the mid-to-high frequency range and the low-frequency range, with the mid-to-high frequency peak occurring at 800 Hz and the low-frequency peak at 40 Hz.

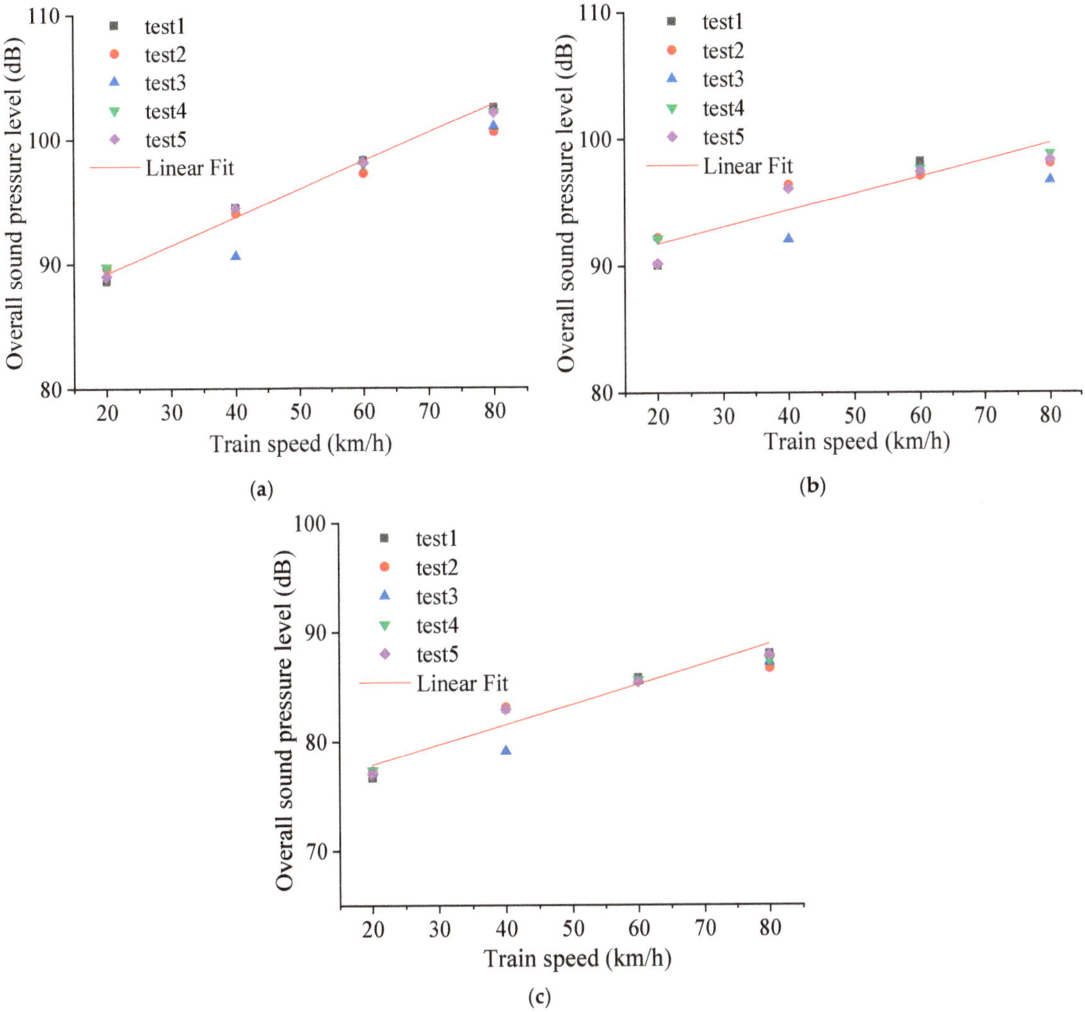

Figure 6. Impact of speed on the linear total sound pressure level of typical noise measurement points: (**a**) measurement point N1; (**b**) measurement point N5; (**c**) measurement point N8.

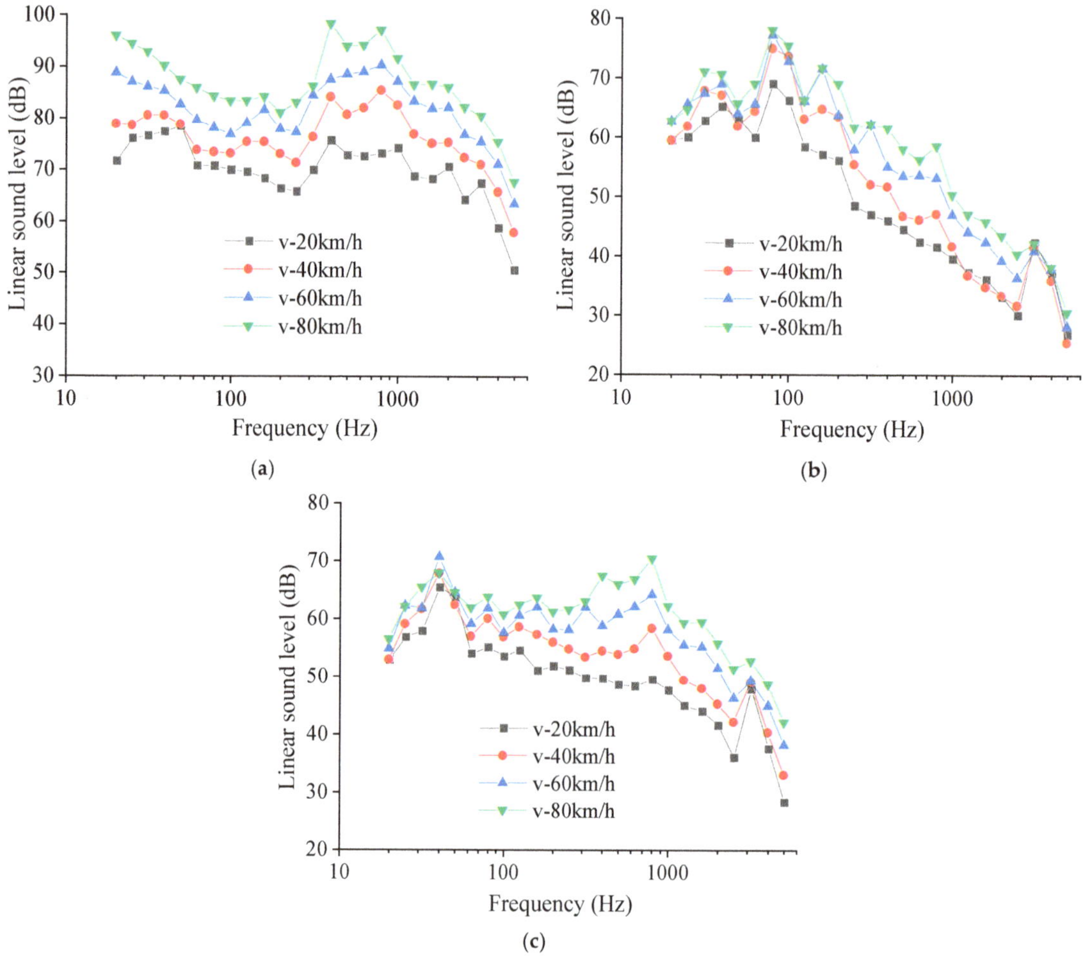

Figure 7. Linear sound-pressure-level spectrum diagram of typical noise measurement points at different speeds/dB: (**a**) measurement point N1′; (**b**) measurement point N5′; (**c**) measurement point N8′.

To further quantitatively analyze the impact of train speed on the noise profile of the vertical sound barrier, the variation in the linear overall sound pressure level with train speed for each typical noise measurement point, N1′, N5′, and N8′ is presented. Noise tests were conducted five times under each speed condition (no aircraft passed by, and dog barking sounds were detected during the test), as shown in Figure 8. From Figure 8, it can be observed that the linear overall sound pressure levels obtained from the noise tests conducted under the same speed condition at each typical measurement point are quite similar, indicating the validity of the noise test data. When the train speed varies from 20 km/h to 80 km/h, there is a strong linear correlation between the linear overall sound pressure level and the train speed for noise measurement points N1′ and N8′, whereas the linear correlation is weaker for measurement point N5′, which is closer to the bottom of the bridge. The noise at each typical measurement point increases with the increase in train speed. For every 10 km/h increase in speed, the linear overall sound pressure levels at measurement points N1′, N5′, and N8′ increase by approximately 3.0 dB, 1.5 dB, and 1.6 dB, respectively. Comparatively, wheel–rail noise increases more rapidly with speed than the structural noise.

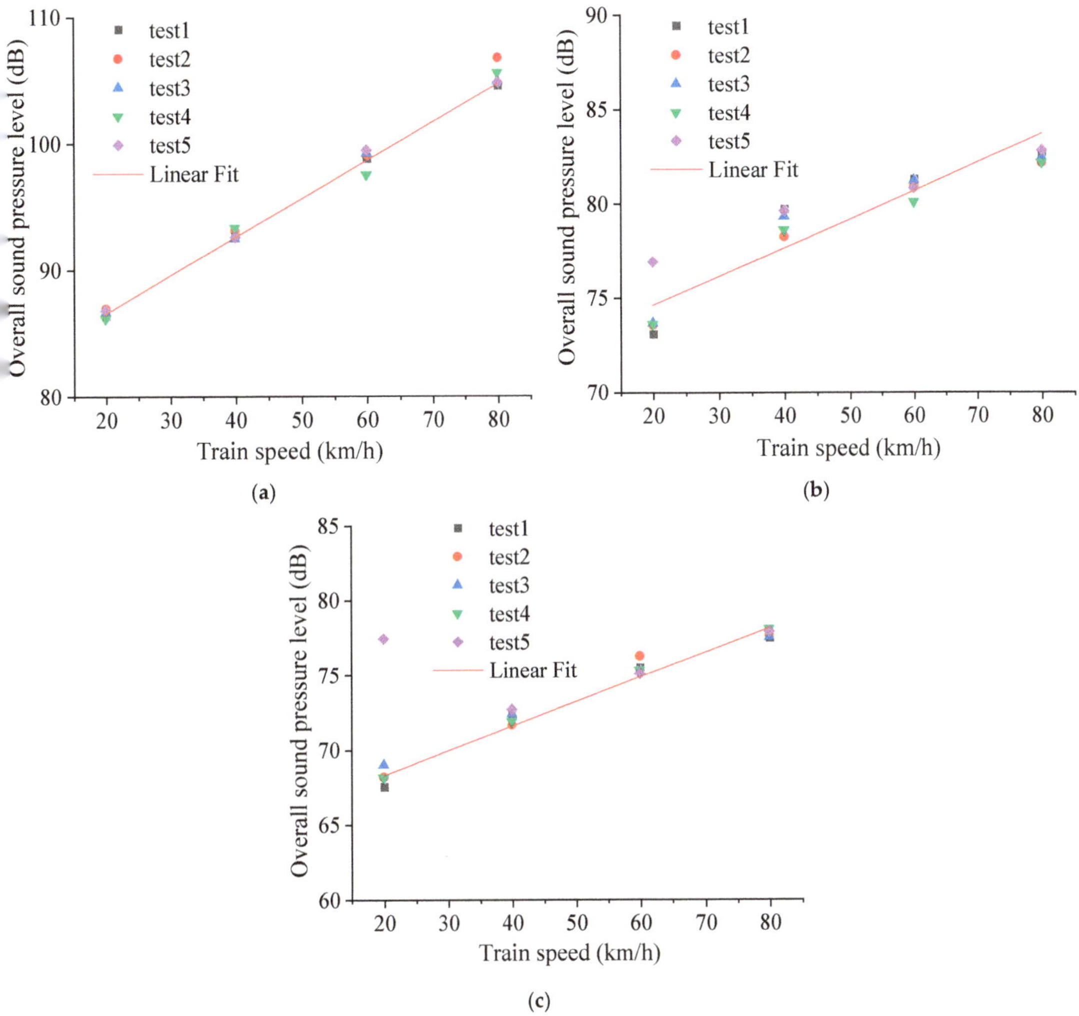

Figure 8. Impact of speed on the linear total sound pressure level of typical noise measurement points: (**a**) measurement point N1′; (**b**) measurement point N5′; (**c**) measurement point N8′.

3.2. Analysis of Sound Source Characteristics in Both Sections

The track-side noise measurement points N1 to N3 and N1′ to N3′ in both sections are located close to the outer rail and are primarily influenced by wheel–rail noise. Meanwhile, the under-beam measurement points N5, N6, N5′, and N6′, situated 0.3 m below the beam and 1.2 m above the ground level, respectively, are mainly affected by the bridge's structural noise. To investigate the amplitude–frequency characteristics of the sound sources in both sections as the train passes through the viaduct, the noise spectrum curves for each typical measurement point at a speed of 80 km/h are provided, along with the total sound pressure level values for these points, as illustrated in Figure 9 and detailed in Table 1.

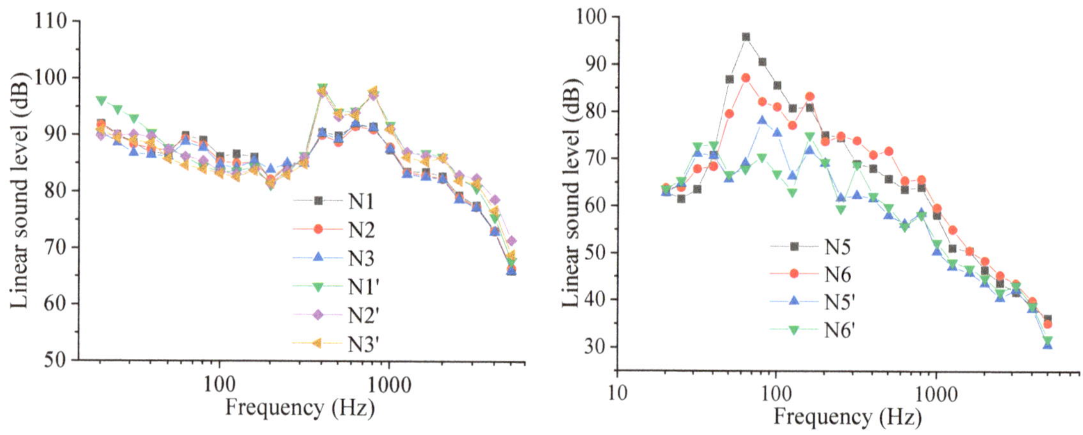

(a) (b)

Figure 9. Spectrum curve diagram of track-side and under-beam noise measurement points: (**a**) wheel–rail noise; (**b**) bridge-structure noise.

Table 1. Total sound pressure level of track-side and under-beam noise measurement points (dB).

Section	Measurement Point Number	Total Sound Pressure Level (dB)
Section Without a Sound Barrier	N1	101.58
	N2	101.21
	N3	100.92
	N5	97.97
	N6	91.13
Section with a Vertical Sound Barrier	N1′	105.24
	N2′	103.99
	N3′	104.00
	N5′	82.40
	N6′	80.93

From Figure 9, it can be observed that the linear sound pressure level curves for the noise measurement points N1 to N3 and N1′ to N3′ exhibit some differences in the low-frequency range between 50 and 100 Hz. However, the peak noise frequencies for both sets of points occur between 300 and 1000 Hz, the peak linear sound pressure levels for points N1 to N3 are 91.7 dB, 91.5 dB, and 92 dB, respectively, while for points N1′ to N3′, they are 98.48 dB, 97.49 dB, and 97.82 dB, respectively. This indicates that the linear sound pressure level amplitudes at points N1 to N3 are lower than those at points N1′ to N3′. Similarly, the linear sound pressure level curves for the noise measurement points N5, N6, N5′, and N6′ also show some differences in the low-frequency range between 40 and 200 Hz. The peak noise frequencies for points N5 and N6 occur at 63 Hz, while for points N5′ and N6′, they occur at 80 Hz and 160 Hz, respectively. The peak linear sound pressure levels for points N5 and N6 are 95.90 dB and 87.15 dB, respectively, whereas for points N5′ and N6′, they are 78.06 dB and 74.95 dB, respectively. This suggests that the linear sound pressure level amplitudes at points N5 and N6 are higher than those at points N5′ and N6′.

From Table 1, it can be seen that the total sound pressure levels at the track-side noise measurement points N1 to N3 are 3 to 4 dB lower than those at points N1′ to N3′. In contrast, the total sound pressure levels at the under-beam noise measurement points N5 and N6 are 10 to 16 dB higher than those at points N5′ and N6′.

In summary, there is a difference in the sound source intensity between the two sections, with the section without a sound barrier exhibiting a greater sound source intensity.

3.3. Analysis of Noise Reduction Effect of the Sound Barrier at Different Speeds

From Section 3.2, it is understood that the sound source intensities differ between the section without a sound barrier and the section with a vertical sound barrier; therefore, it is not appropriate to directly compare the sound pressure levels at the corresponding measurement points of the two sections to study the noise reduction effect of the sound barrier at different speeds. To address this, this section first calculates the transmission loss from the track-side noise measurement point N1 (N1′) to the noise measurement points located 7.5 m and 25 m from the centerline of the track at different speeds. Then, by comparing the transmission losses of the two sections, the noise reduction effect of the vertical sound barrier at different speeds is analyzed.

Figures 10 and 11 present the transmission loss spectrum curves for noise measurement points N1-N8, N9, N13, and N14 in the cross-section without a sound barrier, as well as points N1′-N8′, N9′, N13′, and N14′ in the cross-section with a vertical sound barrier, under different train speed conditions. From Figures 10 and 11, it can be observed that the transmission loss spectrum curves for noise measurement points N1 (N1′), in both the cross-section without a sound barrier and the cross-section with a vertical sound barrier, located 7.5 m and 25 m from the track centerline, show similar trends with respect to frequency. As the train speed increases, the peak transmission loss values for both cross-sections also increase. At a speed of 20 km/h, the peak transmission loss frequency for the cross-section without a sound barrier occurs at a high frequency of 3150 Hz. However, at speeds of 40 km/h, 60 km/h, and 80 km/h, the peak transmission loss frequency shifts to a low frequency of 20 Hz. For the cross-section with a vertical sound barrier, the peak transmission loss frequency occurs at high frequencies between 2000 Hz and 2500 Hz at speeds of 20 km/h and 40 km/h. At speeds of 60 km/h and 80 km/h, the peak transmission loss frequency shifts to a low frequency of 20 Hz.

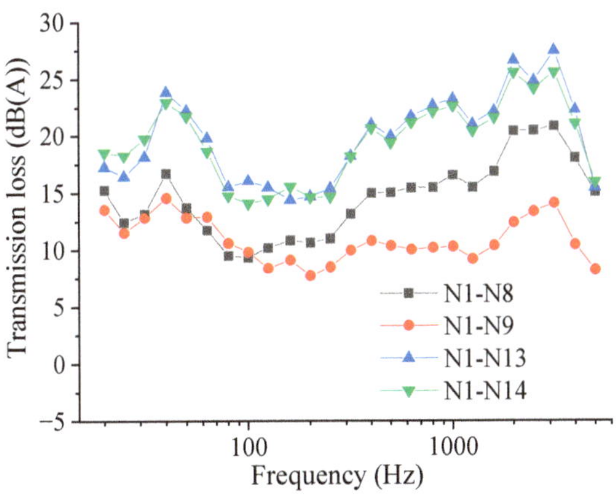

(a)

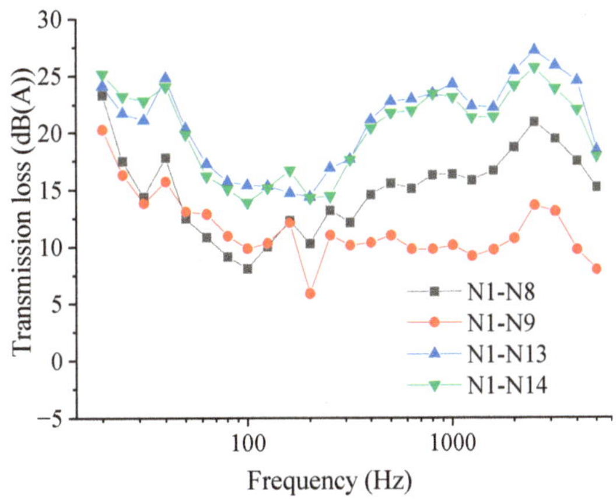

(b)

Figure 10. *Cont.*

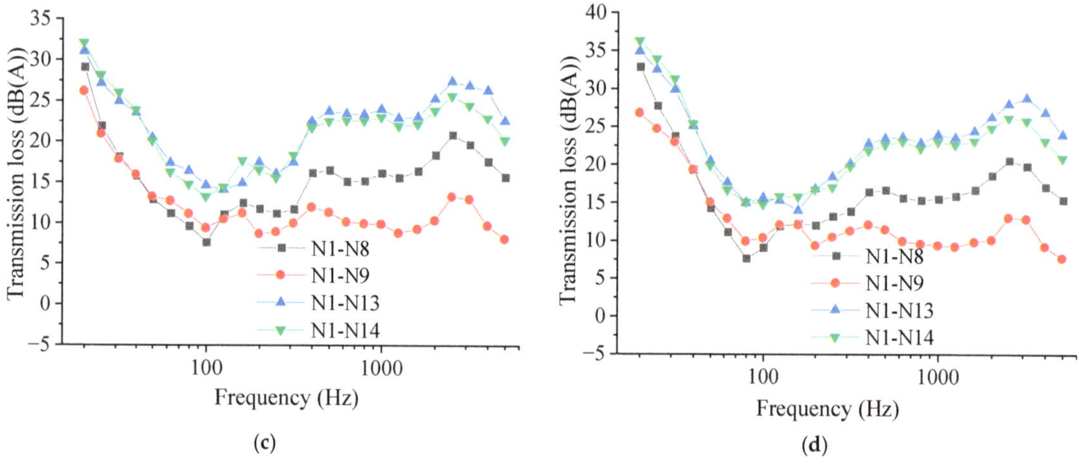

Figure 10. Transmission loss spectrum of the section without a sound barrier: (**a**) 20 km/h; (**b**) 40 km/h; (**c**) 60 km/h; (**d**) 80 km/h.

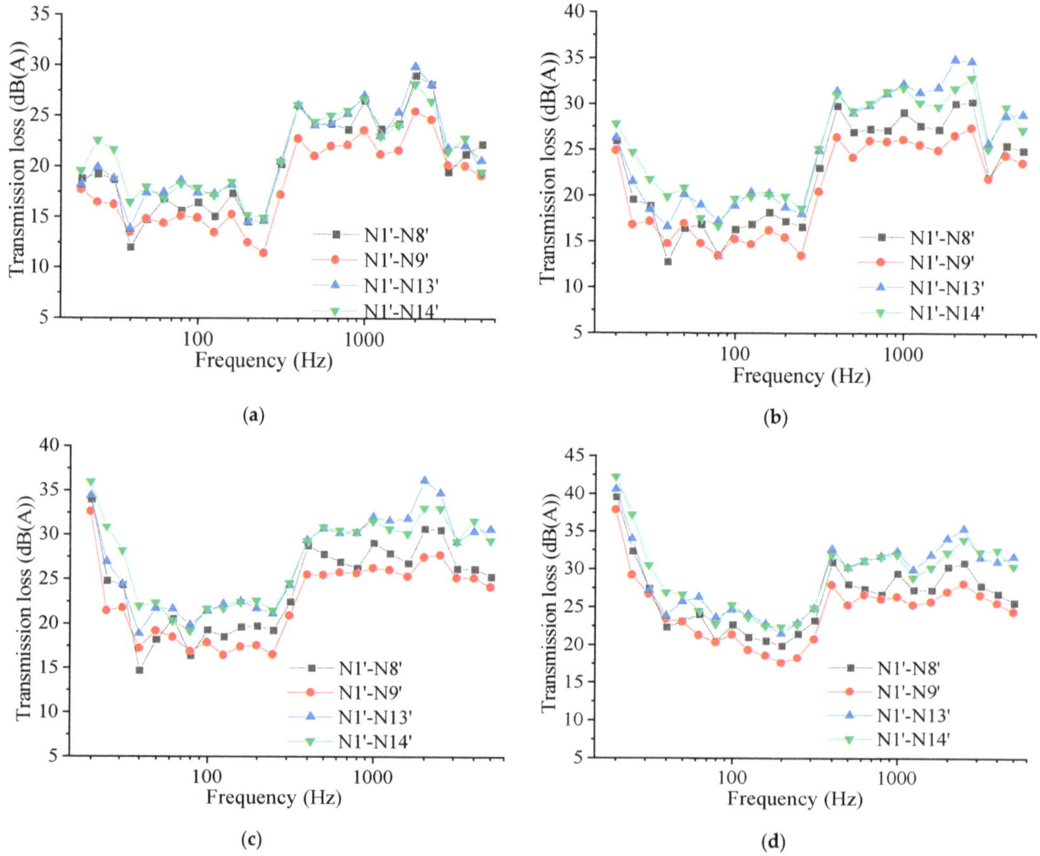

Figure 11. Transmission loss spectrum of the section with a vertical sound barrier: (**a**) 20 km/h; (**b**) 40 km/h; (**c**) 60 km/h; (**d**) 80 km/h.

Figure 12 presents the insertion loss spectra for noise measurement points N8, N9, N13, and N14 under different train speed conditions. From Figure 12, the following observations can be made: The insertion loss spectra curves exhibit similar trends with respect to frequency across different train speeds. The insertion loss initially increases and then decreases as the train speed increases, with the rate of decrease slowing down as the speed increases. In the frequency range of 20~63 Hz, the insertion loss values can be either positive or negative under different speed conditions; however, in the frequency range of 63~5000 Hz, the insertion loss values are predominantly positive, indicating that the vertical sound barrier effectively reduces mid-to-high-frequency noise but may slightly increase noise in the low-frequency range below 63 Hz. This phenomenon is likely due to the vibration of the vertical sound barrier itself, which radiates secondary structural noise when the train passes over the elevated bridge.

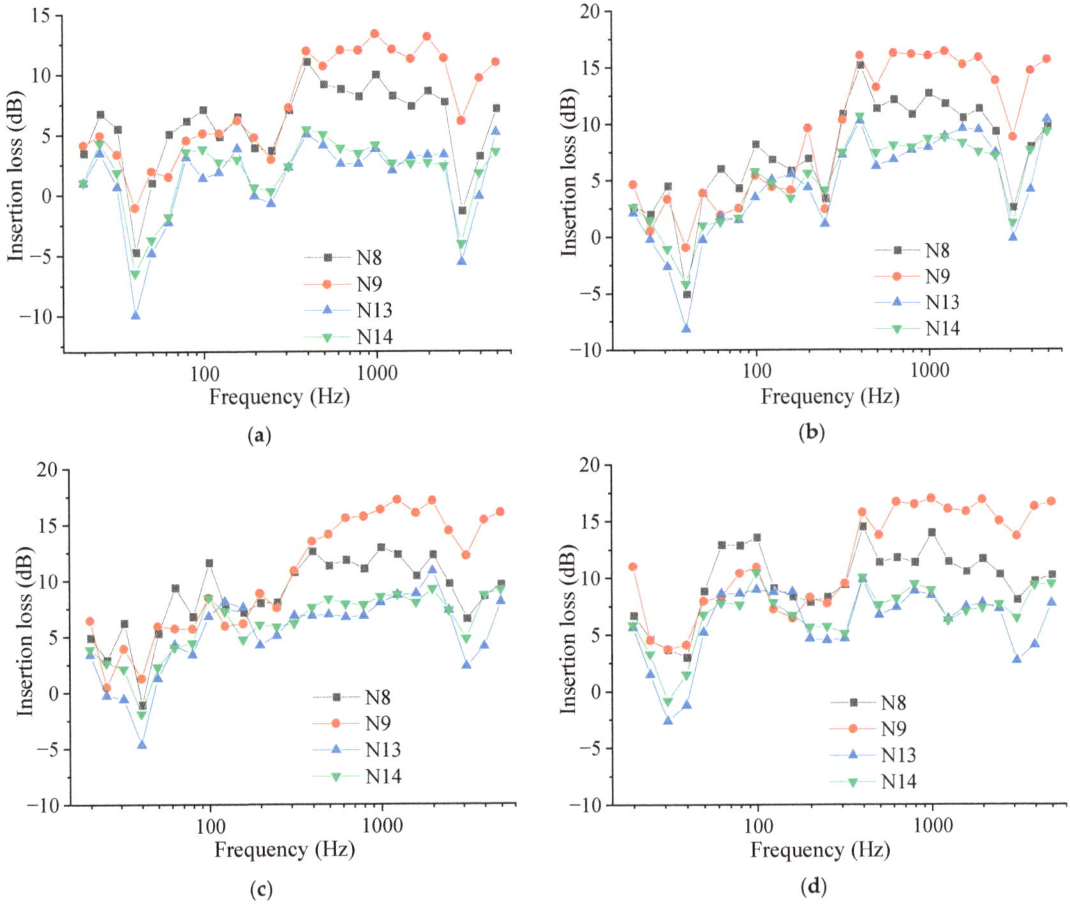

Figure 12. Insertion loss spectrum: (**a**) 20 km/h; (**b**) 40 km/h; (**c**) 60 km/h; (**d**) 80 km/h.

To further quantitatively analyze the noise reduction effect of the vertical sound barrier at different train speeds, Figure 13 presents the insertion loss bar charts for noise measurement points N8, N9, N13, and N14 under different speed conditions. From Figure 13, the following conclusions can be drawn: When the train operates at a speed of 20 km/h, the insertion loss at measurement points located 7.5 m from the outer track centerline ranges from 6.5 to 9.0 dB, while at points 25 m from the outer track centerline, the insertion loss ranges from 1.5 to 2.5 dB. When the train operates at a speed of 40 km/h, the insertion loss at measurement points located 7.5 m from the outer track centerline ranges from 8.5 to 10.5 dB, while at points 25 m from the outer track centerline, the insertion loss ranges from 6.0 to 6.5 dB. When the train operates at a speed of 60 km/h, the insertion loss at measurement points located 7.5 m from the outer track centerline ranges from 7.5 to 9.5 dB, while at points 25 m from the outer track centerline, the insertion loss ranges from 5.5 to 6.0 dB. When the train operates at a speed of 80 km/h, the insertion loss at measurement points located 7.5 m from the outer track centerline ranges from 7.5 to 10.2 dB, while at points 25 m from the outer track centerline, the insertion loss ranges from 5.0 to 6.0 dB.

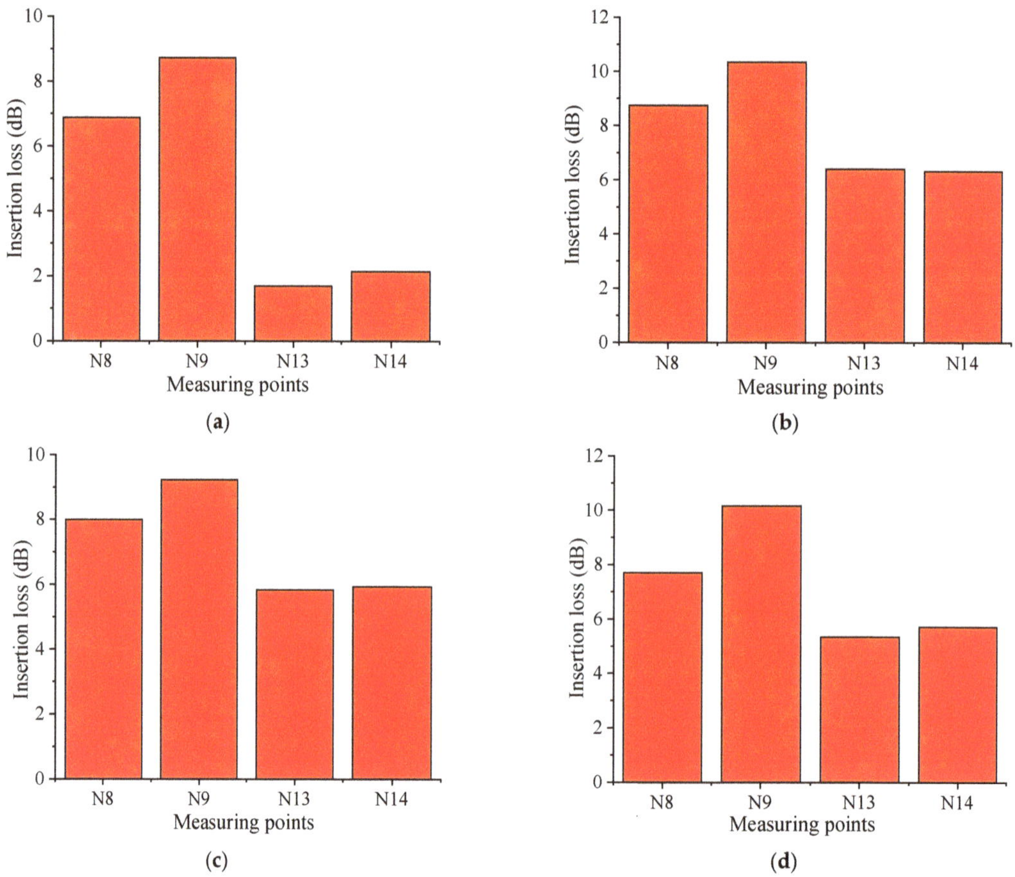

Figure 13. Insertion loss bar chart: (**a**) 20 km/h; (**b**) 40 km/h; (**c**) 60 km/h; (**d**) 80 km/h.

4. Conclusions

This paper focuses on two sections of a viaduct in an urban rail transit system: one without a sound barrier and one with a vertical sound barrier. Through on-site testing, this study evaluates track-side noise, under-beam noise, and beam-side noise for both sections. After data processing and calculations, acoustic characteristics such as the sound-pressure-level spectrum, transmission loss spectrum, insertion loss spectrum, and total sound pressure level at different speed conditions for both sections were obtained. The influence of speed on the noise of both sections was analyzed, and the noise reduction effect of the vertical sound barrier at different speeds was studied. The main conclusions of this paper are as follows:

(1) Regardless of the presence of a vertical sound barrier, the peak frequency of the linear sound pressure level for the track-side noise measurement point N1 (N1′), which is significantly affected by wheel–rail noise, is in the mid-to-high frequency range. The peak frequency for the under-beam noise measurement point N5 (N5′), which is greatly influenced by bridge structural noise, is in the low-frequency range. The beam-side noise measurement point N8 (N8′), which is affected by both wheel–rail noise and secondary structural noise, has peak frequencies in both the mid-to-high and low-frequency ranges.
(2) When train speed varies between 20 to 80 km/h under both sound barrier and non-barrier configurations, comparatively, the overall sound pressure levels at the track-side and bridge-side measurement points demonstrate stronger linear correlations with train speed than those observed at the under-bridge noise measurement points.
(3) In the section without a sound barrier, for every 10 km/h increase in speed, the linear total sound pressure level at noise measurement points N1, N5, and N8 increases by approximately 2.3 dB, 1.3 dB, and 1.8 dB, respectively. In the section with a vertical sound barrier, for every 10 km/h increase in speed, the linear total sound pressure level at noise measurement points N1′, N5′, and N8′ increases by approximately 3.0 dB, 1.5 dB, and 1.6 dB, respectively.
(4) The vertical sound barrier has an effective noise reduction effect on mid-to-high frequency noise, but there is an increase in noise in the low-frequency range between 20~63 Hz, possibly due to the self-vibration of the sound barrier caused by the train passing over the viaduct, which radiates some secondary structural noise.
(5) At speeds of 20 km/h, 40 km/h, 60 km/h, and 80 km/h, the insertion loss at each noise measurement point located 7.5 m from the outer track centerline ranges from 6.5 to approximately 9.0 dB, 8.5 to 10.5 dB, 7.5 to 9.5 dB, and 7.5 to 10.2 dB, respectively. At 25 m from the outer track centerline, the insertion loss ranges from 1.5 to approximately 2.5 dB, 6.0 to 6.5 dB, 5.5 to 6.0 dB, and 5.0 to 6.0 dB, respectively.
(6) The noise reduction performance of vertical sound barriers exhibits an initial increase, followed by a gradual decrease with rising train speeds, with the rate of decrease diminishing progressively. This phenomenon may be attributed to enhanced low-frequency structure-borne noise radiation induced by wheel–rail contact roughness excitation at elevated speeds, which partially offsets the effectiveness of high-frequency noise attenuation.

This research will provide a reference and basis for setting speed limits in rail transit sections equipped with sound barriers. In the future, theoretical analysis, numerical simulation, and experimental research will be combined to conduct more in-depth studies, so as to provide scientific references and basis for the design and selection of sound barriers.

Author Contributions: Conceptualization, L.S. and Q.L.; methodology, L.S.; software, Y.Z.; formal analysis, Y.Z.; writing—original draft preparation, L.S. and Y.Z.; writing—review and editing, Y.L. and R.B.; project administration, L.S. and Q.L.; funding acquisition, L.S. and Q.L. All authors have read and agreed to the published version of the manuscript.

Funding: This research study was funded by the National Natural Science Foundation of China, grant numbers 52378450 and 52372328; the China Postdoctoral Science Foundation, grant numbers 2023T160214 and 2023M731077; the Natural Science Foundation of Jiangxi Province, grant numbers 20232BAB204087 and 20224BAB214074; and the Postgraduate Innovation Special Fund Project of Jiangxi Province, grant Number YC2023-S541.

Data Availability Statement: The data presented in this study are available upon request from the corresponding author.

Conflicts of Interest: The authors declare no conflicts of interest.

References

1. Hou, X.F.; Feng, C.; Yan, H.M.; Zuo, C. Overview of urban rail transit operations in China's mainland in 2024. *Urban Rapid Rail Transit.* **2025**, *38*, 13–19.
2. Qiu, Y.; Zheng, B.K.; Jiang, B.L.; Jiang, S.; Zou, C. Effect of non-structural components on over-track building vibrations induced by train operations on concrete floor. *Int. J. Struct. Stab. Dyn.* **2025**. [CrossRef]
3. Zhang, X.Y.; Thompson, D.J.; Quaranta, E.; Squicciarini, G. An engineering model for the prediction of the sound radiation from a railway track. *J. Sound Vib.* **2019**, *461*, 114921. [CrossRef]
4. Sheng, X.Z.; Cheng, G.; Thompson, D.J.; Ge, S. Research progress on wheel-rail noise prediction models. *J. Traffic Transp. Eng* **2021**, *21*, 20–38.
5. Zou, C.; Li, X.; He, C.; Zhou, S.H. An efficient method for estimating building dynamic response due to train operations in tunnel considering transmission path from source to receiver. *Comput. Struct.* **2024**, *305*, 107555. [CrossRef]
6. Li, X.; Zheng, B.; Chen, Y.; Zou, C. A hybrid methodology for estimating train-induced rigid foundation building vibrations. *Constr. Build. Mater.* **2025**, *460*, 139852. [CrossRef]
7. Tao, Z.; Zhang, D.; Tu, D.; He, L.; Zou, C. Prediction of train-induced ground-borne vibration transmission considering parametric uncertainties. *Probabilistic Eng. Mech.* **2025**, *79*, 103731. [CrossRef]
8. Song, X.D.; Wu, D.J.; Li, Q.; Botteldooren, D. Structure-borne low-frequency noise from multi-span bridges: A prediction method and spatial distribution. *J. Sound Vib.* **2016**, *367*, 114–128. [CrossRef]
9. He, Y.P.; Zhang, Y.; Yao, Y.Y.; He, Y.L.; Sheng, X.S. Review on the prediction and control of structural vibration and noise in buildings caused by rail transit. *Buildings* **2023**, *13*, 2310. [CrossRef]
10. Shi, J.W.; Zhang, J.Y.; Li, T. Numerical investigation on aerodynamic noise source identification and far-field noise characteristics of the high-speed train bogie region. *Acoust. Aust.* **2024**, *52*, 375–391. [CrossRef]
11. Wang, B.; Zhang, W.; Wang, F.; Zhao, Q.J.; Cao, C.K.; Gao, Y. Experimental investigation on aerodynamic and noise characteristics of Fenestron. *Chin. J. Aeronaut.* **2023**, *36*, 88–101. [CrossRef]
12. Kang, H.M.; Kim, C.W.; Cho, T.H.; Kim, D.H.; Yoon, S.H.; Kwon, H.B. Investigation for the aerodynamic characteristics of high speed train pantograph with cover. *J. Comput. Fluids Eng.* **2012**, *17*, 18–24. [CrossRef]
13. Li, R.P.; Zhou, N.; Zhang, W.H.; Mei, G.M.; Chen, Z.B. Calculation and analysis of pantograph aerodynamic uplift force. *J. China Railw. Soc.* **2012**, *34*, 26–32.
14. He, J.L.; Wang, Q.; Jiang, W.K. Analysis of the elevated urban rail transit noise. *Urban Rail Transit Res.* **2007**, *10*, 57–60. [CrossRef]
15. Jin, J.Y.; Zhu, C.Y.; Wu, R.; Liu, Y.M.; Li, M. Comparative noise reduction effect of sound barrier based on statistical energy analysis. *J. Comput. Methods Sci. Eng.* **2021**, *21*, 737–745. [CrossRef]
16. Lee, J.; Lee, H.; Kim, I. Reduction effects of shaped noise barrier by composition with adsorbed filter panel. *Int. J. Smart Hom* **2016**, *10*, 193–202. [CrossRef]
17. Kim, H.I.; Ahn, S.H.; Park, B.J.; Kim, S.G.; Lee, J.H.; Ko, J.H. The analysis of effect of surface shape and pattern of noise barrier panel on the noise reduction. *J. Korean Soc. Environ. Eng.* **2012**, *34*, 835–839. [CrossRef]
18. Li, Q.T.; Liu, Y.; Luo, Y.Y. Noise reduction prediction of a fully-enclosed sound barrier in urban rail transit systems considering incoherence of sources. *Tongji Daxue Xuebao/Tongji Univ.* **2023**, *51*, 117–125.
19. Wang, C.L.; Zhang, L.T.; Song, L.Z.; Wu, X.L. Predication and analysis of noise reduction of near-rail sound barriers on suburban railways. *J. Noise Vib. Control.* **2022**, *42*, 187–191+278.
20. Xin, S.Y.; Zhang, S.F.; Wang, X.W. Design and research on noise reduction system of fully enclosed sound barrier for beijing xiong'an intercity railway. *Railw. Stand. Des.* **2022**, *66*, 163–168.
21. Li, X.Z.; Zhao, Q.C.; Zhang, X.; Yang, D.W. Field test and analysis of noise reduction performance of high-speed railway semi-closed sound barriers. *J. Southwest Jiaotong Univ.* **2018**, *53*, 661–669+755.
22. Sun, W.J.; Liu, L.; Yuan, H.L.; Su, Q.P. Influence of top shape on noise reduction effect of high-speed railway noise barrier. *IOP Conf. Ser. Mater. Sci. Eng.* **2019**, *493*, 012043. [CrossRef]
23. Sun, W.J.; Su, Q.P.; Yuan, H.L. Analysis of influencing factors on noise reduction effect of high-speed railway sound barrier. *IOP Conf. Ser. Mater. Sci. Eng.* **2019**, *493*, 012042. [CrossRef]
24. Jung, S.S.; Kim, Y.T.; Cheong, C.U.; Kim, H.C.; Lee, W.S. Noise barrier with top cylindrical tubes. *J. Korean Phys. Soc.* **2006**, *4*, 145–149.
25. Zhang, X.; Liu, R.; Cao, Z.Y.; Wang, X.Y.; Li, X.Z. Acoustic performance of a semi-closed noise barrier installed on a high-speed railway bridge: Measurement and analysis considering actual service conditions. *Measurement* **2019**, *138*, 386–399. [CrossRef]

6. Liu, Y.; Yang, B.; Zhang, P.X.; Zhong, F.Z.; Dong, X.H.; Sun, Y.H. Test and Analysis of Noise Reduction Effect for Different Heights and Distances of Sound Barrier. *J. Adv. Mater. Res.* **2011**, *1269*, 4447–4450. [CrossRef]
7. Song, X.D.; Li, Q. Numerical and experimental study on noise reduction of concrete LRT bridges. *Sci. Total Environ.* **2018**, *643*, 208–224. [CrossRef]
8. *ISO 3095: 2013*; Railway Applications-Acoustics-Measurement of Noise Emitted by Railbound Vehicles. International Organization for Standardization: Geneva, Switzerland, 2013.

buildings

MDPI

Article

Impact of Urban Elevated Complex Roads on Acoustic Environment Quality in Adjacent Areas: A Field Measurement Study

Guangrui Yang [1], Lingshan He [2], Yimin Wang [1,*] and Qilin Liu [3]

[1] School of Civil Engineering and Transportation, South China University of Technology, Guangzhou 510641, China; 202310181797@mail.scut.edu.cn
[2] Guangzhou Urban Planning & Design Survey Research Institute Co., Ltd., Guangzhou 510060, China; zacharyhe@foxmail.com
[3] School of Economics and Management, Beijing Jiaotong University, Beijing 100044, China; 24126674@bjtu.edu.cn
* Correspondence: ctymwang@scut.edu.cn

Abstract: The current focus of urban environmental governance is on the traffic noise pollution caused by road transportation. Elevated complex roads, defined as transportation systems comprising elevated roads and underlying ground-level roads, exhibit unique traffic noise distribution characteristics due to the presence of double-decked roads and viaducts. This study conducted noise measurements at two sections of elevated complex roads in Guangzhou, including assessing noise levels at the road boundaries and examining noise distribution at different distances from roads and building heights. The results show that the horizontal distance attenuation of noise in adjacent areas exhibits no significant difference from that of ground-level roads, but substantial discrepancies exist in vertical height distribution. The under-viaduct space experiences more severe noise pollution than areas above the viaduct height, and the installation of sound barriers alters the spatial distribution trend of traffic noise. Given that installing sound barriers solely on elevated roads is insufficient to improve the acoustic environment, systematic noise mitigation strategies should be developed for elevated composite road systems. Additionally, the study reveals that nighttime noise fluctuations are significantly greater than those during the day, further exacerbating residents' noise annoyance.

Keywords: traffic noise pollution; elevated complex roads; field measurement; distribution characteristics

1. Introduction

Noise pollution has emerged as one of the most pressing environmental issues attributable to road traffic [1]. Recent research in the field of modern medicine has unequivocally established that traffic noise originating from urban roads poses a substantial risk to human health [2,3]. The primary threats associated with traffic noise include an increased risk of cardiovascular diseases [4–6], diabetes, and hypertension [7], as well as the development of depression [8,9], sleep disorders [10,11], and adverse effects on children's cognitive function [12,13]. Furthermore, recent studies have also established a link between road traffic noise and the prevalence of obesity [14].

With the development of cities, the relentless expansion of road networks, aimed at accommodating the increasing volume of traffic, has led to a substantial increase in noise levels in many urban areas, severely disrupting residents' daily lives [15–17]. Especially in densely populated Asian cities, the high density of traffic and towering building patterns make traffic noise a potentially more significant disturbance for residents. Using three-dimensional technology and the ISO 9613-2 method, Brown et al. [18] conducted a traffic noise simulation across Hong Kong. The findings indicated that the percentage of the Hong Kong population exposed to noise levels exceeding 70 dB(A) was comparable to that in Europe. However, in comparison to European cities, a significantly higher percentage of the Hong Kong population was found to be exposed to noise levels between 60 dB(A) and 64 dB(A). Consequently, as the primary source of noise in urban areas [19], road traffic noise has attracted significant attention from city managers.

Accurate assessment of citizens' noise exposure is essential to inform evidence-based decision-making [20]. To this end, countries including the European Union, the United Kingdom, the United States, China, and Germany have successively introduced noise prediction models. These models incorporate both emission and propagation modules, enabling researchers to leverage non-acoustic data to forecast the spatial distribution of road traffic noise [21]. They have been validated for application in urban road networks, yet the noise sources of the validation datasets are predominantly from surface roads that are consistent with the road types utilized in model development [22–26]. The study by Melo et al. [27] proposes that the road noise prediction model is suitable for estimating urban roads with similar traffic characteristics and road layouts. In addition, the spatial structure of buildings in high-density cities also influences the propagation and distribution of noise [28]. Huang et al. [29] conducted field tests on noise in high-rise buildings along highways and discovered that the FHWA model underestimated the height of the maximum noise level on high-rise buildings compared to actual conditions. Therefore, conducting on-site tests for traffic noise in typical areas with special road types and exploring the distribution of noise, is significant for scientifically assessing the impact of traffic noise in high-density urban and designing noise reduction measures.

Elevated roads represent the primary form of urban transportation expansion aimed at enhancing traffic efficiency. In major cities, elevated roads are often constructed above ground-level roads parallel to their alignment, thus forming an elevated complex road system. The complex arrangement of double-story noise emission sources and viaducts gives rise to alterations in traffic noise propagation patterns [30]. However, no studies have systematically addressed noise propagation in elevated complex road systems. Furthermore, noise barriers are effective noise reduction measures for elevated roads [31], but their application in elevated complex roads still lacks post-evaluation.

The main objective of this study is to deeply investigate the impact of traffic noise generated by elevated complex roads on the sound field distribution in adjacent areas through field tests and to assess the effectiveness of noise mitigation measures. Additionally, this study discusses the temporal variations in traffic noise during different periods of urban traffic operation. Through this study, we hope to offer valuable insights into noise monitoring assessment and reduction strategies for existing road structures of this kind, while also serving as a reference for the construction of elevated roads in other regions, particularly in emerging countries that are experiencing rapid growth in motor vehicle traffic and urban development.

2. Methods

2.1. Location

The analysis data was collected in the vicinity of elevated complex roads in Guangzhou. The city boasts a permanent resident population density of 2532 individuals per square kilometer, with the central urban area reaching a staggering 22,592 people per square kilometer. Alongside this dense population, Guangzhou has constructed an extensive network of 15,017 km of roads and 2234 viaducts [32]. This robust road transportation system, coupled with the dense population distribution, has significantly increased the exposure risk of citizens to traffic noise [33–35].

In the urban core of Guangzhou, two elevated complex road sections were selected as test areas, Test Area #1 and Test Area #2. The selection criteria for the test area are as follows: the primary noise source in the area is road noise generated by traffic flow on elevated complex roads, and the area contains a high-rise building that is affected by noise interference. Additionally, there are no other obvious obstacles between the building and the road except for the sound barrier to be studied.

The ground-level road under the viaduct in Test Area #1 was built in 1974 and serves as a key arterial road. It has four lanes in both directions (17 m wide), serving as a vital link between districts and primarily fulfilling traffic functions. The elevated road, the first urban viaduct in Mainland China, was built in 1987 and stands 5.5 m high. It has two lanes in both directions (9 m wide). Measurement points are set at the junction of the ramp and the main road, where the ramp is 4 m wide.

To analyze the impact of traffic noise at different heights, a road-facing building was selected for noise measurement in each test area: Building #1 (for Test Area #1) and Building #2 (for Test Area #2). Building #1 is a six-story residential building (2.6 m per floor). Floors 1 and 2 are below the viaduct, floors 4 to 6 are above it, and floor 3 is level with the viaduct structure. Building #1 is 7 m from the ground-level road and 3 m from the elevated road. An overview of Test Area #1 is shown in Figure 1.

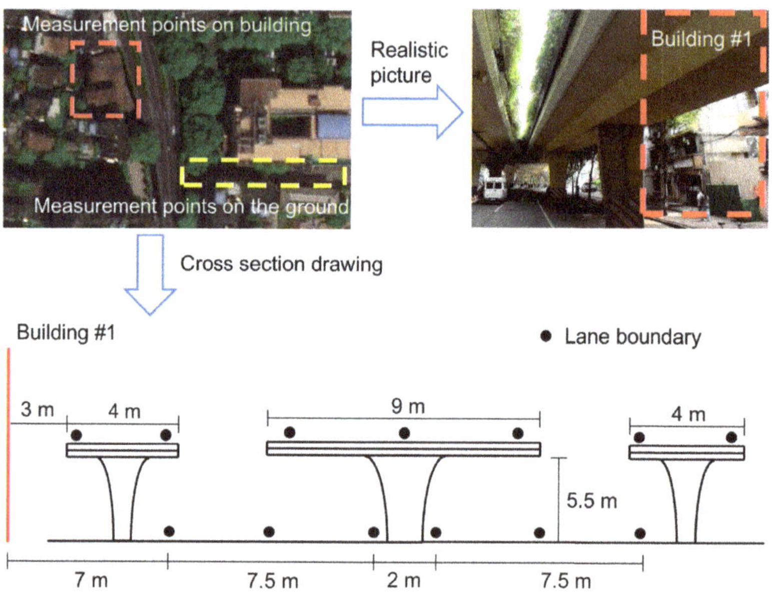

Figure 1. Geographical location of Test Area #1 (satellite image source: Baidu maps).

The ground-level road of Test Area #2 is categorized as an arterial road in the urban road hierarchy. It has eight lanes in both directions and is 33 m wide. The elevated road above is an expressway, designed to ensure smooth traffic flow and enhance urban transportation efficiency. The expressway has eight lanes in both directions (32 m wide), with 3-meter-high arc-shaped top vertical noise barriers on both sides. The viaduct in Test Area #2 is a separate elevated structure, with varying heights for different driving directions: 14 m for east-to-west traffic and 10 m for west-to-east. The road surfaces in both test areas are paved with asphalt.

Building #2 is an eight-story commercial hotel. The first floor is 3.5 m high, and the remaining floors are 2.8 m each. Floors 1 to 5 are located below the viaduct, floors 7 and 8 are above the viaduct, and floor 6 is at the same elevation as the viaduct structure. Building #2 is 8 m from the ground-level road and 7 m from the elevated road. An overview of Test Area #2 is shown in Figure 2.

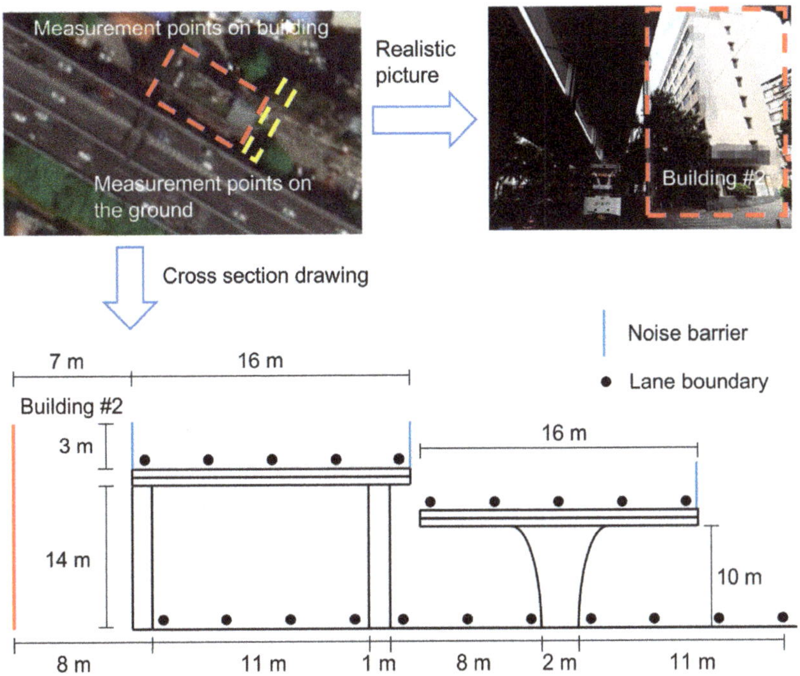

Figure 2. Geographical location of Test Area #2 (satellite image source: Baidu maps).

2.2. Measurement Scheme

Equipment used in the measurement process includes the RION NL-52A (precision sound level meter for noise signal acquisition, manufactured by RION Co., Ltd., Tokyo, Japan) and the RION NC-75 (acoustic calibrator for calibrating precision sound level meters, manufactured by RION Co., Ltd., Tokyo, Japan). They fulfill the requirements of the International Electrotechnical Commission for Class 1 sound level meters (IEC 61672-1:2013) and Class 1 sound calibrators (IEC 60942:2017).

A total of 19 measurement points were deployed across the two test areas to monitor traffic noise levels at varying distances from the roads and different building heights, aiming to analyze the noise attenuation characteristics with distance and variation patterns with height. In Test Area #1, five points were positioned at 0.2 m, 10 m, 20 m, 40 m, and 60 m from the ground-level road, while another five points were set up on the 1st, 2nd, 3rd, 4th, and 6th floors of Building #1. In Test Area #2, four points were arranged at 0.2 m, 5 m, 10 m, and 20 m from the ground-level road, with five additional points on the 1st, 3rd, 4th, 6th, and 8th floors of Building #2. Except for the noise barrier in Test Area #2, no obstructions exist between the measurement points and the roads, or between the measurement points themselves, that could impede noise propagation. It should be noted that the points on Building #1 were located on the road-facing facade, whereas those on Building #2 were positioned outside the side windows, 12 m from the ground-level road and 13 m from the elevated road. During the measurements, the sound level meter was mounted on 1.5 m high tripods, with a 1 m clearance from walls or windows. The microphone was oriented towards the road. The layout of the measurement points is shown in Figure 3.

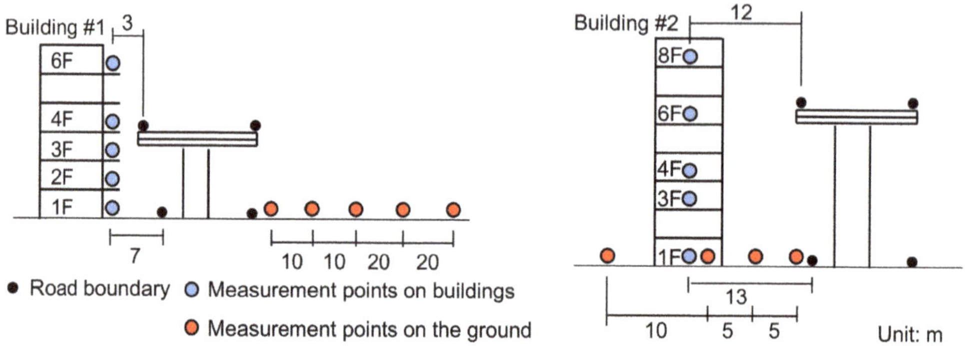

(a) Test Area #1 (b) Test Area #2

Figure 3. Layout diagram of measurement points. (**a**) shows the layout of Test Area #1; (**b**) shows the layout of Test Area #2.

To reflect the impact of traffic flow variations on noise, measurements were conducted in each test area across three time periods: 14:30–16:00 (Period 1, daytime off-peak), 17:30–19:00 (Period 2, daytime peak), and 22:00–23:30 (Period 3, nighttime). Measurements were performed in February–March under clear weather (no rain, snow, thunder, or lightning), with temperatures ranging from 20.4 °C to 27.3 °C, humidity between 40% and 68%, and wind speeds below 2.5 m/s.

Continuous measurements are critical in traffic noise studies. In accordance with Chinese standards, each measurement session lasted 20 min to capture noise levels and fluctuations under stable traffic flow conditions [36]. To ensure simultaneous measurements at all points reflecting noise distribution, multiple operators each controlled an instrument. Additionally, warning signs were placed outside the test areas to exclude construction noise, human activity noise, and other extraneous noise sources that could affect results.

The equivalent continuous A-weighted sound pressure level (L_{Aeq}), maximum sound level (L_{max}), percentile sound levels (L_{10}, L_{50}, L_{90}), and spectral information (obtained through 1/3 octave band analysis) are discussed in this study. These indicators were derived by analyzing the 20 min WAV files collected at each measurement point using the Waveform Analysis Software AS-70 (Version 2.2.0.19) developed by RION Corporation.

Videos were recorded simultaneously during the noise measurement to statistically analyze the traffic flow characteristics within the measurement period.

3. Results

3.1. *Characteristics of Traffic Noise in Urban Actual Traffic Operations*

The temporal variation and frequency components of road traffic noise are analyzed in this section. Table 1 presents the acoustic indicators of noise measured at 20 cm from the road boundary. The variation in noise levels over time is illustrated in Figure 4, where the time-domain graph is plotted based on the sound pressure levels (SPLs) recorded at 100-millisecond intervals.

Table 1. Noise indicators at road boundaries.

Test Area	Period	L_{Aeq} (dB(A))	L_{max} (dB(A))	L_{10} (dB(A))	L_{50} (dB(A))	L_{90} (dB(A))
Area #1	Period 1	67.8	80.2	70.2	67.1	63.5
	Period 2	66.9	92	68.6	65.4	62.7
	Period 3	66.1	93.2	68.9	63.1	57.7
Area #2	Period 1	77.8	96.3	79.2	76.6	74.6
	Period 2	76	92.9	77.5	75.4	73.3
	Period 3	79.9	98.6	82.6	78.1	74.7

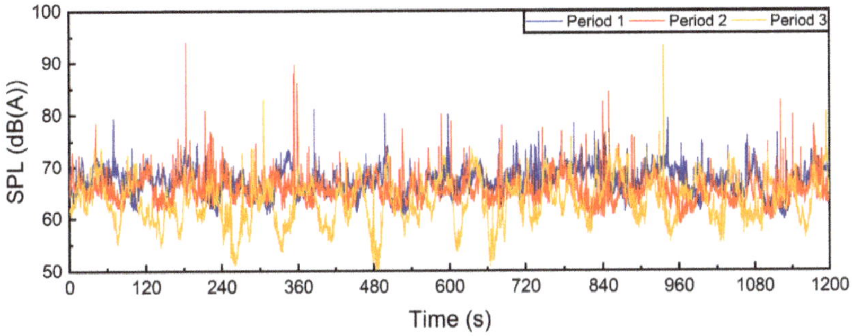

(**a**) Test Area #1

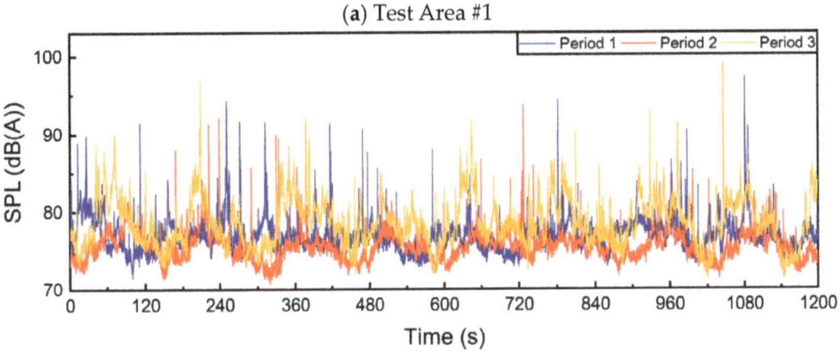

(**b**) Test Area #2

Figure 4. Traffic noise level variations at road boundaries. (**a**) shows the variations of Test Area #1; (**b**) shows the variations of Test Area #2.

Road noise, characterized by significant fluctuation, stands out as a non-steady source of disturbance. Notably, in both test areas, the noise level fluctuation during period 3 exceeds that of periods 1 and 2, and there are reasons for this variation. In Test Area 1, the nighttime background noise level was lower, and the reduction in traffic flow (from 536 in Period 1 and 272 in Period 2 to 224 veh/20min in Period 3) rendered the noise generated by passing vehicles more prominent. This suggests that, even if the L_{Aeq} of nighttime noise is lower, the more pronounced fluctuation may still cause greater annoyance to people [37]. Conversely, in Test Area #2, although the total traffic volume at night decreased compared to daytime (from 1101 in Period 1 and 1012 in Period 2 to 716 in Period 3), the noise fluctuation and level at night were higher. This is because large vehicles with a loading capacity of over 7 tons, which generate significantly louder noise, saw a dramatic increase at night (from 5 in Periods 1 and 2 to 134 veh/20min in Period 3). This surge was due to local government policies prohibiting large vehicles from entering the urban area during the day, leading to their concentrated operation at night. From the perspective of noise indicators, this fluctuation level can be reflected by the difference between the L_{90} and L_{10} values. In test area #1, the daytime noise difference ranged from 6.7 to 5.9 dB(A), whereas during the night, this difference significantly escalated to 11.2 dB(A). Similarly, in test area #2, the daytime noise fluctuation spanned from 4.6 to 4.2 dB(A), yet at night, it witnessed a rise to 7.9 dB(A).

Vehicle horn honking is the main cause of noise peaks, with honking noise levels between 80.2 dB(A) and 98.6 dB(A). Additionally, pneumatic braking of large vehicles can also cause sudden increases in sound pressure levels.

The frequency distribution of traffic noise over the 20-minute duration is shown in Figure 5. To identify the main frequency ranges of the noise, it is necessary to analyze the relative energy contribution of each frequency component, so the data were first normalized. Equation (1) is the formula for SPL superposition based on the principle of energy superposition, which can calculate the total equivalent SPL (L_{eq}) (i.e., the result of energy superposition in the frequency domain) by superimposing the equivalent SPLs (L_n) of each frequency component over the 20 min. Conversely, using Equation (2), the proportion of energy of each frequency in the total energy (P_n) can be calculated based on the total L_{eq} and the L_n of each frequency, and these calculation results are further illustrated in Figure 5.

$$L_{eq} = 10\lg\left(\sum 10^{L_n/10}\right) \quad (1)$$

$$P_n = 10^{L_n/10}/10^{L_{eq}/10} \times 100\% \quad (2)$$

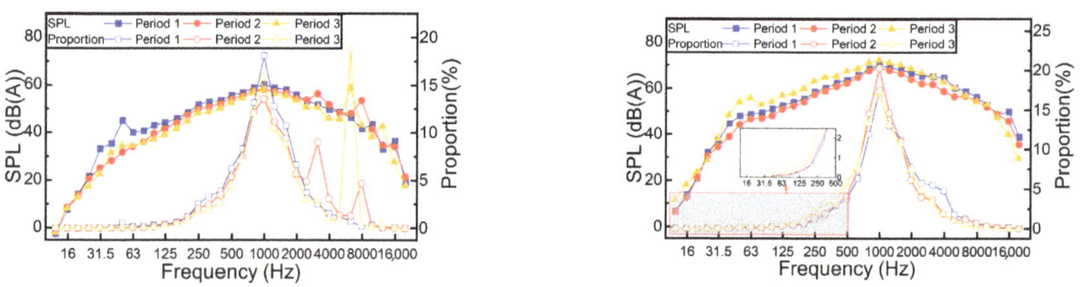

(a) Test Area #1

(b) Test Area #2

Figure 5. Frequency distribution of traffic noise. (**a**) shows the result of Test Area #1; (**b**) shows the result of Test Area #2.

The frequency distribution of traffic noise at the roadside beneath elevated complex road sections closely matches that observed in other urban road types [38]. When assessed under A-weighting conditions, traffic noise predominantly falls within the frequency range of 250 to 4000 Hz, with each frequency contributing over 1% to the total sound pressure level. Notably, a peak in noise is evident at 1000 Hz, making a significant contribution to the total sound pressure level, accounting for a range of 13.6% to 20.3%. The increase in test area #1 during period 3 at 6300 Hz was attributed to sharp noises emitted by bicycle brakes near the testing site.

The low-frequency noise characteristics in the two test areas exhibit differences. Specifically, in test area #2, the sound pressure level contributions within the frequency range of 12.5 to 125 Hz are noticeably higher compared to test area #1. Additionally, during period 3 in test area #2, there is a significant increase in sound pressure level contributions in the frequency range of 12.5 to 500 Hz compared to periods 1 and 2. Conversely, the contributions in the higher frequency range of 6300 to 20,000 Hz are lower during period

3 compared to the earlier periods. The increasing number of large vehicles has resulted in a growth in low-frequency components of traffic noise, a change that is noticeable to human ears. This growth is likely attributed to the more prominent impact of powertrain and structural noises originating from large vehicles, compared to small vehicles.

3.2. Noise Distance Attenuation Under Elevated Complex Roads

Figure 6 illustrates the decrement in traffic noise L_{Aeq} values as the distance from the road escalates. The road boundary serves as the baseline reference point. The specific L_{Aeq} values obtained from the measurements are provided in Appendix A, including the data used in Sections 3.2 and 3.3. Given the prevalence of daytime human activities, it is impractical to ensure the absence of additional noise sources in the vicinity of the measurement points, which can lead to the contamination of certain measurement outcomes. Specifically, the readings recorded in Test Area #1 during period 2, precisely at 40 m and 60 m away from the road, were deemed anomalous and excluded from further analyses.

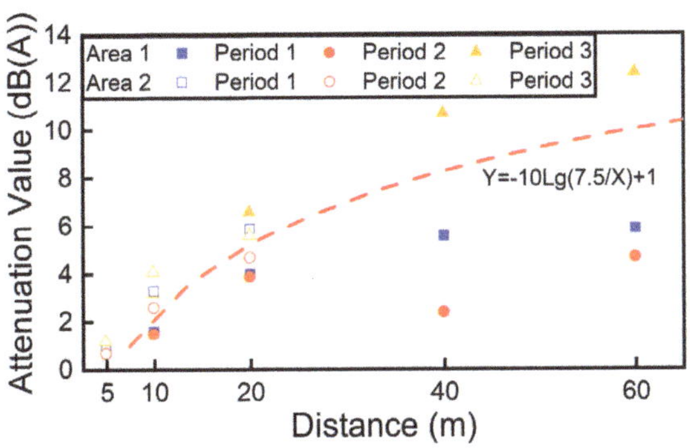

Figure 6. The field-measured attenuation of traffic noise with different distances.

Equation (3) serves as the prediction model for noise distance attenuation, specified in China's acoustic environmental guidelines, which applies to hourly traffic volumes of 300 vehicles or more per hour. The model is adapted from the FHWA prediction model, and its coefficients are calibrated based on Chinese measured data [39]. The horizontal distance between the reference point used in this equation and the noise source is 7.5 m, and '*r*' denotes the distance from the prediction point to the noise source. In practical use, the centerline of the outermost lane is typically taken as the noise emission line. Given that Chinese urban arterial roads and expressways often have lane widths of 3.5 m or 3.75 m, the position 5 m away from the road is approximately considered the reference point. According to the measured data, traffic noise attenuates by approximately 1 dB(A) at this 5 m distance. Therefore, when comparing the predicted values of Equation (3) with measured attenuation values at other distances, a correction of 1 dB(A) is applied, as indicated by the red dashed line in Figure 6.

$$\Delta_D = 10\lg(7.5/r) \tag{3}$$

Table 2 presents the differences between the actual noise distance attenuation values shown in Figure 6 and the predicted values from Equation (3). The results indicate that within 20 m, the difference is less than 2 dB(A); within 40 m, the difference is less than 3 dB(A). Therefore, Equation (3) can meet the requirements for predicting the distance attenuation of traffic noise under elevated complex roads, and it is more accurate at shorter distances.

Table 2. Comparison of the predicted values and actual measurements.

Difference (dB(A))	Distance (m)			
	10	20	40	60
Minimum	−0.7	−1.4	−1.7	−4.1
Maximum	1.9	1.3	2.4	2.4

The measurement results of period 3 were utilized to conduct a spectral analysis, from which the corresponding attenuation values were calculated (as shown in Figure 7). In the primary frequency range of traffic noise, spanning from 250 to 4000 Hz, the observed attenuation values across various frequencies generally align with those of the sound pressure level. Overall, the rate of attenuation is faster in the high-frequency range compared to the low-frequency range.

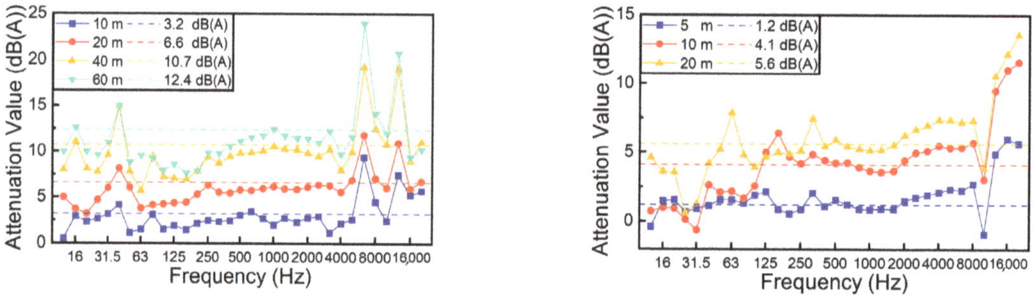

(**a**) Test Area #1 (**b**) Test Area #2

Figure 7. Attenuation of SPL in frequency bands of traffic noise with different distances (period 3). (**a**) shows the result of Test Area #1; (**b**) shows the result of Test Area #2.

3.3. The Variation in Traffic Noise with Height

As depicted in Figure 8, the distribution of traffic noise for Building #1 and Building #2 reveals significant noise impacts, particularly during nighttime. Building #1 surpasses the standard limit (70 dB(A) daytime and 55 dB(A) nighttime [36]) by a range of 9.8 to 12.3 dB(A), whereas Building #2 exceeds it by a margin of 18.8 to 20.7 dB(A). Notably, despite the installation of noise barriers, noise levels at heights above the viaduct for building #2 still exceed the limit, suggesting the need for more effective noise reduction measures.

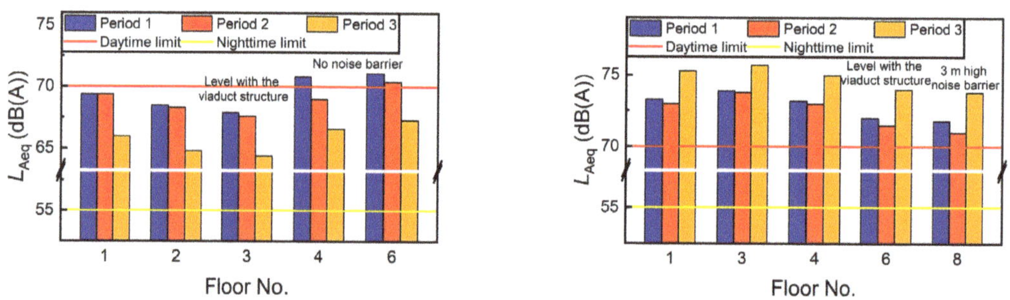

(**a**) Building #1 (**b**) Building #2

Figure 8. Measured vertical distribution of traffic noise at various buildings' floors. (**a**) shows the distribution of Building #1; (**b**) shows the distribution of Building #2.

The pattern of traffic noise distribution differs between the two buildings, which is related to the height of the viaduct and the arrangement of noise barriers. Specifically, for building #1, the traffic noise diminishes initially as one ascends the floors, reaching its nadir at the level corresponding to the viaduct's height. Then, as the floor level continues to rise, the noise level begins to increase. Conversely, building #2 experiences a rise and subsequent decline in traffic noise with increasing floor levels, peaking at the third floor.

Extensive research has concurred that traffic noise exhibits a trend of initial increase followed by a decrease with ascending height [40], stemming from the ground effect and variations in the propagation medium. The variation in noise levels above building #1's viaduct indicates that this pattern is not limited to ground-level roads, but also extends to the noise generated by vehicles on elevated roads. At the same time, the test results for building #2 indicate that the configuration of noise barriers has modified the pattern of traffic noise's variation with height. The specific causes of this phenomenon will be elaborated in Section 4.2 of the subsequent discussion.

Changes in noise levels within the under-viaduct height range are also noteworthy. Contrary to the expected trend where noise level increases with proximity to the elevated road, the mid-height under the viaduct experiences more intense traffic noise. This is partly because the noise from the ground-level road, as previously mentioned, first increases with height, and partly due to the noise reflection from the viaduct underside.

Figure 9 exhibits the frequency distribution of traffic noise across various heights during period 3. Within the primary frequency spectrum of traffic noise, ranging from 250 to 4000 Hz, the SPL variations across different frequency bands exhibit a consistent height-dependent pattern with the L_{Aeq} trend in Figure 8. Additionally, in contrast to the bridge-related low-frequency noise phenomena studied by other researchers [41,42], floors at the same elevation as the viaduct structure do not exhibit a significant rise in low-frequency noise. The lack of increase in emissions can be attributed to the relatively lower driving speeds and the lightweight characteristics of vehicles traveling on urban roads.

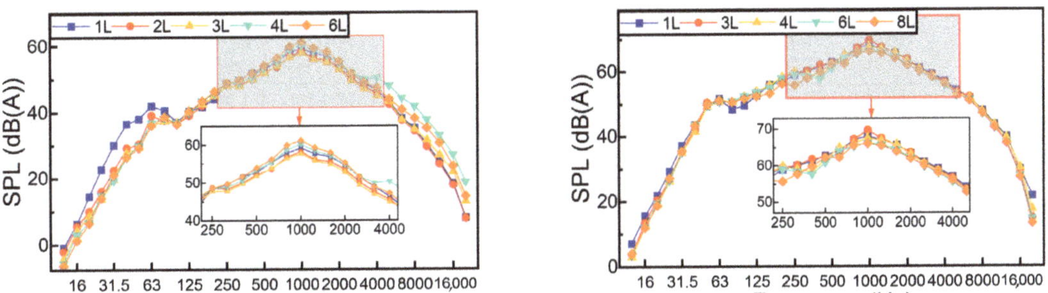

(a) Building #1 (b) Building #2

Figure 9. Frequency distribution of traffic noise at different floors of buildings (period 3). (**a**) shows the frequency distribution of Building #1; (**b**) shows the frequency distribution of Building #2.

4. Discussion

4.1. How Can We Better Address Residential Noise Nuisance at Night?

According to the complaint results of the Guangzhou Municipal Transportation Bureau, complaints about traffic noise are concentrated during the period from 22:00 to 7:00. The residents' desire for a low-noise environment during their nighttime rest and the implementation of stricter nighttime noise limits are considered as contributing factors to the concentration of complaints at night. However, our measurement data may provide an alternative explanation. Despite the reduced traffic volume at night, the sound pressure level from passing vehicles remains largely unchanged. Consequently, with a decreased level of background noise, traffic noise with similar daytime sound pressure levels becomes more noticeable. This significant fluctuation in noise is more likely to annoy residents, reduce sleep quality, and even affect children's cognitive development [37,43,44]. One possible solution is to enforce different speed limits during the day and night, as vehicle speed significantly affects the noise level generated by passing vehicles [45]. This approach, however, requires further research by considering factors such as road capacity and traffic flow, to ensure that traffic noise is reduced while maintaining smooth and safe road operations.

To ensure the safety and smooth flow of urban road traffic, many Chinese cities designate areas and time restrictions for the passage of trucks in urban areas. The periods for prohibiting truck traffic are usually during morning and evening rush hours or the daytime. However, this approach has detrimentally impacted the nighttime acoustic environment. In test area #2, although the number of small vehicles has halved at night, the number of large vehicles soared from 5 to 134 veh/20 min, resulting in a significant increase in nighttime noise levels by between 2.1 and 3.9 dB(A). In addition, more obvious low-frequency noise has been found at night when the number of large vehicles increases. Compared with medium and high-frequency noise, low-frequency noise is easier to propagate, and it is more likely to make residents feel fatigued, dizzy, emotionally volatile, and other discomforts, aggravate the annoyance suffered by residents, and reduce speech clarity [46–48].

Therefore, due to the high noise level of large vehicles, their tendency to generate low-frequency noise, and the more pronounced noise fluctuations they cause, urban freight traffic planning needs to consider traffic noise and its interference with residential areas. It is essential to demarcate the scope of densely populated residential areas, as this can maximize the avoidance of affecting residents when designing truck-prohibited routes. Freight terminals and distribution loading/unloading points with heavy truck traffic are recommended to be located away from densely populated residential areas, and special freight corridors should be designed. Ma et al. [49] designed a lane restriction system to separate trucks from passenger vehicles, which is considered to have better traffic efficiency and reduce environmental costs. Although this system does not take noise into account, it can provide a design idea for freight corridors. Emerging technologies can be used to supplement cargo transportation in urban areas and reduce noise impacts, such as combining trucks with drones for package delivery [50,51].

Furthermore, as trucks travel at lower speeds in urban areas, their powertrain noise becomes the primary noise source. This makes the noise advantage of electric trucks even more prominent. Under low-speed conditions below 50 km/h, the main noise source of electric trucks is tire/road noise, in contrast to traditional diesel trucks. Electric trucks have the potential to significantly reduce noise by over 8 dB(A) [52]. Currently, new energy vehicles (NEVs) are developing rapidly, but their applications remain primarily focused on small and medium-sized passenger cars and public transport. There is still a need for technological R&D and policy formulation to promote the upgrading and replacement of conventional trucks.

4.2. *Improvements Needed for Noise Prediction Methods in Adjacent Areas of Elevated Complex Roads*

The construction of elevated roads is an important infrastructure measure to promote urban economic growth, but it may have adverse impacts on residents' quality of life [53]. Hadi Baaj et al. [54] used the FHWA prediction model to forecast the noise impact of elevated complex roads. The results showed that the newly built elevated roads would increase vehicle speeds, thus causing greater noise pollution. Notably, this study only considered the noise levels at ground height (1.5 m above the ground), without accounting for the variation in noise with height. Wu et al. [55] used the CadnaA software to predict the three-dimensional distribution of noise from individual elevated roads and the influence of village morphology on noise propagation. For the three-dimensional distribution of pollutants in the vicinity of elevated complex roads, Lu et al. [56] analyzed the diffusion patterns of air pollutants such as PM_1 and CO based on measured data and found a bimodal distribution in the vertical profile. Nonetheless, this study did not focus on noise distribution, and our study fills this gap.

In Section 3.2, the study compared the differences between the measured and predicted values of distance attenuation for road traffic noise on the horizontal plane. Within a range of 40 m, existing prediction methods can meet practical needs with an error within 3 dB(A). However, for the noise levels at different floors of buildings, the existing road noise prediction models struggle to accurately predict the vertical distribution of noise in the areas adjacent to elevated complex roads.

In China's road noise prediction model, the noise levels at prediction points are obtained by superimposing the calculation results of the two roads, without accounting for noise reflection from the overpass bottom, which leads to inaccurate predictions. This is exemplified by our simulation using CadnaA (Version 2021 MR1) software for double-deck road noise distribution: the predicted noise level in the under-viaduct space increased continuously with height, which is inconsistent with the noise results shown in Figure 8. Noise reflection is a complex propagation process, where the viaduct height (height of the reflection surface), the width of the ground-level road (distribution of sound sources), the relative position between the viaduct and the ground-level road (relative position between the reflection surface and sound sources), and the distance between the building and the viaduct/ground-level road all affect the noise distribution on the building. Although our survey results did not provide a specific calculation method, they proved that the maximum noise height appears in the middle of the under-viaduct height, rather than increasing continuously with height.

Another factor to consider is that for the above-viaduct space of elevated complex roads without sound barriers, road noise first surges and then continues to increase slightly with height, as shown in the test results of the 3rd, 4th, and 6th floors in Figure 8a. The initial surge is due to the influence of traffic on the viaduct, while the subsequent slight increase occurs because the guardrails on both sides of the elevated road are higher than the equivalent sound source height of small vehicles, acting as a shorter sound barrier—particularly in this case, plant decorations on the guardrails further enhance the obstructive effect. According to the Huygens–Fresnel principle, the diffraction sound attenuation is related to the sound path difference. The larger the sound path difference, the greater the Fresnel number (N), and the more significant the attenuation effect. Therefore, as the height increases, the attenuation effect of the guardrails gradually decreases, and the noise level gradually increases.

Conversely, in the area with sound barriers, most of the noise in the above-viaduct space comes from the ground-level road. Due to the obstruction of the viaduct, the noise on higher floors is more significantly reduced, as shown with the 6th and 8th floors in Figure 8b. However, it is worth noting that although the height difference between the 8th and 6th floors is 6 m, the noise level difference between them is less than 0.5 dB(A). This is because, as analyzed earlier regarding the impact of sound path differences on barrier attenuation, higher floors, while experiencing less noise from ground-level roads, are subjected to more severe noise from elevated roads. Therefore, given that all measurement points were located within the acoustic shadow zones created by these barriers due to the limitation of floor height, the noise distribution at higher positions requires further investigation.

In summary, to better predict the noise distribution in the areas adjacent to elevated complex roads, it is necessary to improve existing noise prediction models by incorporating considerations of noise reflection from viaducts and the impact of guardrails.

4.3. Comprehensive Design of Multiple Noise Reduction Measures Needed

In major urban areas, noise reduction design is prioritized for the elevated sections of complex roads, as opposed to ground-level roads. So far, Guangzhou's primary urban elevated roads have installed 65,551 m of noise barriers. However, a sole focus on noise reduction design for elevated roads alone may not result in the anticipated reduction in noise levels.

Our test data reveals that despite the presence of noise barriers, the noise level on the road-facing facade of the building still surpasses 70 dB(A). The inadequacy of the barriers' noise reduction capabilities is a contributing factor, but the impact of the roads beneath viaducts also merits consideration. As traffic noise from ground-level roads rises upwards, it is influenced by the viaducts, causing a portion to reflect to the ground, thereby elevating noise levels beneath the viaducts. The maximum noise levels may be recorded within this height range. Meanwhile, the remaining noise continues to propagate upward, affecting floors higher up. Consequently, the design of noise reduction measures for ground-level roads must be taken seriously.

Low-noise asphalt pavement is widely utilized as a noise reduction measure. Practical experience in Guangzhou has confirmed the conclusions of many studies [57–60]. This type of pavement can initially reduce traffic noise by 3 to 8 dB(A) compared to traditional asphalt pavement and by over 10 dB(A) compared to cement concrete pavement. Furthermore, some innovative concrete materials (such as foamed concrete and porous concrete) have been developed [61], which have excellent sound absorption effects. A comprehensive noise reduction design for the under-viaduct space that can be adopted is to use low-noise asphalt pavement on the ground-level road to reduce the sound source level while using sound-absorbing concrete at the bottom of the bridge to reduce noise reflection. Notably, the utilization of solid waste has become a key development direction for emerging road surfaces [62,63]. The development of low-noise pavement materials using recycled materials will further promote the sustainable development of cities [64,65].

For areas with severe noise exceeding standards, noise reduction measures at the receiver (buildings) should also be considered. Acoustic treatment of building facades can effectively reduce indoor noise levels [66]. Naish et al. [67] proposed a design guide to optimize the acoustic treatment of balconies. Eggenschwiler et al. [68] found that using absorptive facades can reduce indoor noise levels, thereby decreasing residents' annoyance. Although slightly rotating the building orientation has a limited noise reduction effect, it can significantly reduce noise interference. In addition, some plant treatment methods applied to building facades can mitigate the impact of road traffic on the acoustic environment [69].

Based on the measured results, this section discusses the comprehensive noise reduction design for elevated complex roads. In addition to road traffic, the subway system is also an important component of urban three-dimensional transportation. Similar to the noise problems caused by roads, the vibration and structure-borne noise problems caused by subways also annoy residents, and some studies are dedicated to the prediction and control of these problems [70,71].

5. Conclusions

This study conducts a comprehensive analysis of the distribution and temporal variations in traffic noise in areas adjacent to elevated complex roads, based on field measurements, and the following conclusions were drawn:

- The distance attenuation of noise from elevated complex roads at ground level is not significantly different from that of general roads. By comparing the measured data with the predicted data, it is demonstrated that the distance attenuation formulas of the Chinese Sound Environment Guidelines and the US FHWA Road Noise Prediction Model remain applicable to elevated complex roads.
- Given the presence of noise barriers on many elevated roads adjacent to residential buildings, it is expected that the maximum noise levels will be concentrated within the height range underneath the viaduct. The reflection of noise from the viaduct structure results in the maximum noise level underneath the viaduct being in the middle of the viaduct height. In cases where there are no noise barriers, the noise level on the viaduct initially shows an increasing trend with height. However, with the installation of noise barriers, this upward trend is reversed and the noise level within the acoustic shadow zone gradually decreases as height increases.
- Within the frequency range of 250 to 4000 Hz, the sound pressure levels exhibit consistent variations with the total sound pressure level. Additionally, due to the lower speeds and lighter weights of urban road vehicles, little low-frequency noise pollution caused by viaduct vibration has been observed.

In our study, the revealed noise distribution pattern updates the understanding of road noise propagation in areas adjacent to elevated complex roads. This unique propagation law helps planners design quieter three-dimensional road traffic systems and adopt more targeted comprehensive noise reduction measures. Meanwhile, the research results provide key references for the improvement direction of future noise prediction models, especially by incorporating the structural parameters of viaducts and the acoustic effects of guardrails. The discussions on nighttime noise fluctuations and low-frequency noise from heavy vehicles are also key points of this paper, through which we hope to inspire more acoustically friendly urban traffic organization strategies. Overall, these insights are particularly applicable to medium and large cities with dense populations, heavy traffic, and urgent needs for three-dimensional road systems, while providing a reference for noise management in other urban environments.

Author Contributions: Conceptualization, G.Y. and Y.W.; methodology, G.Y. and L.H.; formal analysis, G.Y. and L.H.; investigation, G.Y. and Q.L.; data curation, Q.L.; visualization, G.Y. and Q.L.; writing—original draft, G.Y.; writing—review and editing, L.H. and Y.W.; supervision, Y.W.; validation, L.H.; resources, Y.W. All authors have read and agreed to the published version of the manuscript.

Funding: This research was funded by Collaborative Innovation Center for Natural Resources Planning and Marine Technology of Guangzhou grant number No.2023B04J0301 and No.2023B04J0326.

Data Availability Statement: All data used in this study are available from the corresponding author upon request.

Acknowledgments: The authors would like to acknowledge the contribution of Ke Li, Rongjing Zhang, Minghao Xiao, Zixiong Shen and Desi Tu for conducting the in situ measurements and surveys. Many thanks for the support of Academic Specialty Group for Urban Sensing in Chinese Society of Urban Planning.

Conflicts of Interest: Author Lingshan He was employed by the company Guangzhou Urban Planning & Design Survey Research Institute Co., Ltd. The remaining authors declare that the research was conducted in the absence of any commercial or financial relationships that could be construed as a potential conflict of interest.

Appendix A

The Appendix provides the SPL test results and points locations of each measuring point on the horizontal plane at ground height (Table A1 for Section 3.2) and each measuring point on different floors of the building (Table A2 for Section 3.3).

Table A1. SPL test results and locations of horizontal measuring points at ground height.

Test Area	Period	Height of Measurement Point Above Ground (m)	Distance from Road Boundary (m)	L_{Aeq} (dB(A))
Area #1	Period 1	1.5	0.2	67.8
			10	66.2
			20	63.8
			40	62.2
			60	61.9
	Period 2	1.5	0.2	66.9
			10	65.4
			20	63
			40	64.5 (excluded)
			60	62.2 (excluded)
	Period 3	1.5	0.2	66.1
			10	62.9
			20	59.5
			40	55.4
			60	53.7
Area #2	Period 1	1.5	0.2	77.8
			5	76.8
			10	74.4
			20	71.9
	Period 2	1.5	0.2	76
			5	75.3
			10	73.4
			20	71.3
	Period 3	1.5	0.2	79.9
			5	78.7
			10	75.8
			20	74.3

Table A2. SPL test results and locations of floor-wise measuring points.

Test Area	Period	Distance from Road Boundary (m)	Height of Measurement Point Above Ground (m)	L_{Aeq} (dB(A))
Area #1	Period 1	3	1.5 (1L)	69.4
			4.1 (2L)	68.5
			6.7 (3L)	67.9
			9.3 (4L)	70.8
			14.5 (6L)	71.1
	Period 2	3	1.5 (1L)	69.4
			4.1 (2L)	68.3
			6.7 (3L)	67.6
			9.3 (4L)	69
			14.5 (6L)	70.4
	Period 3	3	1.5 (1L)	66
			4.1 (2L)	64.8
			6.7 (3L)	64.4
			9.3 (4L)	66.6
			14.5 (6L)	67.3
Area #2	Period 1	13	1.5 (1L)	73.3
			7.8 (3L)	73.9
			10.6 (4L)	73.2
			16.2 (6L)	72
			21.8 (8L)	71.8
	Period 2	13	1.5 (1L)	73
			7.8 (3L)	73.8
			10.6 (4L)	73
			16.2 (6L)	71.5
			21.8 (8L)	71
	Period 3	13	1.5 (1L)	75.3
			7.8 (3L)	75.7
			10.6 (4L)	75
			16.2 (6L)	74
			21.8 (8L)	73.8

eferences

Phillips, B.B.; Bullock, J.M.; Osborne, J.L.; Gaston, K.J. Spatial extent of road pollution: A notional analysis. *Sci. Total Environ.* **2021**, *773*, 145589. [CrossRef] [PubMed]

Singh, D.; Kumari, N.; Sharma, P. A review of adverse effects of road traffic noise on human health. *Fluct. Noise Lett.* **2018**, *17*, 183001. [CrossRef]

Tobollik, M.; Hintzsche, M.; Wothge, J.; Myck, T.; Plass, D. Burden of disease due to traffic noise in Germany. *Int. J. Environ. Res. Public Health* **2019**, *16*, 2304. [CrossRef]

4. Banerjee, D.; Das, P.P.; Foujdar, A. Association between road traffic noise and prevalence of coronary heart disease. *Environ. Monit. Assess.* **2014**, *186*, 2885–2893. [CrossRef]
5. Yang, W.-T.; Wang, V.-S.; Chang, L.-T.; Chuang, K.-J.; Chuang, H.-C.; Liu, C.-S.; Bao, B.-Y.; Chang, T.-Y. Road Traffic Noise, Air Pollutants, and the Prevalence of Cardiovascular Disease in Taichung, Taiwan. *Int. J. Environ. Res. Public Health* **2018**, *15*, 1707. [CrossRef]
6. Cai, C.; Xu, Y.-N.; Wang, Y.; Wang, Q.-K.; Liu, L. Experimental Study on the Effect of Urban Road Traffic Noise on Heart Rate Variability of Noise-Sensitive People. *Front. Psychol.* **2022**, *12*, 749224. [CrossRef]
7. Shin, S.; Bai, L.; Oiamo, T.H.; Burnett, R.T.; Weichenthal, S.; Jerrett, M.; Kwong, J.C.; Goldberg, M.S.; Copes, R.; Kopp, A.; et al. Association Between Road Traffic Noise and Incidence of Diabetes Mellitus and Hypertension in Toronto, Canada: A Population-Based Cohort Study. *J. Am. Heart Assoc.* **2020**, *9*, e013021. [CrossRef]
8. Lin, J.-Y.; Cheng, W.-J.; Wu, C.-F.; Chang, T.-Y. Associations of Road Traffic Noise and Its Frequency Spectrum with Prevalent Depression in Taichung, Taiwan. *Front. Public Health* **2023**, *11*, 1116345. [CrossRef]
9. Shi, J.; Huang, J.; Guo, M.; Tian, L.; Wang, J.; Wong, T.W.; Webster, C.; Leung, G.M.; Ni, M.Y. Contributions of Residential Traffic Noise to Depression and Mental Wellbeing in Hong Kong: A Prospective Cohort Study. *Environ. Pollut.* **2023**, *338*, 122641. [CrossRef] [PubMed]
10. Sygna, K.; Aasvang, G.M.; Aamodt, G.; Oftedal, B.; Krong, N.H. Road Traffic Noise, Sleep and Mental Health. *Environ. Res.* **2014**, *131*, 17–24. [CrossRef] [PubMed]
11. Smith, M.G.; Younes, M.; Aeschbach, D.; Elmenhorst, E.-M.; Müller, U.; Basner, M. Traffic Noise-Induced Changes in Wake Propensity Measured with the Odds-Ratio Product (ORP). *Sci. Total Environ.* **2022**, *805*, 150191. [CrossRef] [PubMed]
12. van Kempen, E.; Fischer, P.; Janssen, N.; Houthuijs, D.; van Kamp, I.; Stansfeld, S.; Cassee, F. Neurobehavioral Effects of Exposure to Traffic-Related Air Pollution and Transportation Noise in Primary Schoolchildren. *Environ. Res.* **2012**, *115*, 18–25. [CrossRef]
13. Bao, W.-W.; Xue, W.-X.; Jiang, N.; Huang, S.; Zhang, S.-X.; Zhao, Y.; Chen, Y.-C.; Dong, G.-H.; Cai, M.; Chen, Y.-J. Exposure to Road Traffic Noise and Behavioral Problems in Chinese Schoolchildren: A Cross-Sectional Study. *Sci. Total Environ.* **2022**, *837*, 155806. [CrossRef] [PubMed]
14. Cai, Y.; Zijlema, W.L.; Sørgjerd, E.P.; Doiron, D.; de Hoogh, K.; Hodgson, S.; Wolffenbuttel, B.; Gulliver, J.; Hansell, A.L.; Nieuwenhuijsen, M.; et al. Impact of Road Traffic Noise on Obesity Measures: Observational Study of Three European Cohorts. *Environ. Res.* **2020**, *191*, 110013. [CrossRef] [PubMed]
15. Annual Report on Prevention and Control of Noise Pollution in China 2024. Available online: https://www.mee.gov.cn/hjzl/sthjzk/hjzywr/202408/W020240829356736814731.pdf (accessed on 4 June 2025).
16. Faulkner, J.; Murphy, E. Road traffic noise modelling and population exposure estimation using CNOSSOS-EU: Insights from Ireland. *Appl. Acoust.* **2022**, *192*, 108692. [CrossRef]
17. Amoatey, P.; Omidvarbona, H.; Baawain, M.S.; AI-Mayahi, A.; AI-Mamun, A.; AI-Harthy, I. Exposure assessment to road traffic noise levels and health effects in an arid urban area. Environ. *Sci. Pollut. Res.* **2020**, *27*, 35051–35064. [CrossRef]
18. Brown, A.L.; Lam, K.C.; van Kamp, I. Quantification of the exposure and effects of road traffic noise in a dense Asian city: A comparison with western cities. *Environ. Health* **2015**, *14*, 22. [CrossRef]
19. Lee, N.F.; Walker, E.D. Research in action-tourism and its impacts on the environmental soundscape-A community-initiated pilot study. *Environ. Impact Assess. Rev.* **2024**, *105*, 107450. [CrossRef]
20. Huang, B.; Pan, Z.; Wang, G. A methodology to control urban traffic noise under the constraint of environmental capacity: A case study of a double-decision optimization model. *Transp. Res. Part D Transp. Environ.* **2015**, *41*, 257–270. [CrossRef]
21. Guarnaccia, C.; Mascolo, A.; Aumond, P.; Can, A.; Rossi, D. From Early to Recent Models: A Review of the Evolution of Road Traffic and Single Vehicles Noise Emission Modelling. *Curr. Pollut. Rep.* **2024**, *10*, 662–683. [CrossRef]
22. Fiedler, P.E.K.; Zannin, P.H.T. Evaluation of noise pollution in urban traffic hubs—Noise maps and measurements. *Environ. Impact Assess. Rev.* **2015**, *51*, 1–9. [CrossRef]
23. Zhou, Z.; Zhang, M.; Gao, X.; Gao, J.; Kang, J. Analysis of traffic noise spatial distribution characteristics and influencing factors in high-density cities. *Appl. Acoust.* **2024**, *217*, 109838. [CrossRef]
24. Cai, C.; Mak, C.M.; He, X. Analysis of urban road traffic noise exposure of residential buildings in Hong Kong over the past decade. *Noise Health* **2019**, *21*, 142–154.
25. Di, G.; Liu, X.; Lin, Q.; Zheng, Y.; He, L. The relationship between urban combined traffic noise and annoyance: An investigation in Dalian, north of China. *Sci. Total Environ.* **2012**, *432*, 189–194. [CrossRef] [PubMed]
26. Yang, W.; He, J.; He, C.; Cai, M. Evaluation of urban traffic noise pollution based on noise maps. *Transp. Res. Part D Transp. Environ.* **2020**, *87*, 102516. [CrossRef]
27. Melo, R.A.; Pimentel, R.L.; Lacerda, D.M.; Silva, W.M. Applicability of models to estimate traffic noise for urban roads. *J. Environ. Health Sci. Eng.* **2015**, *13*, 83. [CrossRef]
28. Schiff, M.; Hornikx, M.; Forssén, J. Excess attenuation for sound propagation over an urban canyon. *Appl. Acoust.* **2010**, *71*, 510–517. [CrossRef]

29. Huang, B.; Pan, Z.; Liu, Z.; Hou, G.; Yang, H. Acoustic amenity analysis for high-rise building along urban expressway: Modeling traffic noise vertical propagation using neural networks. *Transp. Res. Part D Transp. Environ.* **2017**, *53*, 63–77. [CrossRef]
30. Li, W.; Zhai, J.; Zhu, M. Characteristics and perception evaluation of the soundscapes of public spaces on both sides of the elevated road: A case study in Suzhou, China. *Sustain. Cities Soc.* **2022**, *84*, 103996. [CrossRef]
31. Su, K.; Cao, R.; Wang, Q.; Peng, Z.; He, Y. Prediction of Noise Reduction Effect of Sound Barriers and Evaluation of Noise Annoyance. *Fluct. Noise Lett.* **2025**, *24*, 2550002. [CrossRef]
32. Guangzhou Statistical Yearbook 2024. Available online: https://tjj.gz.gov.cn/datav/admin/home/www_nj/ (accessed on 4 June 2025).
33. Cai, M.; Zou, J.; Xie, J.; Ma, X. Road traffic noise mapping in Guangzhou using GIS and GPS. *Appl. Acoust.* **2015**, *87*, 94–102. [CrossRef]
34. Wang, H.; Chen, H.; Cai, M. Evaluation of an urban traffic Noise-Exposed population based on points of interest and noise maps: The case of Guangzhou. *Environ. Pollut.* **2018**, *239*, 741–750. [CrossRef]
35. Lee, H.M.; Luo, W.; Xie, J.; Lee, H.P. Urban traffic noise mapping using building simplification in the Panyu District of Guangzhou City, China. *Sustainability* **2022**, *14*, 4465. [CrossRef]
36. *GB 3096-2008*; Environmental Quality Standard for Noise. Ministry of Ecology and Environment of the People's Republic of China/General Administration of Quality Supervision, Inspection and Quarantine of the Peoples Republic of China: Beijing, China, 2008.
37. Alsina-Pagès, R.M.; Benocci, R.; Brambilla, G.; Zambon, G. Methods for noise event detection and assessment of the sonic environment by the harmonica index. *Appl. Sci.* **2021**, *11*, 8031. [CrossRef]
38. Wu, J.; Zou, C.; He, S.; Sun, X.; Wang, X.; Yan, Q. Traffic noise exposure of high-rise residential buildings in urban area. *Environ. Sci. Pollut. Res.* **2019**, *26*, 8502–8515. [CrossRef]
39. Yan, X.; Wu, Z.; Wu, Z.; Wang, H. Study on the network acoustics environment effects of traffic management measures by a bilevel programming model. *Sustain. Cities Soc.* **2024**, *101*, 105203. [CrossRef]
40. Qin, X.; Li, Y.; Ma, L.; Zhang, Y. Traffic noise distribution characteristics of high-rise buildings along ultra-wide cross section highway with multiple noise reduction measures. *Environ. Sci. Pollut. Res.* **2024**, *31*, 20601–20620. [CrossRef]
41. Xie, X.; Wu, D.; Zhang, H.; Shen, Y.; Mikio, Y. Low frequency noise radiation from traffic-induced vibration of steel double-box girder bridge. *J. Vib. Control* **2012**, *18*, 373–384. [CrossRef]
42. He, Y.; Zhou, Q.; Sheng, X. Characterizing low-frequency structure-borne noise from multi-span bridges on high-speed railways. *J. Cent. South Univ.* **2024**, *31*, 976–988. [CrossRef]
43. Sanok, S.; Berger, M.; Müller, U.; Schmid, M.; Weidenfeld, S.; Elmenhorst, E.-M.; Aeschbach, D. Road traffic noise impacts sleep continuity in suburban residents: Exposure-response quantification of noise-induced awakenings from vehicle pass-bys at night. *Sci. Total Environ.* **2022**, *817*, 152594. [CrossRef]
44. Foraster, M.; Esnaola, M.; López-Vicente, M.; Rivas, I.; Álvarez-Pedrerol, M.; Persavento, C.; Sebastian-Galles, N.; Pujol, J.; Dadvand, P.; Sunyer, J. Exposure to road traffic noise and cognitive development in schoolchildren in Barcelona, Spain: A population-based cohort study. *PLoS. Med.* **2022**, *19*, e1004001. [CrossRef] [PubMed]
45. Lan, Z.; Cai, M. Dynamic traffic noise maps based on noise monitoring and traffic speed data. *Transp. Res. Part D Transp. Environ.* **2021**, *94*, 102796. [CrossRef]
46. Balážiková, M.; Pačaiová, H.; Tomašková, M. A Proposal for Risk Assessment of Low-Frequency Noise in the Human–Machine–Environment System. *Appl. Sci.* **2023**, *13*, 13321. [CrossRef]
47. Persson Waye, K.; Bengtsson, J.; Rylander, R.; Hucklebridge, F.; Evans, P.; Clow, A. Low frequency noise enhances cortisol among noise sensitive subjects during work performance. *Life Sci.* **2002**, *70*, 745–758. [CrossRef]
48. Berglund, B.; Hassmén, P.; Soames Job, R.F. Sources and effects of low-frequency noise. *J. Acoust. Soc. Am.* **1996**, *99*, 2985–3002. [CrossRef]
49. Ma, J.; Wu, X.; Jiang, J. Lane restriction system to reduce the environmental cost of urban roads. *Transp. Res. Part D Transp. Environ.* **2023**, *115*, 103575. [CrossRef]
50. Qu, X.; Zeng, Z.; Wang, K.; Wang, S. Replacing urban trucks via ground–air cooperation. *Commun. Transp. Res.* **2022**, *2*, 10080. [CrossRef]
51. Wang, Y.; Wang, Z.; Hu, X.; Xue, G.; Guan, X. Truck–drone hybrid routing problem with time-dependent road travel time. *Transp. Res. Part C Emer.* **2022**, *14*, 103901. [CrossRef]
52. Pallas, M.A.; Chatagnon, R.; Lelong, J. Noise emission assessment of a hybrid electric mid-size truck. *Appl. Acoust.* **2014**, *76*, 280–290. [CrossRef]
53. Lak, A.; Amiri, N.; Aghamolaei, R. Urban Elevated Highways in Residential Districts: New Developed Elevated Highways from Residents' Perspectives. *J. Urban Plann. Dev.* **2022**, *148*, 05022006. [CrossRef]
54. Hadi Baaj, M.; El-Fadel, M.; Shazbak, S.M.; Saliby, E. Modeling Noise at Elevated Highways in Urban Areas: A Practical Application. *J. Urban Plan. Dev.* **2001**, *127*, 169–180. [CrossRef]

55. Yu, W.L.; Kang, J. Resistance of Villages to Elevated-Road Traffic Noise. *J. Environ. Plan. Manage.* **2018**, *62*, 492–516. [CrossRef]
56. Lu, K.-F.; He, H.-D.; Wang, H.-W.; Li, X.-B.; Peng, Z.-R. Characterizing temporal and vertical distribution patterns of traffic-emitted pollutants near an elevated expressway in urban residential areas. *Build. Environ.* **2020**, *72*, 106678. [CrossRef]
57. Yuan, M.; Wang, J.; Wang, Y.; Shao, S. Study on noise reduction with paving different low noise pavement materials. *Appl. Sci.* **2021**, *11*, 10273. [CrossRef]
58. Sandberg, U.; Żurek, B.Ś.; Ejsmont, J.A.; Ronowski, G. Tyre/road noise reduction of poroelastic road surface tested in a laboratory. In Proceedings of the Acoustics 2013—Victor Harbor, Victor Harbor, Australia, 17–20 November 2013.
59. Dong, S.; Han, S.; Luo, Y.; Han, X.; Xu, O. Evaluation of tire-pavement noise based on three-dimensional pavement texture characteristics. *Constr. Build. Mater.* **2021**, *306*, 124935. [CrossRef]
60. Fujiwara, T.; Meiarashi, S.; Namikawa, Y.; Hasebe, M. Reduction of equivalent continuous A-weighted sound pressure levels by porous elastic road surfaces. *Appl. Acoust.* **2005**, *66*, 766–778. [CrossRef]
61. Fediuk, R.; Amean, M.; Vatin, N.; Vasilev, Y.; Lesovik, V.; Ozbakkaloglu, T. Acoustic properties of innovative concretes: A review. *Materials* **2021**, *14*, 398. [CrossRef] [PubMed]
62. Sun, X.; Xu, H.; Qin, X.; Zhu, Y.; Jin, J. Cross-scale study on the interaction behaviour of municipal solid waste incineration fly ash-asphalt mortar: A macro-micro approach. *Int. J. Pavement Eng.* **2025**, *26*, 2469114. [CrossRef]
63. Sun, X.; Qin, X.; Liu, Z.; Yin, Y. Damaging effect of fine grinding treatment on the microstructure of polyurea elastomer modifier used in asphalt binder. *Measurement* **2025**, *242 Pt B*, 115984. [CrossRef]
64. Poulikakos, L.D.; Athari, S.; Mikhailenko, P.; Kakar, M.R.; Bueno, M.; Piao, Z.; Pieren, R.; Heutschi, K. Effect of waste materials on acoustical properties of semi-dense asphalt mixtures. *Transp. Res. Part D Transp. Environ.* **2022**, *102*, 103154. [CrossRef]
65. Faßbender, S.; Oeser, M. Investigation of the Reusability of a Polyurethane-Bound Noise-Absorbing Pavement in Terms of Reclaimed Asphalt Pavement. *Materials* **2022**, *15*, 3040. [CrossRef]
66. Secchi, S.; Astolfi, A.; Calosso, G.; Casini, D.; Cellai, G.; Scamoni, F.; Scrosati, C.; Shtrepi, L. Effect of outdoor noise and façade sound insulation on indoor acoustic environment of Italian schools. *Appl. Acoust.* **2017**, *126*, 120–130. [CrossRef]
67. Naish, D.A.; Tan, A.C.; Demirbilek, F.N. Simulating the effect of acoustic treatment types for residential balconies with road traffic noise. *Appl. Acoust.* **2014**, *79*, 131–140. [CrossRef]
68. Eggenschwiler, K.; Heutschi, K.; Taghipour, A.; Pieren, R.; Gisladottir, A.; Schäffer, B. Urban design of inner courtyards and road traffic noise: Influence of façade characteristics and building orientation on perceived noise annoyance. *Build. Environ.* **2022**, *224*, 109526. [CrossRef]
69. Kang, J.; Hornikx, M.; van Renterghem, T.; Smyrnova, Y.; Forssén, J.; Cheal, C.; Botteldooren, D.; Yang, H.-S.; Jeon, J.Y.; Jang, H.S. et al. Vegetation in urban streets, squares, and courtyards. In *Environmental Methods for Transport Noise Reduction*, 1st ed.; Nilsson M., Bengtsson, J., Klaeboe, R., Eds.; CRC Press: London, UK, 2014.
70. Tao, Z.; Zhang, D.; Tu, D.; He, L.; Zou, C. Prediction of train-induced ground-borne vibration transmission considering parametric uncertainties. *Probab. Eng. Mech.* **2025**, *79*, 103731. [CrossRef]
71. Qiu, Y.; Zheng, B.; Jiang, B.; Jiang, S.; Zou, C. Effect of Non-Structural Components on Over-Track Building Vibrations Induced by Train Operations on Concrete Floor. *Int. J. Struct. Stab. Dyn.* **2025**. [CrossRef]

MDPI AG
Grosspeteranlage 5
4052 Basel
Switzerland
Tel.: +41 61 683 77 34

Buildings Editorial Office
E-mail: buildings@mdpi.com
www.mdpi.com/journal/buildings

www.ingramcontent.com/pod-product-compliance
Lightning Source LLC
Chambersburg PA
CBHW061502210926
51850CB00008B/1796
* 9 7 8 3 7 2 5 8 4 9 9 3 2 *